Other Titles in This Series

173 V. Kharlamov, A. Korchagin, G. Polotovskiĭ, and O. Viro, Editors, Topology of Real Algebraic Varieties and Related Topics
172 K. Nomizu, Editor, Selected Papers on Number Theory and Algebraic Geometry
171 L. A. Bunimovich, B. M. Gurevich, and Ya. B. Pesin, Editors, Sinai's Moscow Seminar on Dynamical Systems
170 S. P. Novikov, Editor, Topics in Topology and Mathematical Physics
169 S. G. Gindikin and E. B. Vinberg, Editors, Lie Groups and Lie Algebras: E. B. Dynkin's Seminar
168 V. V. Kozlov, Editor, Dynamical Systems in Classical Mechanics
167 V. V. Lychagin, Editor, The Interplay between Differential Geometry and Differential Equations
166 O. A. Ladyzhenskaya, Editor, Proceedings of the St. Petersburg Mathematical Society, Volume III
165 Yu. Ilyashenko and S. Yakovenko, Editors, Concerning the Hilbert 16th Problem
164 N. N. Uraltseva, Editor, Nonlinear Evolution Equations
163 L. A. Bokut', M. Hazewinkel, and Yu. G. Reshetnyak, Editors, Third Siberian School "Algebra and Analysis"
162 S. G. Gindikin, Editor, Applied Problems of Radon Transform
161 Katsumi Nomizu, Editor, Selected Papers on Analysis, Probability, and Statistics
160 K. Nomizu, Editor, Selected Papers on Number Theory, Algebraic Geometry, and Differential Geometry
159 O. A. Ladyzhenskaya, Editor, Proceedings of the St. Petersburg Mathematical Society, Volume II
158 A. K. Kelmans, Editor, Selected Topics in Discrete Mathematics: Proceedings of the Moscow Discrete Mathematics Seminar, 1972–1990
157 M. Sh. Birman, Editor, Wave Propagation. Scattering Theory
156 V. N. Gerasimov, N. G. Nesterenko, and A. I. Valitskas, Three Papers on Algebras and Their Representations
155 O. A. Ladyzhenskaya and A. M. Vershik, Editors, Proceedings of the St. Petersburg Mathematical Society, Volume I
154 V. A. Artamonov et al., Selected Papers in K-Theory
153 S. G. Gindikin, Editor, Singularity Theory and Some Problems of Functional Analysis
152 H. Draškovičová et al., Ordered Sets and Lattices II
151 I. A. Aleksandrov, L. A. Bokut', and Yu. G. Reshetnyak, Editors, Second Siberian Winter School "Algebra and Analysis"
150 S. G. Gindikin, Editor, Spectral Theory of Operators
149 V. S. Afraĭmovich et al., Thirteen Papers in Algebra, Functional Analysis, Topology, and Probability, Translated from the Russian
148 A. D. Aleksandrov, O. V. Belegradek, L. A. Bokut', and Yu. L. Ershov, Editors, First Siberian Winter School "Algebra and Analysis"
147 I. G. Bashmakova et al., Nine Papers from the International Congress of Mathematicians, 1986
146 L. A. Aĭzenberg et al., Fifteen Papers in Complex Analysis
145 S. G. Dalalyan et al., Eight Papers Translated from the Russian
144 S. D. Berman et al., Thirteen Papers Translated from the Russian
143 V. A. Belonogov et al., Eight Papers Translated from the Russian
142 M. B. Abalovich et al., Ten Papers Translated from the Russian
141 H. Draškovičová et al., Ordered Sets and Lattices
140 V. I. Bernik et al., Eleven Papers Translated from the Russian
139 A. Ya. Aĭzenshtat et al., Nineteen Papers on Algebraic Semigroups
138 I. V. Kovalishina and V. P. Potapov, Seven Papers Translated from the Russian
137 V. I. Arnol'd et al., Fourteen Papers Translated from the Russian
136 L. A. Aksent'ev et al., Fourteen Papers Translated from the Russian

(*Continued in the back of this publication*)

*Dedicated to the memory of
Dmitriĭ Andreevich Gudkov (1918–1992)*

American Mathematical Society

TRANSLATIONS

Series 2 • Volume 173

Advances in the Mathematical Sciences – 29

(*Formerly Advances in Soviet Mathematics*)

Topology of Real Algebraic Varieties and Related Topics

V. Kharlamov
A. Korchagin
G. Polotovskiĭ
O. Viro
Editors

American Mathematical Society
Providence, Rhode Island

ADVANCES IN THE MATHEMATICAL SCIENCES
EDITORIAL COMMITTEE

V. I. ARNOLD

S. G. GINDIKIN

V. P. MASLOV

Translation edited by A. B. Sossinsky

1991 *Mathematics Subject Classification.* Primary 14Hxx, 14Pxx, 57Mxx, 57R45; Secondary 05A15.

ABSTRACT. The volume is dedicated to the memory of the Russian mathematician D. A. Gudkov. It contains papers written by his friends, students, and collaborators, devoted mainly to the areas where D. A. Gudkov made important contributions. The main topic is the topology of real algebraic varieties. In particular, several papers include new results on the topology of real plane algebraic curves (the Hilbert 16th problem).

For graduate students and researchers working in algebraic geometry and algebraic topology.

Library of Congress Card Number 91-640741
ISBN 0-8218-0555-X
ISSN 0065-9290

Copying and reprinting. Material in this book may be reproduced by any means for educational and scientific purposes without fee or permission with the exception of reproduction by services that collect fees for delivery of documents and provided that the customary acknowledgment of the source is given. This consent does not extend to other kinds of copying for general distribution, for advertising or promotional purposes, or for resale. Requests for permission for commercial use of material should be addressed to the Assistant to the Publisher, American Mathematical Society, P. O. Box 6248, Providence, Rhode Island 02940-6248. Requests can also be made by e-mail to reprint-permission@ams.org.

Excluded from these provisions is material in articles for which the author holds copyright. In such cases, requests for permission to use or reprint should be addressed directly to the author(s). (Copyright ownership is indicated in the notice in the lower right-hand corner of the first page of each article.)

© Copyright 1996 by the American Mathematical Society. All rights reserved.
The American Mathematical Society retains all rights
except those granted to the United States Government.
Printed in the United States of America.

∞ The paper used in this book is acid-free and falls within the guidelines
established to ensure permanence and durability.
♻ Printed on recycled paper.

10 9 8 7 6 5 4 3 2 1   01 00 99 98 97 96

Contents

Dmitriĭ Andreevich Gudkov
GRIGORIĬ M. POLOTOVSKIĬ   1

Recollection of D. A. Gudkov
EVGENIĬ I. GORDON   11

Remarks on the Enumeration of Plane Curves
V. I. ARNOLD   17

On Diagrams of Configurations of 7 Skew Lines of $\mathbb{R}^3$
ALBERTO BOROBIA and VLADIMIR F. MAZUROVSKIĬ   33

Smoothing Isolated Singularities on Real Algebraic Surfaces
BENOÎT CHEVALLIER   41

Quadratic Transformations $\mathbb{R}P^2 \to \mathbb{R}P^2$
ALEXANDER DEGTYAREV   61

Real Algebraic Curves and Link Cobordism. II
PATRICK GILMER   73

Morsifications of Rational Functions
V. V. GORYUNOV   85

Real Algebraic Curves with Real Cusps
ILIA ITENBERG and EUGENIĬ SHUSTIN   97

Towards the Maximal Number of Components of a Nonsingular Surface of Degree 5 in $\mathbb{R}P^3$
V. KHARLAMOV and I. ITENBERG   111

Stable Equivalence of Real Projective Configurations
SERGEĬ I. KHASHIN and VLADIMIR F. MAZUROVSKIĬ   119

Smoothing of 6-Fold Singular Points and Constructions of 9th Degree M-Curves
A. B. KORCHAGIN   141

Automaton Model of Relations Between Two Countries
MARK KUSHELMAN   157

Classification of Curves of Degree 6 Decomposing into a Product of M-Curves in General Position
T. V. KUZMENKO and G. M. POLOTOVSKIĬ   165

Spinors and Differentials of Real Algebraic Curves
S. M. NATANZON   179

On the Topological Classification of Real Enriques Surfaces. I
VYACHESLAV V. NIKULIN   187

Critical Points of Real Polynomials, Subdivisions of Newton Polyhedra and Topology of Real Algebraic Hypersurfaces
EVGENIĬ SHUSTIN   203

On a Nonlinear Boundary Value Problem of Mathematical Physics
G. A. UTKIN   225

Generic Immersions of the Circle to Surfaces and the Complex Topology of Real Algebraic Curves
OLEG VIRO   231

Stratified Spaces of Real Algebraic Curves of Bidegree $(m, 1)$ and $(m, 2)$ on a Hyperboloid
V. I. ZVONILOV   253

Dmitriĭ Andreevich Gudkov

Grigoriĭ M. Polotovskiĭ

§I. Curriculum vitae

Dmitriĭ Andreevich Gudkov was born on May 18, 1918, in the city of Vologda. His father Andreĭ Fedorovich, a land-surveyor, was mobilized when World War I began, became an officer in the Red Army, and disappeared in 1919. His mother Nina Pavlovna (née Chekalova) was a physician. She was a very well educated person.

In 1926 Nina Pavlovna and Dima came to Nizhniĭ Novgorod, and then lived in a village called "Memory of the Paris Commune" on the bank of the Volga river near the city. Here Nina Pavlovna was married for the second time and in 1931 Dima's brother Kostya was born. The second marriage was not happy and did not last long, so Dima was a sort of a father and nurse to his brother.

In 1935 the family returned to Gor′kiĭ (the new name for Nizhniĭ Novgorod), where Dima graduated from high school and entered the Department of Physics and Mathematics of Gor′kiĭ State University. He graduated from the University in 1941 and was immediately drafted. For a short time he studied in an artillery school, and from October 1941 until the end of the war he was at the front line. Dmitriĭ Andreevich took part in the attack and seizure of Berlin. He was rewarded with several medals. During the war Dmitriĭ Andreevich applied for membership in the Communist Party,[1] but was only accepted in 1948.

In February of 1946 Dmitriĭ Andreevich came back to Gor′kiĭ State University and worked there until the end of his life. Here is the short list of positions that he occupied:

- 1946 — Assistant at the Chair of Algebra and Analysis.
- 1948 — Post-graduate student, then (1952) assistant, then (1954) docent (associate professor) at the Chair of Mathematical Analysis.
- 1961 — Head of the Chair of Mathematics at the Radiophysical Department (full professor since 1971).
- 1978–1988 — Head of the Chair of Geometry and Higher Algebra.
- 1988–1992 — Full professor at this Chair.

1991 *Mathematics Subject Classification*. Primary 01A70.

The author is grateful to V. Kharlamov and E. Shustin for discussions and useful remarks.

[1] Many years later he related: "Whenever I made an attempt to restore justice, I would be told: 'You, Gudkov, are not a member of the Party, so you should keep silent.' I understood that as a nonparty member I could not achieve anything."

©1996, American Mathematical Society

In 1953 Dmitriĭ Andreevich married. His wife Nataliya Vasil′evna graduated from the Physics Department of Gor′kiĭ State University, where she met him as a student in one of his exercise groups. Later their children, Yuriĭ and Alexandra, graduated from the same Department.

On March 13, 1992 Dmitriĭ Andreevich Gudkov died, several days after a heart attack.

§II. Mathematical works and their development

Dmitriĭ Andreevich wrote in his memoirs[2] about the war years: "I began to like mathematics already in the 6th grade.[3] This was due to the influence of a remarkable teacher, Pyotr Mikhailovich Bezelev." And further: "At first I wanted to be an engineer. In the 10th grade I decided firmly to be a mathematician and even told to my friends that I will become a Professor. Why did I like mathematics? I think the reason was in my character. Mathematics gives a person more independence than any other field of science."

Dmitriĭ Andreevich did very well in school and graduated from the University with honors. But his own mathematical investigations began much later: the war had started.

When Gudkov returned from the army to the University, a large group of physicists and mathematicians, Andronov's scientific school, was actively working in Gor′kiĭ. A. A. Andronov, the physicist, had a very wide range of scientific interests. In particular, he was interested in some purely mathematical problems. He had noticed the analogy between the topological structure of dynamical systems and the topology of algebraic curves. In 1948 he conjectured that the elaboration of the theory of bifurcations for algebraic curves on the basis of the notions of "robustness" and "degrees of nonroughness", which were at first defined for dynamical systems,[4] would be fruitful for the classification of algebraic curves as well. At that time A. A. Andronov and his colleague A. G. Meyer proposed this problem to Gudkov.

A little bit later, in the spring of 1950, I. G. Petrovskiĭ, the author of the remarkable paper "On the topology of real plane algebraic curves",[5] advised Gudkov to concentrate his efforts on the problem of classifying nonsingular curves of degree 6. It is well known that this task was included by D. Hilbert in the 16th problem of his famous list.[6]

It turned out that all the mathematical activity of Dmitriĭ Andreevich was to be centered around the above-mentioned problems. His main results in the general theory of bifurcations of algebraic curves are contained in [2, 5–8].[7] Gudkov's results about 6th degree curves exerted a strong influence on the further development of real algebraic geometry, so it is necessary to describe this topic in more detail.

The technique for solving the problem of the arrangements of ovals for 6th degree curves was already elaborated in its main features by Gudkov in his Kandidat's

---

[2]*Artillery Technician*, in the book "Those who defended the motherland", Gor′kiĭ, 1991, pp. 142–147. (Russian)

[3]At the same time Dmitriĭ Andreevich became interested in chess. This interest persisted for all his life. For an amateur he played extremely well.

[4]See A. A. Andronov and L. S. Pontryagin, *Robust systems*, Dokl. Akad. Nauk SSSR **14** (1937), no. 5, 247–252. (Russian)

[5]Ann. of Math. **39** (1938), no. 1, 187–209.

[6]D. Hilbert, *Mathematische Probleme*, Arch. Math. Phys. 3 Reiche, Bd. 1 (1901), 44–63, 213–237. (German)

[7]See the list of Gudkov's publications in §III below.

(=Ph.D.) thesis [1]. It is based on bifurcation theory and consists of investigating transformations of the curve under continuous and not necessarily small changes of its coefficients. The very idea of such an approach belongs to D. Hilbert. His students G. Kahn[8] and K. Löbenstein[9] tried to prove the nonexistence of a 6th degree curve with scheme 11 (i.e., consisting of 11 ovals one outside the other). Then K. Rohn[10] tried to prove the nonexistence of 6th degree curves with schemes 11 and $\frac{10}{1}$ by the same method.[11] But the proofs in the above-mentioned papers contain essential gaps. Gudkov wrote: "*Without a classification by degrees of nonrobustness, it is practically impossible to look into all the numerous logically possible complicated situations* (of a curve under changes of its coefficients, G. P.)" ([23, p. 44]). But: "*K. Rohn made an important contribution to the development of Hilbert's idea*" (ibid) and "*We consider Rohn's works to be very significant*" ([16, p. 153]). Proceeding from this assessment, Gudkov called (in [18, 23]) the above-mentioned approach the *Hilbert–Rohn method*.

So, the Hilbert–Rohn method was elaborated practically in [1], but the classification of nonsingular curves of degree 6 in [1] and [3] contained mistakes; in particular, in [1] there was an incorrect proof[12] of Hilbert's conjecture about the nonexistence of curves with scheme $\frac{5}{1}5$. The correction of all the mistakes required 15 more years of intensive work, which was summarized in the series of papers [11–17] and in Gudkov's Doctorate thesis [18].

The main results of Gudkov's investigations may be classified in the following way:
 (i) the elaboration of the Hilbert–Rohn method;
 (ii) the isotopy classification of nonsingular plane projective 6th degree curves;
 (iii) the discovery of a congruence-type restriction[13] to $2k$-degree M-curves:[14]

$$\chi(B_+) \equiv k^2 \pmod 8;$$

here $\chi(B_+)$ is the Euler characteristic of the oriented "half" of the complement to the curve (i.e., B_+ is the union of all components of the complement on which the polynomial determining the curve has the same sign and B_+ is orientable[15]);
 (iv) the development of deformation theory for real algebraic curves;
 (v) the classification of 4th degree plane curves.

We already spoke about (i). Let us discuss the other items.

Item (ii) means that one of the specific questions by Hilbert into his 16th problem was finally answered. It turned out that Hilbert's conjecture about the nonexistence of a curve with the scheme $\frac{5}{1}5$ was wrong. It is interesting to remark that the first proof of this fact in [18] was extraordinarily complicated. It takes up 28 pages of text, is a

---

[8] *Eine allgemeine Methode zur Untersuchung der Gestalten algebraischer Kurven*, Inaugural Dissertation, Göttingen, 1909. (German).

[9] *Ueber den Satz, dass ebene algebraische Kurve 6 Ordnung mit 11 sich ein ander ausschliessenden Ovalen nicht existiert*, Ibid, 1910.

[10] *Die ebene Kurve 6 Ordnung mit elf Ovalen*, Berichte über die Verhandl **63** (1911), 540–555. (German); *Die Maximalzahl und Anordnung der Ovale bei der ebenen Kurve 6 Ordnung und bei der Fläche 4 Ordnung*, Math. Ann. **73** (1913), 177–229. (German).

[11] In Gudkov's notation, $\frac{m}{1}n$ is the scheme consisting of $m + n$ ovals one outside of each other plus one more oval which encircles m of these ovals.

[12] This incorrect proof was not published.

[13] All the previously known restrictions on the topology of a curve had the form of inequalities.

[14] This term is due to I. G. Petrovskiĭ, see the definition below.

[15] Gudkov's initial formulation used another terminology, see below.

"pure existence proof", and was obtained by means of a combination of the Hilbert–Rohn method with quadratic transformations.[16] To my knowledge, Gudkov was the first who applied quadratic transformations to the investigation of the topology of nonsingular curves.[17]

The congruence in item (iii) gave the impact which took real algebraic geometry out of its "period of stagnation".

It is worth noting that until 1970 Gudkov worked essentially alone,[18] although I. G. Petrovskiĭ, O. A. Oleĭnik, E. A. Leontovich-Andronova, and (after sixteen years) V. V. Morozov showed interest in his work.

The situation changed drastically after the discovery of congruence (iii). Gudkov had found it as a periodicity law in his list of realized schemes of nonsingular 6th degree curves. The test that he carried out showed that the congruence is true for all even degree curves constructible by methods known at the time. Gudkov understood the importance of this congruence, but on the other hand he understood that he did not have enough experience in topology to prove it in a short time. So he announced it as a conjecture, first in private conversations and in talks at seminars, and then in [**19**].

The reaction appeared very soon: in 1971 V. I. Arnold published a paper[19] with a proof of "half" of this conjecture (i.e., modulo 4); this opened a new period of investigations in real algebraic geometry. In his answer to the question: "Which of your results do you consider the most important?" in the interview granted to the "*Gazette des Mathématiciens*",[20] V. I. Arnold said in particular:

*I recall that it was I. G. Petrovskiĭ, the Rector of Moscow University and the founder of the theory of real algebraic curves, who asked me to read D. A. Gudkov's thesis ([**18**]). Gudkov had found the answer to the question of Hilbert's 16th problem about the mutual dispositions of the ovals of real algebraic curves of degree 6 in the projective plane. In this very difficult thesis that I had not read before, I was struck by a modulo 8 congruence conjectured by Gudkov:*[21]

$$p - m \equiv k^2 \pmod{8},$$

where p is the number of ovals of a smooth curve of degree 2k "contained inside" an even number of ovals, and m is that contained inside an odd number of ovals, provided that the total number of ovals reaches its maximum value. As shown by Harnack, this maximum value equals $g + 1$, where $g = (n - 1)(n - 2)/2$ is the genus of the curve. Curves of degree n with $g + 1$ ovals exist for any n; they were called M-curves (M from maximum) by Petrovskiĭ.

Gudkov had found all the configurations of 11 ovals of the 6th degree M-curves (see Figure 1).

---

[16]Shortly after D. A. Gudkov suggested significantly simpler *constructions* of curves having this scheme, see [**19, 21, 23**].

[17]Later quadratic transformations were effectively used in other investigations on the 16th Hilbert problem. In particular, O. Ya. Viro suggested the use of quadratic transformations of another sort ("hyperbolisms"), by means of which he, A. B. Korchagin, E. I. Shustin, and others obtained many important results.

[18]On the one hand, this may be explained by the difficulty of the problem, and on the other hand, by the wrong (as it is clear now) opinion that the problem is too exotic and distant from the mainstream of mathematical development.

[19]*Arrangement of ovals of real plane algebraic curves, involutions of four-dimensional manifolds and the arithmetic of integer quadratic forms*, Funktsional. Anal. i Prilozhen. **5** (1971), no. 3; English transl., Functional Anal. Appl. **5** (1971), no. 3, 169–175.

[20]no. 52, Avril 1992.

[21]Actually Gudkov's Thesis does not contain this conjecture.

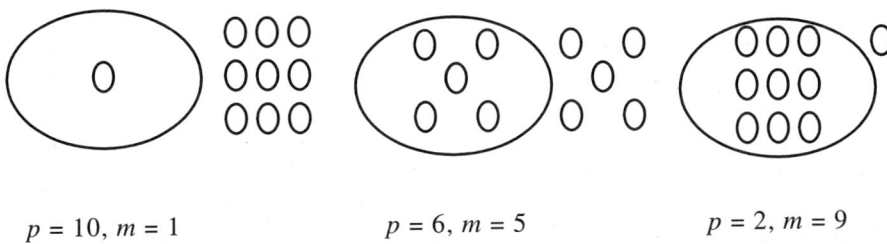

$p = 10, m = 1$ $p = 6, m = 5$ $p = 2, m = 9$

FIGURE 1. Eleven ovals of M-curves

Gudkov's congruence has been confirmed by all the M-curves known until now. However, there seemed to be no relation between the configuration of ovals in the projective plane and arithmetic.

V. I. Arnold discovered this relation, and "investigations in the topology of real algebraic varieties flowed into the mainstream of differential topology" ([**23**, p. 5]).

Shortly afterwards, V. A. Rokhlin[22] proved Gudkov's conjecture completely and obtained sweeping generalizations of the congruence in (iii). Since then the stream of works on real algebraic geometry has not been interrupted. V. A. Rokhlin attracted the attention of a group of his students (O. Ya. Viro, V. M. Kharlamov, T. Fiedler, V. I. Zvonilov, N. M. Mishachev, and others) to this subject. Gudkov's new students (G. M. Polotovskiĭ, E. I. Shustin, A. B. Korchagin, G. F. Nebukina) began their work. The interest in this problem was regenerated in the West (G. Wilson, A. Marin, J.-J. Risler, and others). Now there are many surveys[23] in real algebraic geometry (the paper [**23**] was one of the first) in which the development of the subject and new approaches can be traced. Below we shall mention only the works of Gudkov himself and their immediate development.

In (iv), we are referring to Gudkov's generalizations of Brusotti's theorem[24] about the independence of smoothings of singularities of a plane curve having nondegenerate double points only. At first this was done for plane curves with cusps ([**5, 8**]),[25] then for curves on quadrics ([**26**]). It should be noted that these results were rediscovered by other authors.[26]

---

[22]*Congruences modulo 16 in Hilbert's 16th problem*, Funktsional. Anal. i Prilozhen. **6** (1972), no. 4; English. transl., Functional Anal. Appl. **6** (1972), no. 4, 301–306.

[23]G. Wilson, *Hilbert's sixteenth problem*, Topology **17** (1978), no. 1, 53–73; N. A'Campo, *Sur la 1ère partie du 16e problème de Hilbert*, Seminaire Bourbaki, 31e année, 1978–1979, no. 537; V. A. Rokhlin, *Complex topological characteristics of real algebraic curves*, Uspekhi Mat. Nauk **33** (1978), no. 5, 77–89: V. I. Arnold and O. A. Oleĭnik, *Topology of real algebraic varieties*, Vestnik Moskov. Univ. Ser. I Mat. Mekh. (1979), no. 6, 7–17; V. M. Kharlamov, *Topology of real algebraic varieties*, Petrovskiĭ I. G. Selected Papers. Systems of Partial Differential Equations. Algebraic Geometry, Moscow, "Nauka", 1986, pp. 465–493; O. Ya. Viro, *Progress in the topology of real algebraic varieties over the last six years*, Russian Math. Surveys **41** (1986), no. 3, 55–82; O. Ya. Viro, *Real algebraic plane curves: constructions with controlled topology*, Leningrad Math. J. **1** (1990), 1059–1134.

[24]L. Brusotti, *Sulla "piccola variazione" di una curva piana algebrica reali*, Rend. Rom. Acc. Lincei (5) **30** (1921), 375–379.

[25]A similar result for complex curves was known, probably, to F. Severi (see O. Zariski, *Algebraic surfaces*, 2nd ed., Heidelberg, Springer-Verlag, 1971).

[26]A. Nobile, *On families of singular plane projective curves*, Ann. Mat. Pura Appl. (4) **138** (1984), 341–378; M. Gradolato, E. Mezzetti, *Curves with nodes, cusps and ordinary triple points*, Ann. Univ. Ferrara Sez. VII **31** (1985), 23–47; M. Lindner, *Über Mannigfaltigkeiten ebener Kurven mit Singularitäten*, Arch. Math. **28** (1977), 603–610.

The problem mentioned in (v) concerns the detailed investigation of 4th degree curves. The isotopy classification of plane nonsingular curves of degrees $\leqslant 5$ is not difficult, but this is not so for curves with singularities. At the beginning of the 18th century, Isaac Newton knew all about plane cubics. As the result of the investigations of D. A. Gudkov and his followers and students, now we have approximately the same amount of knowledge about plane quartics. Namely, in [10, 51, 62] the algebraic-topological[27] classification of such curves was obtained, in [37, 38, 43–49, 52–60] the classification of the forms of curves was obtained,[28] and in [53, 61, 62] the stratification of the space of coefficients was studied.

Let us mention some other works of Gudkov and their development.

(1) Gudkov's student G. A. Utkin investigated the classification of nonsingular projective surfaces. This question was mentioned by D. Hilbert in the 16th problem too. In Utkin's papers of 1967–1969, the results and methods of [11–18] are essentially used. (The topological classification of nonsingular 4th degree surfaces was finished by V. M. Kharlamov in 1976. Later V. M. Kharlamov and V. V. Nikulin obtained finer classifications.)

(2) Using the Hilbert–Rohn method and quadratic transformations, in 1979 G. M. Polotovskiĭ obtained the classification of 6th degree curves that decompose into a product of two nonsingular cofactors in general position. Polotovskiĭ's results in turn found different applications, including Kharlamov's investigations (mentioned in (1)) and Gudkov's works about curves on quadrics (see (5) below).

(3) In 1973 D. A. Gudkov and A. D. Krakhnov [22], and simultaneously and independently V. M. Kharlamov, proved analogs of the congruence in (iii) for the cases of $(M - 1)$-manifolds and $(M - 1)$-pairs. For an $(M - 1)$-curve of degree $2k$, this congruence can be written as

$$p - m \equiv (k^2 \pm 1) \pmod{8}.$$

(4) Using results mentioned in (2) and in (iv), D. A. Gudkov obtained ([27, 28]) the classification of nonsingular 8th degree curves on the hyperboloid. Let us note that Hilbert investigated this problem too. At present V. I. Zvonilov, G. B. Mikhalkin, and S. Matsuoka continue studying curves on the hyperboloid.

(5) In 1984 E. I. Shustin significantly developed the Hilbert–Rohn method. In particular, he constructed and applied new versions of this method to the classification of smoothings of some types of singularities and obtained new sufficient conditions for the independence of smoothings of singular points (compare to (iv)).

(6) What collection of singularities does an nth degree curve admit? Papers [32–36, 40, 41] are devoted to this problem for $n = 5$. In particular, in [32] a 5th degree curve with five (the maximal possible number) real cusps is constructed. (Investigation of this question is being continued by E. I. Shustin and I. V. Itenberg.)

(7) A special place in the activity of Gudkov is occupied by the investigation of the initial period of N. I. Lobachevskiĭ's biography. Note that Gudkov worked in the history of mathematics professionally. He elaborated and for many years delivered a two-semester course of lectures on the history of mathematics (after A. G. Meyer, for 35 years the history of mathematics was not taught in Gor′kiĭ University).

---

[27] Finer than isotopy.

[28] Here the arrangement of the points of inflection on the branches is taken into consideration.

Gudkov's interest in Lobachevskiĭ's biography had an "inherited" character too: he aspired to finish the investigations that A. A. Andronov and his group had undertaken in 1948–1956. On the basis of his own long-term investigations, in particular in the archives of Gor′kiĭ, Leningrad, and other cities (in 1986–1991), Gudkov uncovered a number of new documents and came to the conclusion that the old conjecture[29] asserting that N. I. Lobachevskiĭ and his brothers were in fact the sons of S. S. Shebarshin (and not of Ivan Lobachevskiĭ) is true. His arguments were presented in detail in the nice monograph [63] published posthumously.

§III. The list of scientific works of D. A. Gudkov

1. D. A. Gudkov, *The ascertainment of all existing types of nonsingular plane projective curves of the 6th order with real coefficients*, Master's Thesis, Gor′kiĭ, 1952, pp. 1–172. (Russian)
2. _____, *On the space of coefficients of plane algebraic curves of nth order*, Dokl. Akad. Nauk SSSR **98** (1954), no. 3, 337–340. (Russian)
3. _____, *The complete topological classification of nonsingular real algebraic curves of the 6th order in the real projective plane*, Dokl. Akad. Nauk SSSR **98** (1954), no. 4, 521–524. (Russian)
4. _____, *On the topology of plane real curves of the 6th order*, Proc. Third All-Union Math. Congr. (Moscow, 1956), vol. 1, Izdat. Akad. Nauk SSSR, Moscow, 1956, p. 149. (Russian)
5. _____, *Bifurcations of simple double points and cusps of real plane algebraic curves*, Dokl. Akad. Nauk SSSR **142** (1962), no. 5, 990–993. (Russian)
6. _____, *Variability of simple double points of real plane curves*, Dokl. Akad. Nauk SSSR **142** (1962), no. 6, 1233–1235. (Russian)
7. _____, *On certain questions in the topology of plane algebraic curves*, Mat. Sb. **58 (100)** (1962), no. 1, 95–127. (Russian)
8. _____, *On the ideas of roughness and degrees of nonroughness for plane algebraic curves*, Mat. Sb. **67 (109)** (1965), no. 4, 481–527. (Russian)
9. _____, *On qualitative methods in the topology of plane algebraic curves*, Abstracts of short scientific reports of Intern. Congr. Math., Section 10, 1966, p. 12. (Russian)
10. D. A. Gudkov, G. A. Utkin, and M. L. Tai, *The complete classification of the nondecomposing curves of the 4th order*, Mat. Sb. **69 (111)** (1966), no. 2, 222–256. (Russian)
11. D. A. Gudkov, *Certain theorems on curves of order m*, Uch. Zap. Gor′kov. Univ. **87** (1969), 5–13; English transl. in Amer. Math. Soc. Transl. Ser. 2 **112** (1978).
12. _____, *Ovals of sixth order curves*, Uch. Zap. Gor′kov. Univ. **87** (1969), 14–20; English transl. in Amer. Math. Soc. Transl. Ser. 2 **112** (1978).
13. _____, *Systems of k points in general position and algebraic curves of different orders*, Uch. Zap. Gor′kov. Univ. **87** (1969), 21–58; English transl. in Amer. Math. Soc. Transl. Ser. 2 **112** (1978).
14. _____, *Properties of rough spaces of sixth order curves with k singular points*, Uch. Zap. Gor′kov. Univ. **87** (1969), 59–85; English transl. in Amer. Math. Soc. Transl. Ser. 2 **112** (1978).
15. _____, *The change of the topology of a sixth order curve under a continuous change of its coefficients*, Uch. Zap. Gor′kov. Univ. **87** (1969), 86–117; English transl. in Amer. Math. Soc. Transl. Ser. 2 **112** (1978).
16. _____, *Complete topological classification of the arrangement of ovals of a 6th degree curve in the projective plane*, Uch. Zap. Gor′kov. Gos. Univ. **87** (1969), Gor′kiĭ, 118–153; English transl., D. A. Gudkov and G. A. Utkin, *Nine papers on Hilbert's 16th problem*, Amer. Math. Soc. Transl. Ser. 2 **112** (1978).
17. _____, *Position of the circuits of a curve of sixth order*, Dokl. Akad. Nauk SSSR **185** (1969), no. 2, 260–263; English transl. in Soviet Math. Dokl. **10** (1969), no. 2, 332–335.
18. _____, *On the topology of plane algebraic curves*, Doctor's Thesis, Gor′kiĭ, 1969, pp. 1–351. (Russian)
19. _____, *Construction of a new series of M-curves*, Dokl. Akad. Nauk SSSR **200** (1971), no. 6, 1269–1272; English transl., Soviet Math. Dokl. **12** (1971), no. 5, 1559–1563.
20. _____, *On the topology of real algebraic varieties*, Abstracts of Sixth All-Union Topological Conf. (Tbilisi, 1972), pp. 43–44. (Russian)
21. _____, *The construction of 6th order curve of the type $\frac{5}{1}5$*, Izv. Vyssh. Uchebn. Zaved. Mat. **130** (1973), no. 3, 28–36; English transl. in Soviet Math. (Iz. VUZ) (1973).

---

[29]Put forward in 1929 by I. I. Vishnevskiĭ, the archivist of Nizhniĭ Novgorod.

22. D. A. Gudkov and A. D. Krakhnov, *On the periodicity of the Euler characteristic of real algebraic $(M-1)$-manifolds*, Funktsional. Anal. i Prilozhen. **7** (1973), no. 2, 15–19; English transl., Functional Anal. Appl. **7** (1973), no. 2, 82–98.
23. D. A. Gudkov, *The topology of real projective algebraic varieties*, Uspekhi Mat. Nauk **29** (1974), no. 4, 3–79; English transl., Russian Math. Surveys **29** (1974), no. 4, 1–79.
24. _____, *Letter to Editor*, Uspekhi Mat. Nauk **30** (1975), no. 4, 300; English transl. in Russian Math. Surveys **30** (1975), no. 4.
25. _____, *The real algebraic variety*, Mathematical Encyclopedia, vol. 2, Moscow, 1979, pp. 70–73; English transl., Kluwer, Amsterdam. (Russian)
26. _____, *The Brusotti theorem for curves on a second order surface*, Uspekhi Mat. Nauk **33** (1979), no. 4, 159–160; English transl. in Russian Math. Surveys **33** (1979), no. 4.
27. _____, *On the topology of algebraic curves on a hyperboloid*, Abstracts of reports in Intern. Topological Conf., Moscow, 1979, pp. 36. (Russian)
28. _____, *On the topology of algebraic curves on a hyperboloid*, Uspekhi Mat. Nauk **34** (1979), no. 6, 26–32; English transl. in Russian Math. Surveys **34** (1979), no. 6.
29. _____, *Generalized Brusotti's theorem for curves on second order surfaces*, Funktsional. Anal. i Prilozhen. **14** (1980), no. 1, 20–24; English transl. in Functional Anal. Appl. **14** (1980), no. 1.
30. D. A. Gudkov and A. E. Usachev, *Nonsingular curves of lowest orders on a hyperboloid*, Methods of the Qualitative Theory of Differential Equations (E. A. Leontovich-Andronova, ed.), Gor′kiĭ State Univ., Gor′kiĭ, 1980, pp. 96–103. (Russian)
31. D. A. Gudkov and E. I. Shustin, *A classification of nonsingular curves of order eight on an ellipsoid*, Methods of the Qualitative Theory of Differential Equations (E. A. Leontovich-Andronova, ed.), Gor′kiĭ State Univ., Gor′kiĭ, 1980, pp. 104–107. (Russian)
32. D. A. Gudkov, *On a fifth-order curve with five cusps*, Funktsional. Anal. i Prilozhen. **16** (1982), no. 3, 54–55; English transl., Functional Anal. Appl. **16** (1982), no. 3, 201–202.
33. D. A. Gudkov and E. I. Shustin, *Invariants of singular points and 5th order curves*, Uspekhi Mat. Nauk **37** (1982), no. 4, 94–95; English transl. in Russian Math. Surveys **37** (1982), no. 4.
34. D. A. Gudkov and L. V. Golubina, *The complete classification of two-point collections of singular points of unicursal curves of fifth order*, Deposited in VINITI, No. 2819-82Dep, 1–11. (Russian)
35. _____, *On the classification of collections of singular points of unicursal plane curves of fifth order*, Sbornik "Differentsial′nye i Integral′nye Uravneniya", Gor′kovskiĭ Universitet, Gor′kiĭ, 1982, pp. 126–132. (Russian)
36. D. A. Gudkov, L. V. Golubina, L. G. Kubrina, and A. V. Zarodova, *Classification of triplets of singular points of unicursal curves of 5th order*, Methods of the Qualitative Theory of Differential Equations (E. A. Leontovich-Andronova, ed.), Gor′kiĭ State Univ., Gor′kiĭ, 1982, pp. 123–134; English transl. in Selecta Math. Soviet. **7** (1988), no. 2, 183–189.
37. D. A. Gudkov, *Points of inflection and double tangents of fourth order curves*. I, Deposited in VINITI, No. 4207-82Dep, 1–9. (Russian)
38. D. A. Gudkov, N. A. Kirsanova, and G. F. Nebukina, *Points of inflection and double tangents of fourth order curves*. II, Deposited in VINITI, No. 17-83Dep, 1–14. (Russian)
39. D. A. Gudkov and E. I. Shustin, *On the intersection of close algebraic curves*, Abstracts of Leningrad Intern. Topological Conf., Leningrad, 1982, p. 58. (Russian)
40. D. A. Gudkov and L. V. Golubina, *Classification of four-point collections of singular points of unicursal curves of 5th order*, Deposited in VINITI, No. 4558-83Dep, 1–9. (Russian)
41. D. A. Gudkov, A. M. Kiselev, and N. K. Komleva, *Classification of five- and six-point collections of singular points of unicursal curves of 5th order*, Deposited in VINITI, No. 5437-83Dep, 1–10. (Russian)
42. D. A. Gudkov and E. I. Shustin, *On the intersection of close algebraic curves*, Lecture Notes in Math., vol. 1060, Springer-Verlag, Berlin, 1984, pp. 278–289.
43. D. A. Gudkov and G. F. Nebukina, *Double tangents and points of inflection of 4th order curves*, Uspekhi Mat. Nauk **39** (1984), no. 4, 112–113; English transl. in Russian Math. Surveys **39** (1984), no. 4.
44. _____, *Points of inflection and double tangents of fourth order curves*. III, Deposited in VINITI, Nc. 704-84Dep, 1–18. (Russian)
45. _____, *Points of inflection and double tangents of fourth order curves*. IV, Deposited in VINITI, No. 6708-B85, 1–23. (Russian)
46. _____, *Points of inflection and double tangents of fourth order curves*. V, Deposited in VINITI, No. 6709-B85, 1–17. (Russian)
47. _____, *Points of inflection and double tangents of fourth order curves*. VI, Deposited in VINITI, No. 6710-B85, 1–26. (Russian)

48. _____, *Points of inflection and double tangents of fourth order curves.* VII, Deposited in VINITI, No. 6711-B85, 1–15. (Russian)
49. _____, *Types and forms of 4th order curves with imaginary singular points*, Uspekhi Mat. Nauk **40** (1985), no. 5, 211; English transl. in Russian Math. Surveys **40** (1985), no. 5.
50. D. A. Gudkov and N. K. Komleva, *On the calculation of invariants of a singular point of an algebraic curve*, Differential and Integral Equations, Gor'kiĭ State Univ., Gor'kiĭ, 1985, pp. 84–86. (Russian)
51. D. A. Gudkov and G. F. Nebukina, *Real curves of fourth order with imaginary singular points*, Deposited in VINITI, No. 1108-B86, 1–22. (Russian)
52. D. A. Gudkov, *Real curves of 4th order: the survey of results*, Abstracts of Baku Intern. Topological Conf., Part II, Baku, 1987, pp. 90. (Russian)
53. D. A. Gudkov and G. M. Polotovskiĭ, *Stratification of the space of the 4th order curves. The joining of stratums*, Uspekhi Mat. Nauk **42** (1987), no. 4, 152; English transl. in Russian Math. Surveys **42** (1987), no. 4.
54. D. A. Gudkov, G. F. Nebukina, and T. I. Tetneva, *Special forms of fourth order curves with imaginary singular points*, Deposited in VINITI, No. 4374-B88, 1–18. (Russian)
55. D. A. Gudkov, *Plane real projective quartic curves*, Lecture Notes in Math., vol. 1346, Springer-Verlag, Berlin, 1988, pp. 341–347.
56. _____, *Special forms of fourth order curves. Part* 1, Deposited in VINITI, No. 9208-B88, 1–36. (Russian)
57. _____, *Special forms of the fourth order curves. Part* 2, Deposited in VINITI, No. 9207-B88, 1–57. (Russian)
58. _____, *Special forms of the fourth order curves. Part* 3, Deposited in VINITI, No. 6435-B89, 1–67. (Russian)
59. _____, *Special forms of the fourth order curves. Part* 4, Deposited in VINITI, No. 1239-B90, 1–55. (Russian)
60. _____, *Special forms of the fourth order curves. Part* 5, Deposited in VINITI, No. 3847-B90, 1–30. (Russian)
61. D. A. Gudkov and G. M. Polotovskiĭ, *Stratification of the space of plane algebraic 4th order curves with respect to algebraic-topological types.* I, Deposited in VINITI, No. 5600-B87, 1–55. (Russian)
62. _____, *Stratification of the space of plane algebraic 4th order curves with respect to algebraic-topological types.* II, Deposited in VINITI, No. 6331-B90, 1–33. (Russian)
63. D. A. Gudkov, *N. I. Lobachevskiĭ. Mysteries of a biography*, Izdat. Nizhegorodskogo Univ., Nizhniĭ Novgorod, 1992. (Russian)

Translated by THE AUTHOR

Recollection of D. A. Gudkov

Evgeniĭ I. Gordon

Once, not long before Dmitriĭ Andreevich died, as we were strolling out of the University, he told me, "I understood that what we have in this country is fascism back in 1938."

"How did that occur, Dmitriĭ Andreevich?"

"You see, at that time I had a lot of respect for Bukharin. It is now that I realize that he was no better than the others, but at that time I greatly respected him. So, when during the trial the newspapers were writing that he had ordered to add crushed glass to butter, I understood this kind of lie was a manifestation of fascism, and that one had to be very cautious and not speak frankly with anybody."

This and many other conversations of that time revealed many new features in Dmitriĭ Andreevich's character to me. I understood that all his life was given to consciously opposing fascism, fighting it the way he thought necessary.

He was not a fighter against the political system, nor was he dissident. I do not remember him ever saying anything against the regime. Even when he would come to our house in the evening to read *samizdat* (my parents always had something "fresh"), he would always leave without saying a word about what he had read, even though he read everything, some things for several evenings. Although to some extent this was in contrast with his absolutely fearless attitude towards the University heads, this made me, then a young man, distressed and uneasy. Now I understand that Dmitriĭ Andreevich thought his goal to be different: he tried to preserve around him a spirit of decency and honesty, to save science from decay and corruption. In the last decades, the fascism that Dmitriĭ Andreevich spoke about was already rotting and poisoning everything around, including science and culture. Provinces were suffering most; in the capitals there was still a critical mass of real scientists.[1]

This was felt very strongly in Gor'kiĭ University, especially in the field of mathematics. For many years the leading scientists, now the pride of the University, were badgered by mediocrities occupying the leading administrative or Party posts. In 1951, after his pedagogical activity had been "discussed" for a long time (actually, this was the result of a series of political denunciations), professor A. G. Meyer, Gudkov's scientific supervisor, a brilliant teacher and a favorite of the students, died of a stroke. In 1959, professor A. G. Sigalov, a prominent mathematician who contributed much

---

1991 *Mathematics Subject Classification.* Primary 01A70.

[1] In some provinces the situation was also different, depending on how zealous local authorities were.

to the investigation of the 19th, 20th, and 23rd Hilbert problems and created a well-known school in the calculus of variations, was forced to leave the University after several anti-Semitic provocations.

I think, though, that anti-Semitism was not the main motive in this case. Other but equally vicious means were used against A. G. Meyer, who was not a Jew (his ancestors were Germans, in fact one of them was described by M. Yu. Lermontov in "The Hero of Our Time"). He was accused of "political and ideological mistakes" in his lectures on the history of mathematics, reprimanded for his divorce, etc. The administrators did not disdain any means. Here is one of the scenes that Dmitriĭ Andreevich told me about concerning this badgering. During a meeting of the Chair of Mathematical Analysis presided by Meyer, two members of the Chair suddenly entered, completely drunk, stamping their feet. They sat down and began to discuss their own problems out loud. Meyer demanded that they should leave, but they rudely refused to go. After this had been going on for some time, Meyer canceled the meeting. Next day he was officially reprimanded for "disruption of the Chair meeting".

Dmitriĭ Andreevich opined that those members had been ordered to do so by the Party. I doubt it was a direct order: it was the atmosphere, in which such actions were encouraged by the authorities. Dmitriĭ Andreevich, however, thought that those who organized this campaign were consciously badgering Meyer, driving him to death (Meyer's serious hypertension problem being quite well-known).

Certainly, those people had their own purely mercantile aims. After the real scientists and teachers were driven out from the Department of Mechanics and Mathematics, they were the ones who became the heads of the Chairs, and were doing nothing but zealously striving to get rid of gifted young people, who were a constant potential threat to them.

That mathematics at the faculty was not totally destroyed was to a considerable extent due to D. A. Gudkov. Of course, he succeeded only after the situation became more tolerable, after the "Krushchev thaw", which became apparent at the University when Korshunov, a competent researcher in chemistry and honest man, became its Rector in 1960 and made great efforts to help the mathematicians who were still working there. He actively supported the opening of a Department of Computational Mathematics and Cybernetics in 1963, and the Chair of Mathematics at the Department of Radiophysics in 1961. Gudkov was proposed as its head.

Dmitriĭ Andreevich's own scientific interests were far from physics. From the very start, he thought that his work at the Department of Radiophysics was not permanent, and envisaged a rapid return to the Department of Mechanics and Mathematics. His main idea was to preserve mathematical science in the University. The matter was that the atmosphere at the Department of Radiophysics, which was founded 15 years before by Academician A. A. Andronov and Professor M. T. Grekhova, was much better. I think that this was due to several reasons. Since radiophysicists worked for important defense projects, they were subjected to much less ideological pressure. The very fact that they took part in military research was considered as evidence of their loyalty, and that, actually, was true. Besides, in these projects real results were required (at least at that time), and they could not be obtained just by being devoted to the Party. This fact, and, certainly, personalities of the organizers of that Department, made it possible to gather a considerable number of efficient and honest specialists there.

Having become the head of the Chair of Mathematics, Dmitriĭ Andreevich began to develop the mathematics syllabus for physicists, with all his thoroughness. Being a very broad-minded person, ready to work in close contact with physicists and to take into account their interests and demands, but without lowering the high level of mathematical culture (and that was not easy, even at the Radiophysics Department), he succeeded in raising the level of mathematical education to a considerable height, and that was felt long after Dmitriĭ Andreevich came back to the Department of Mechanics and Mathematics.

Being the head of the Chair, Dmitriĭ Andreevich thought it his duty to protect his collaborators and provide suitable working conditions for those whose work was efficient and corresponded to the high scientific level of the Chair. He was always firm in getting rid of persons who did not meet the requirements of the University, and his actions in this case were always open and well-grounded. Sometimes he helped those expelled to find other employment, better suited to their abilities.

Owing to his honesty, adherence to principles, his feeling of equity and justice, openness and benevolence, and his scientific qualifications, he was greatly respected by members of his Chair. This did not stop them, however, from arguing with him about various problems, since Dmitriĭ Andreevich was very easy-going.

Thus, the Chair had its own specific atmosphere, quite unusual for a standard Soviet University chair, that of mutual benevolence, keenness on pedagogical and scientific activities, and an atmosphere of calm confidence. The members of the Chair felt protected by D. A. Gudkov. He was always fearless in defending justice and used his reputation in academic circles and favorable (for the Party and administration) personal particulars: he was ethnically Russian, a member of the Party, and a war veteran.[2]

D. A. Gudkov became a member of the Party after returnig from the army and left it as soon as it was possible (in 1991).

Dmitriĭ Andreevich always told the truth courageously and publicly, not caring about personalities in everything that did not concern politics. This was, of course, a very rare thing and greatly irritated the authorities. Here is one episode that characterizes his behavior; I was also involved. When I was still an undergraduate in 1971, Dmitriĭ Andreevich decided to hire me as a member of his Chair. At that time there were firm instructions for the Party leaders (though, or course, not in writing) not to give employment to Jews, especially as teachers, since such employment in Soviet society was always considered closely connected with ideology.[3]

That was why the heads of the University did not give their permission for half a year. Finally Dmitriĭ Andreevich gained it, but not before he asked many times for explanations. At that time, many people tried to employ ethnic Jews. As a rule, they were told meaningfully, "Unfortunately, we cannot do it at present. You understand ...." Most of the Soviet people would catch the meaning on the uptake and leave silently. Those who were not so "quick-minded" were explained directly that there is an instruction "from above" not to give employment to Jews. Nobody would dare to protest against it in public.

But as for Dmitriĭ Andreevich, the authorities were afraid to tell him the real reason, although he knew it for sure. They just doubted that he would keep silent. And he would not have kept silent, he would have spoken against it at an open Party meeting, and they would have had to answer, which was not so easy, since officially,

---

[2] In the Soviet Union it was a necessary (and quite often a sufficient) condition for a successful career to have at least the first two of these particulars in one's CV.

[3] To be completely fair it should be noted that there was no racial discrimination in the admission of students to the Gor′kiĭ University.

the Party proclaimed "proletarian internationalism", equality and brotherhood of all nations, and so on, and so forth.

At that time only a handful of people had the courage for such outspokenness. Dmitriĭ Andreevich could not permit himself to speak out against the regime openly, not in any form: this meant that he would have given the authorities a weapon, and, in the long run, that he would have been deprived of the opportunity to protect people and protect science at the University, which he considered, as I have mentioned already, his main goal. Besides, he evidently thought such acts to be absolutely useless self-denunciations. In this connection I would like to recall another episode.

Once, in the middle of the 70ths, a group of Moscow Jewish "otkazniks", before the beginning of a seminar on human rights held illegally in Moscow, issued a questionnaire and distributed it among Soviet ethnical Jews. It contained mostly innocent questions, like "Would you like your children to study Hebrew?" or "Would you like Jewish cafés to open?", etc. Some such questionnaires reached Gor'kiĭ. Among the people who filled them in, there were two students of the Radiophysics Department, the Kovner sisters, whose mother also worked at the department. The KGB made up a loud political process out of this. Local papers wrote about a "Zionist plot". The sisters were expelled from the University, and their mother was forbidden to teach, after a huge meeting of teachers and students of the Radiophysics Department. Dmitriĭ Andreevich, evidently, was afraid that I would come to this meeting and speak in their defense (though, frankly, my rather "loyal" behavior gave no grounds for such fears), and two days before the meeting he called me up and told me, almost literally, the following words, "Zhenya, I ask you not to attend the meeting. If you go there, you will speak up, and that is absolutely senseless and won't help them. Please understand that this meeting is a provocation and is especially organized to reveal those who side with them."

Having been granted this deliverance, I did not attend the meeting, but Dmitriĭ Andreevich went there, and even had to mutter a few neutral words that could be treated as disapproval: obviously, he and other members of the Party had been bullied into saying something there.

Certainly, one can assess this episode differently. I think that there will be many, especially among those who were close to dissidents, who would condemn Dmitriĭ Andreevich.

But not all honest people were dissidents, and not only dissidents were honest and courageous citizens, people whose behavior greatly contributed to the preservation of culture and science in our country. Certainly, they had to compromise, and to decide to what extent was always a very difficult problem; everyone solved it for himself. As far as I know, that episode was the greatest compromise that Dmitriĭ Andreevich permitted himself to make. For this scene to be understood more clearly, it should be said that the Kovners, by that time, had already applied for emigration to Israel. The permission was given unusually fast (1976), and Dmitriĭ Andreevich was absolutely right in his assessment of the main purpose of the meeting.

Though Dmitriĭ Andreevich did not have the possibility of saying the truth, he, at least in my memory, never said a word of falsehood. The impossibility of telling the truth publicly was very hard for him. In general, he was a social person. His inner freedom was not enough for him, but he missed the possibility to express his opinion openly very much.

This was especially clear in his last years, when Dmitriĭ Andreevich, earlier than many others, began to speak directly and openly about what he thought and had always thought about the regime. Everyone could see how it pleased him. I should say that Dmitriĭ Andreevich was delighted by the changes that had occurred, by the fact that fascism had been destroyed. And he paid absolutely no attention to new material difficulties, which of course were not as harsh as those under which he had led the greater part of his life. The only thing that troubled him was that many good mathematicians, his colleagues and collaborators, were leaving the country. Certainly, he did not disapprove of anyone's decision, and treated it with complete understanding, but it was surely hard for him.

It should also be said that during the Brezhnev "stagnation period" Dmitriĭ Andreevich's behavior also caused constant disapproval by the authorities. I remember that in 1983, during Gudkov's re-election as Head of the Chair of Geometry and Algebra, A. G. Ugodchikov, then the Rector of the University, began to put forward various obstacles. The election was postponed several times; Gudkov was said to have failed to fulfill certain queer requirements (like the introduction of computers to education, though they were then practically unavailable) and his scientific activity was described as lacking practical applications, etc. Nevertheless at first everything was going all right. Dmitriĭ Andreevich later recalled that all the trouble started at a meeting of the University Council, when the Rector had just returned from a meeting of the regional Party committee and suddenly attacked Gudkov with great irritation, this giving rise to all the subsequent developments.

Before then the relationship between Gudkov and the Rector had always been quite normal, so Dmitriĭ Andreevich put forward the following version of what had happened. Shortly before that, he had been to Leningrad at a topological conference. There he went to the Party committee of Leningrad University and began, "as a communist", to demand giving re-employment to Vladimir Abramovich Rokhlin, who had been suddenly pensioned off. Since high Party spheres had something to do with this (as far as I know, Rokhlin gave only a satisfactory grade at an examination to the daughter of G. V. Romanov, then the First Secretary of the Leningrad regional Party committee, later a member of the Politbureau), the local communists reacted violently and began to shout at Gudkov, undoubtedly having first found out who he was. Dmitriĭ Andreevich also answered in sharp words. Probably, only owing to the fact that Academician A. V. Gaponov-Grekhov defended him at the Gor'kiĭ regional Party committee (Professor M. A. Miller, at my request, asked him to do so, while Dmitriĭ Andreevich was obstinate in not contacting anyone, and tried to stop me from doing anything about this), Gudkov stayed at the head of the Chair.

Such collisions with the authorities were constant and were detrimental to the force and health of Gudkov. I remember that once, when there were only two of us in the Chair room, a woman came in (she was probably from the War Museum) and asked Dmitriĭ Andreevich, as a war veteran, to make a speech for students on an anniversary of the Great Victory. Dmitriĭ Andreevich answered to her in a tired and sad voice, "You know, I am not in the mood to make speeches. I am a peaceful man. It's been forty years since the war ended and it seems to me that in these forty years I have been much more useful than during the war. And I have been badgered all these years. And only before Victory Day do they remember how good I am and ask me to tell about the war. As you like, I can make a speech, but I will tell everything I know and everything I think about it." That did not suit the woman, and she left disappointed.

Dmitriĭ Andreevich worked at the Radiophysics Department for 17 years, until 1978. All this time he stayed in the course of everything that was going on at the

Department of Mechanics and Mathematics, and always would come to help. For example, when a group of young talented geometers had trouble there, they were at once adopted by the Radiophysics Department, and later came back, together with Gudkov. Actually, it was Dmitriĭ Andreevich who made it possible to form a strong team of mathematicians, students of Professors A. G. Sigalov and V. I. Plotnikov, at the Chair of Mathematical Physics. When he became the head of the Chair of Geometry and Algebra, Dmitriĭ Andreevich actively supported the algebraists who had worked there already, in particular M. I. Kuznetsov, to whom Dmitriĭ Andreevich entrusted the Chair in 1988.

These notes give my impressions only about one side of Dmitriĭ Andreevich Gudkov's activities. I could tell about many other things. He was a very broad-minded, delightful, and simply very kind person. I think that I was very lucky to know him for more than twenty years.

Translated by THE AUTHOR

Remarks on the Enumeration of Plane Curves

V. I. Arnold

In memory of Professor D. A. Gudkov

ABSTRACT. Combinatorial problems related to the classification of the generic immersions of a circle into the plane are discussed. They are reduced to the study of special functions on vertices of trees and these special functions are enumerated for the simplest trees. Tables of $1 + 2 + 6 + 19 + 76$ curves on the sphere having $1, \ldots, 5$ double points and of irreducible curves with up to 7 double points are presented.

The set of theorems formulated below is a by-product of the enumeration of the generic immersions of a circle to a plane up to diffeomorphisms of the circle and of the plane (see [1]). They might be formulated in a purely combinatorial way as statements about words in the symmetric group, mostly in the group of permutations of four elements. However I shall present these theorems in their initial form, which is simpler to perceive.

§1. Hangable pendants

The description of the image of a generic immersion of a circle into the plane starts from the decomposition of this image into the exterior contour ("*cactus*") and the rest, hanging inside the loops of the exterior contour.

DEFINITION. A *pendant* is the pair formed by a generic immersion of a circle into the plane and by n generic *distinguished points* on its exterior contour.

In the sequel equivalences are homeomorphisms of the plane.

Any pendant may be transformed by a homeomorphism of the plane into an equivalent pendant, inscribed into the unit disk in such a way that the distinguished points lie exactly on the boundary circle of the disk. The union of the inscribed curve with the ascribed boundary circle have only double point singularities (up to plane homeomorphisms). Hence it is equivalent to the image of a generic immersion of a disjoint union of circles into the plane.

The union curve is unique (up to a homeomorphism of the plane).

1991 *Mathematics Subject Classification*. Primary 53A04; Secondary 05A15.

The decomposition of the union curve into the images of the circles is unique. For instance, the number of immersed circles is unambiguously defined by the initial pendant. The union curve is called irreducible if this number is equal to one.

DEFINITION. A pendant is *hangable* if its union with the ascribed circle is irreducible, i.e., if this union is equivalent to the image of a generic immersion of a circle into the plane (the equivalence being a homeomorphism of the plane).

EXAMPLES. The curve "8" with two distinguished points is hangable iff the distinguished points are separated by the double points (Figure 1). The curve "○" with n distinguished points is hangable iff the number of points is odd. The three pendants at the right in Figure 1 are not hangable.

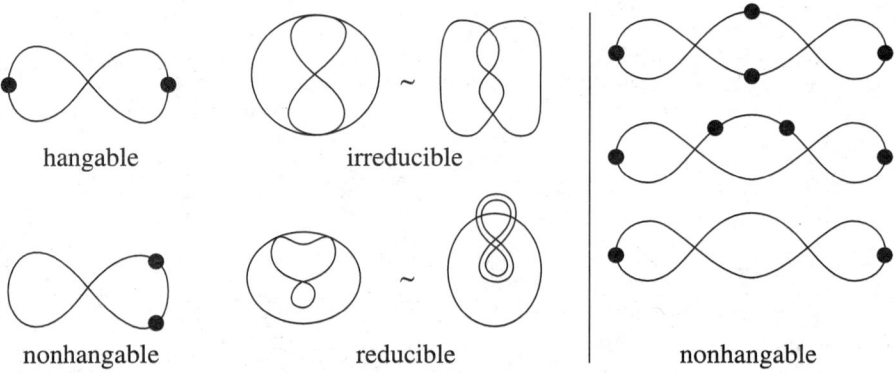

FIGURE 1. Hangable and nonhangable pendants

REMARK. The property of being hangable is topological. Hence the double points of the immersions are sometimes shown (in Figure 1 and in the sequel) as points of self-tangency (being in fact crossings, topologically equivalent to "×").

Now we start to investigate which pendants are hangable and which are not.

§2. The two elementary moves

The first move is the elimination of two neighboring distinguished points.

LEMMA 1. *A pendant containing two distinguished points (which are not separated by a distinguished or double point on the exterior contour) is hangable if and only if the pendant obtained from it by deleting these two points is hangable.*

PROOF. See Figure 2.

The second move is the jump of a distinguished point over a double point.

DEFINITION. A double point of a generic immersion of a circle into the plane is called *separating* if the two loops into which the curve is divided by this point do not intersect each other.

EXAMPLES. The double points of the "8" curve and of the other curve having one double point are separating, those of the trefoil curve are not.

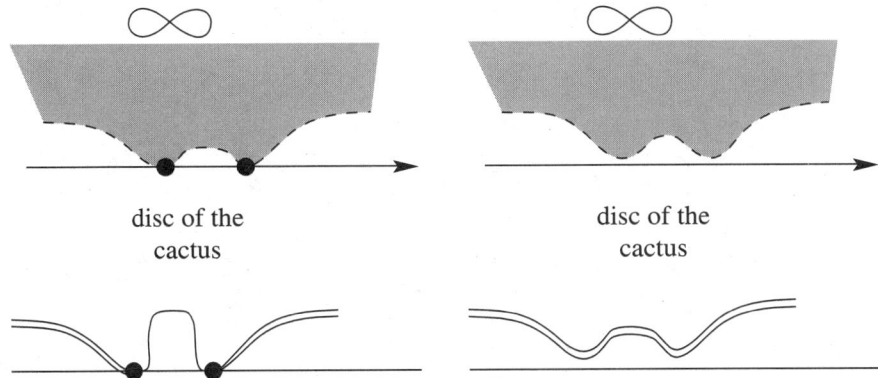

FIGURE 2. Cancellation of two distinguished points

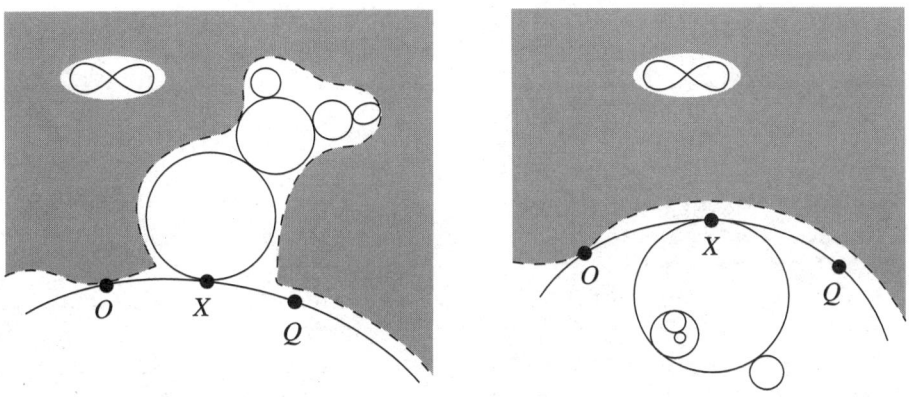

FIGURE 3. Jump of a distinguished point over a separating point

Consider a pendant having a separating point X (Figure 3). Suppose that it has a distinguished point O, which is not separated from the separating point X by any other double point or distinguished point.

In this case both end-parts of the loop containing the distinguished point O near the separating double point X belong to the exterior contour of the pendant (Figure 3). Choose a new point Q on the other end of the loop (Q is neither distinguished nor double and there are no double or distinguished points between Q and X).

THEOREM 1. *The pendant containing the distinguished point O is hangable if and only if the pendant obtained from it by the deletion of the distinguished point O and by the inclusion of a new distinguished point Q is hangable.*

In other words, a distinguished point of a pendant may jump over a separating point without disturbing the hangability of the pendant.

PROOF. Suppose that the pendant is hangable and choose the orientation on the union curve (according to an orientation of the unique circle immersed onto the union curve). We shall use the representation in which the exterior contour of the pendant and the ascribed circle forming the union curve are tangent at the distinguished points.

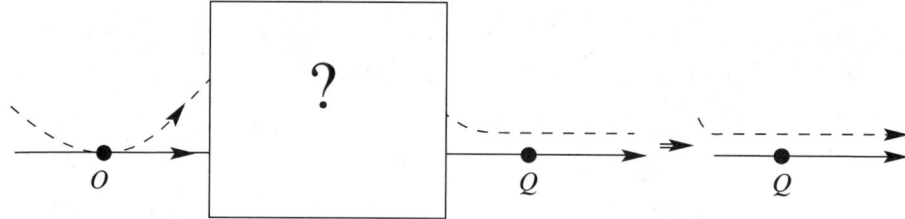

FIGURE 4. Parallelism at O implies parallelism at Q

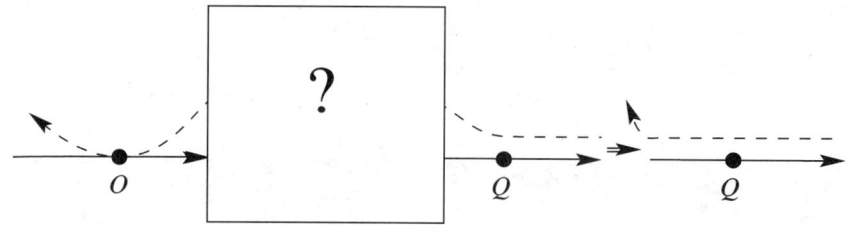

FIGURE 5. Antiparallelism at O implies antiparallelism at Q

There are two possibilities:

(i) *The orientations of the two branches of the union curve crossing at O are equal* (Figure 4), *say from O to X.*

In this case the orientations of both branches passing near Q are also parallel (and directed from X to Q if the first orientation was from O to X).

Indeed, we can extend the second loop in a box (a disk, whose boundary will intersect the first loop and the described circle transversally) near O and near Q, and exactly once.

If the curve enters the box twice (near O), it should leave it twice near Q.

(ii) *The orientations at O are opposite* (Figure 5).

Then they are opposite near Q, too (by the same "box argument").

In case (i), cover the union curve, with the exception of the neighborhoods of O and Q, by two boxes (one containing one of the loops and the other containing the other loop (Figure 6)).

The behavior of the union curve inside the boxes is not known. However there are two branches entering each box and two leaving it, hence each box defines a permutation of two elements. We denote by P_i the permutation corresponding to the ith box. The neighborhoods of O and Q also define such permutations P_O and P_Q: transposition at O, a trivial one at Q. The total monodromy is $M = P_1 P_Q P_2 P_O$.

The hangability condition in these notations reads: M is nontrivial.

But the group of permutations of two elements is commutative. Hence if M is nontrivial (respectively trivial), it will remain nontrivial (respectively trivial) when one permutes P_Q and P_O (making the permutation at the neighborhood of O trivial and at the neighborhood of Q nontrivial).

Thus in (i) the irreducibility of the union curve does not change when the distinguished point O jumps across the separating point X.

In case (ii), represent the jump of the distinguished point O to the new position Q as the product of two moves: first remove O and then create Q.

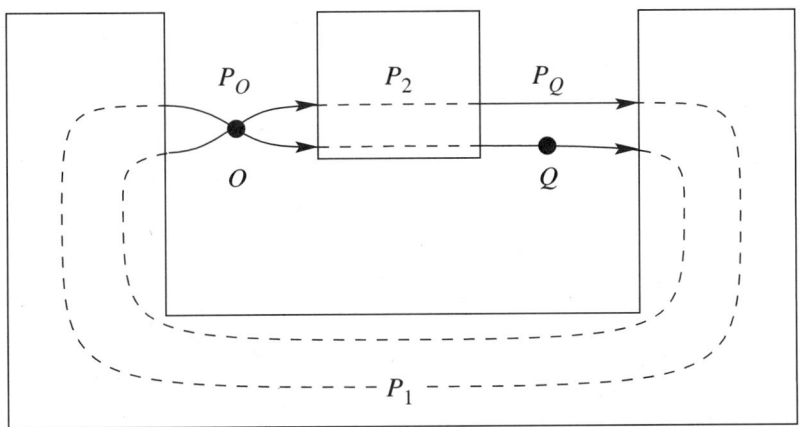

FIGURE 6. The two boxes connected by four segments

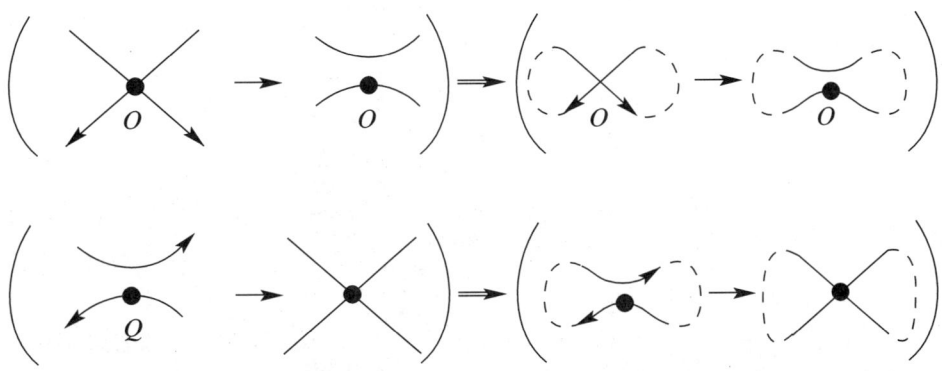

FIGURE 7. The removal of O and the creation of Q

Each of the two operations does not change the irreducibility of the union curve in case (ii) (see Figure 7). Theorem 1 is thus proved.

§3. Cacti and tree-like curves

DEFINITION. A generic immersion of a circle into the plane is *tree-like* if all its double points are separating (Figure 8). A plane curve is *tree-like* if it is equivalent to the image of a tree-like immersion (by a homeomorphism of the plane).

EXAMPLE. The exterior contour of any generic immersion of a circle into the plane is a tree-like curve.

DEFINITION. A *circular cactus* is the boundary of a contractible union of a finite set of closed disks in the plane, tangent exteriorly. A *cactus* is any plane curve equivalent to a circular cactus (a curve which may be transformed into a circular cactus by a homeomorphism of the plane).

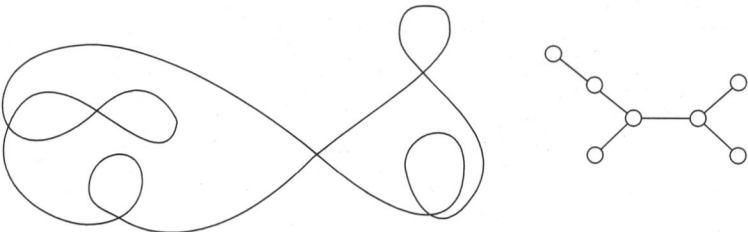

FIGURE 8. A tree-like curve

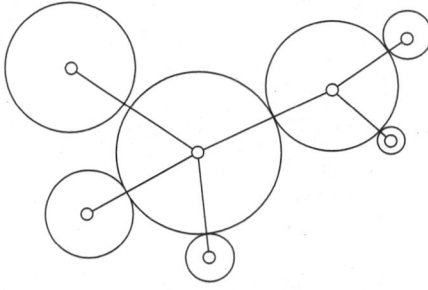

FIGURE 9. A circular cactus

REMARK. The centers of the disks and the segments joining the centers of the tangent disks of a circular cactus form a tree (Figure 9).

Any cactus is a tree-like curve.

EXAMPLE. The exterior contour of any generic immersion of a circle into the plane is a cactus. We shall call it the *boundary cactus of the immersion*.

REMARK. The boundary cactus structure of the exterior contour does not contain information on those double points of the initial curve which do not correspond to the tangency of the circles of the cactus.

THEOREM 2. *Whether a tree-like curve with n distinguished points on the exterior contour is hangable or not depends only on the parities of the numbers of distinguished points on each disk of its boundary cactus.*

PROOF. Using Lemma 1 and Theorem 1, we can transform configurations of distinguished points to each other provided that the parities of the numbers of the points on each disk are equal: Theorem 1 allows us to jump over the double points along a disk of the boundary cactus, and Lemma 1 kills pairs of neighboring points.

EXAMPLE. If the pendant is a cactus, its hangability depends only on the parities of the numbers of the distinguished points on its disks.

REMARK. Theorem 2 (with its proof) holds for more general classes of curves:
(i) for all curves for which all the double points on the exterior contour are separating;
(ii) it is even sufficient that every loop leaving the exterior contour at a double point never meets it before the return to the same double point.

§4. Hangable pendants of type A_n

Consider a plane tree and a function with values 0 or 1 defined on its vertices.

EXAMPLE. A cactus-pendant defines such a function: it associates to each disk the parity of the number of distinguished points on its boundary.

DEFINITION. This function will be called the *characteristic* of the cactus-pendant.

DEFINITION. A characteristic is *good* if the corresponding cactus-pendants are hangable (and *bad* in the opposite case).

It follows from §3 that the property of being hangable or not for a cactus-pendant depends only on its characteristic.

EXAMPLE. The characteristics (1) and $(1, 1)$ on trees with one or two vertices are good, while $(1, 1, 1)$ is bad (Figure 10).

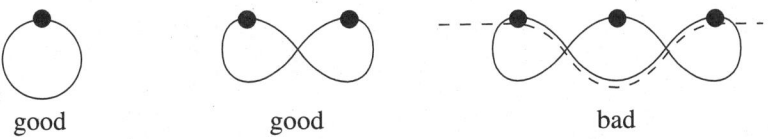

FIGURE 10. Good and bad characteristics

The same arguments show that the characteristic $(1, 1, \ldots, 1)$ on the tree $A_n = \cdot - \cdot - \cdot \cdots - \cdot$ with n vertices is bad if and only if $n = 3k$.

From now on I consider the characteristic of the tree A_n, whose vertices are labeled by the numbers $(1, 2, \ldots, n)$ (in this order) once and for all. The goal is the description and the enumeration of the good characteristics. The enumeration of good and bad characteristics on nonoriented A_n-trees also follows from the results presented below.

A characteristic on the tree A_n is a sequence of n binary digits.

DEFINITION. The *character* τ of a characteristic on the tree A_n is the mod 3 residue of the number
$$\tilde{\tau} = \beta_1 - \beta_2 + \beta_3 - \cdots,$$
where the integers β_i are defined by the following construction. In the characteristic delete the maximal subsequences of even numbers of zeros. The resulting shorter sequence is formed by the series of 1's separated by the isolated zeros. The numbers β_i are the numbers of 1's in the subsequent series (reading the characteristics from the left). The zeros at the ends of a characteristics do not change the numbers β_i.

EXAMPLE. $\tau(1, 1, 1, 0) = 0$, $\tau(1, 0, 1, 0) = 0$, $\tau(0, 1, 1, 0) = 2$.

THEOREM 3. *A pendant cactus whose tree is A_n is hangable if and only if its character is different from zero.*

PROOF. We may represent the cactus as a horizontal chain of disks formed by two threads intersecting at $n - 1$ points. We may choose the distinguished points at the upper part of some disks (Theorem 1 of §2). Then the ascribed circle will consist of two threads, one higher and the other lower than the cactus, connected with the upper part of the cactus at the distinguished points and with each other at the left and at the

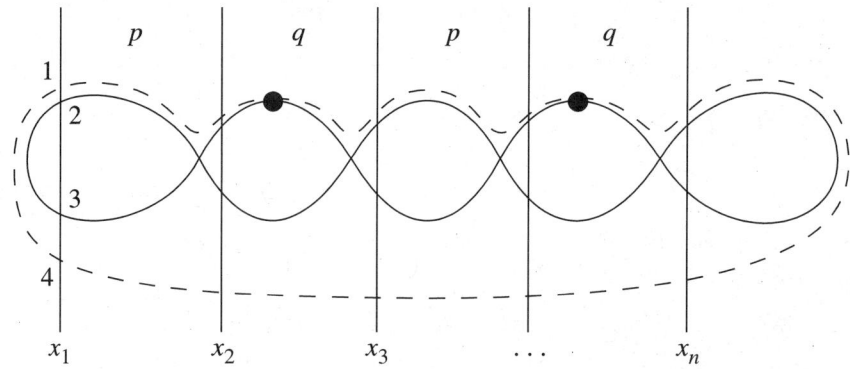

FIGURE 11. Construction of a permutation from a pendant cactus

right end (Figure 11). We may suppose that there are no distinguished points on the left most and on the right most disks.

The generic vertical lines, $x = $ const, crossing the cactus have four common points with its union with the ascribed curve. Denote these points by $(1, 2, 3, 4)$ $(y_1 > y_2 > y_3 > y_4)$ for every vertical line. Choose for every disk of the cactus a generic vertical line, crossing it near the left most point.

Now we can associate to each disk, except the right most one, a permutation of three elements $(1, 2, 3)$, continuing every thread from the line $x = x_i$ to the line $x = x_{i+1}$ making a crossing at each double or distinguished point.

If the disk contains no distinguished point, the permutation is

$$p = \text{the transposition of 2 and 3.}$$

If the disk contains one distinguished point, the permutation is

$$q = \text{the cyclic permutation } (1, 2, 3) \to (3, 2, 1).$$

Consider the product of all the permutations (from x_1 to x_2, from x_2 to x_3, ..., from x_{n-1} to x_n). The pendant cactus is hangable if and only if this product of permutations does not send the thread 1 to 1. Indeed, the threads 2 and 3 are connected at the left end, as also are 1 and 4. Hence if the left 1 is sent at the right end to 2 or 3, then 2 or 3 of the left end will be sent to 1 at the right end. Hence the threads 2 and 3 at the left end are continuations of the 1 of the right end and hence are connected with 4 at the right end, hence with 4 at the left end, hence with 1 at the left end and thus all 4 threads form one irreducible immersion with crossings at all double points.

If the left 1 is sent to the right 1, then the immersion is reducible ($(1, 4)$ and $(2, 3)$ belong to different circles).

Since $p^2 = 1$ and $q^3 = 1$, we may reduce the cactus, deleting consecutive pairs of disks with no distinguished points and consecutive triples of disks containing one distinguished point; these operations do not change the hangability. Now the product permutation takes the form $p^\omega q^{\beta_k} p q^{\beta_k - 1} p \cdots p q^{\beta_1} p^\alpha$, where α and ω are 0 or 1, and $\beta_i = 1$ or 2.

Since p leaves 1 fixed, the p's at the left end at the right end are not important. Since $pqp = q^2$ and $pq^2p = q$ (a reflection of a triangle reverses the direction of its rotation) our product may be written in the form

$$p^\omega \cdots (pq^{\beta_4}p)q^{\beta_3}(pq^{\beta_2}p)q^{\beta_1}p^\alpha = p^\omega q^{\beta_1-\beta_2+\beta_3-\cdots}p^\alpha = p^\omega q^\tau p^\alpha.$$

This permutation leaves 1 fixed if and only if q^τ does, i.e., if and only if $\tau = 3k$. This proves the theorem.

§5. Enumeration of good characteristics

DEFINITION. A function $\{1,\ldots,n\} \to \{0,1\}$ is called *good* if its character does not vanish, and *bad* if it vanishes (the character of such a function is defined in §4; I switch from "characteristics" to "function" to emphasize the independence of the notion of goodness of any immersions, pendants, cacti, etc.).

EXAMPLE. Among the 8 characteristics of length 3 there exist 5 good ones, namely

$$(1\ 1\ 0),\ (1\ 0\ 0),\ (0\ 1\ 1),\ (0\ 1\ 0),\ (0\ 0\ 1).$$

Among the 16 characteristics of length 4 there exist 5 bad ones, namely

$$(1\ 1\ 1\ 0),\ (1\ 0\ 1\ 0),\ (0\ 1\ 1\ 1),\ (0\ 1\ 0\ 1),\ (0\ 0\ 0\ 0).$$

THEOREM 4. *The numbers $G(n)$ and $B(n)$ of good and of bad characteristics depend on n the following way*:

n	1	2	3	4	5	6	7	...
$G(n)$	1	3	5	11	21	43	85	...
$B(n)$	1	1	3	5	11	21	43	...

In particular, the following identity holds:

$$G(n) = B(n+1).$$

REMARK. Since $G(n) + B(n) = 2^n$, the theorem implies the formula

$$B(n) = (2^n - (-1)^n)/3.$$

THEOREM 5. *A bijection between the sets of bad characteristics of length $n+1$ and of good ones of length n is given by the following operation.*

Choose any pair of neighboring places in the sequences of 0 and 1 of length $n+1$ (say, the pair consisting of the ith and of the $(i+1)$th elements).

Replace the pair by 1 if the values of the characteristics at these two places are equal and by 0 otherwise.

EXAMPLE. $(i, i+1) = (1,2)$, $n=4$:

$$(1\ 1\ 1\ 0) \to (1\ 1\ 0),\quad (1\ 0\ 1\ 0) \to (0\ 1\ 0),\quad (0\ 1\ 1\ 1) \to (0\ 1\ 1),$$
$$(0\ 1\ 0\ 1) \to (0\ 0\ 1),\quad (0\ 0\ 0\ 0) \to (1\ 0\ 0).$$

The proofs of Theorems 5–8 are elementary and are left to the reader.

To enumerate the nonoriented characteristics, it is sufficient to enumerate the symmetric good and the symmetric bad characteristics. Denote their numbers by $SG(n)$ and $SB(n)$.

EXAMPLE. Among the four symmetric characteristics of length 4, exactly one is bad, namely (0 0 0 0). Thus $SB(4) = 1$.

THEOREM 6. *The numbers of good and of bad symmetric characteristics depends on n the following way*:

n	1	2	3	4	5	6	7	8	9	...
$SG(n)$	1	1	1	3	3	5	5	11	11	...
$SB(n)$	1	1	3	1	5	3	11	5	21	...

In particular, the following identities hold:

$$SG(2n) = G(n), \quad SG(2n-1) = B(n),$$
$$SB(2n) = B(n), \quad SB(2n-1) = G(n).$$

THEOREM 7. *A bijection between the set of good (respectively bad) characteristics of length n and the set of good (respectively bad) symmetric characteristics of length $2n$ is given by the reflection operation.*

EXAMPLE. The 3 bad characteristics (among the 8 characteristics of length 3) are

$$(1\ 1\ 1),\ (1\ 0\ 1),\ (0\ 0\ 0).$$

The reflection transforms them into the 3 characteristics of length 6

$$(1\ 1\ 1\ 1\ 1\ 1),\ (1\ 0\ 1\ 1\ 0\ 1),\ (0\ 0\ 0\ 0\ 0\ 0).$$

These are all the bad characteristics among the 8 symmetric characteristics of length 6.

THEOREM 8. *A bijection between the set of bad characteristics of length n and the set of good symmetric characteristics of length $2n-1$ is given by the reflection followed by the replacement of the two central digits by one digit, namely* 1.

EXAMPLE. The 3 bad characteristics among the 8 characteristics of length 3 are transformed into the 3 good symmetric characteristics among the 8 symmetric characteristics of length 5, namely

$$(1\ 1\ 1) \to (1\ 1\ 1\ 1\ 1), \quad (1\ 0\ 1) \to (1\ 0\ 1\ 0\ 1), \quad (0\ 0\ 0) \to (0\ 0\ 1\ 0\ 0).$$

§6. Applications of the enumeration of curves

A *curve* below means a generic immersion of a circle into the sphere or into the plane, whose only singularities are n double points.

First consider the classification of spherical curves. Fix the component of the complement to the curve having the maximal number m of sides. Put ∞ into this component. The spherical curve is then represented as a plane curve, for which the unbounded component of the complement has the maximal numbers of sides, m.

The n double points divide the curve into $2n$ parts, hence $m \leqslant 2n$. If $m = 2N - k$, then the curve consists of the exterior contour, which is a cactus, and of the interior parts, formed by k segments, which hang in the disks of the cactus. These interior parts are connected with the exterior contour at d distinguished points. Thus the equality $m = 2v - 2 + d$ holds, where v is the number of disks of the cactus (of the vertices of its tree).

THEOREM. *The curves with $m = 2n$ are enumerated by the plane trees consisting of $n + 1$ vertices.*

EXAMPLE. There exist 6 plane trees with 6 vertices, shown in Figure 12. They enumerate the 6 curves having $m = 2n = 10$.

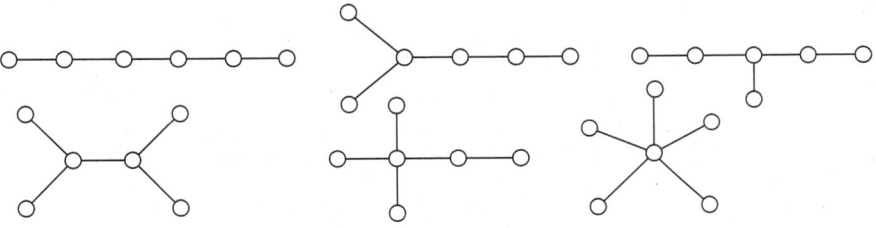

FIGURE 12. Enumeration of curves having $m = 2n = 10$ exterior edges

PROOF. Since $k = 0$, the curve having $m = 2n$ is a cactus with no distinguished points, coinciding with its exterior contour.

THEOREM. *The curves with $m = 2n - 1$ are enumerated by the plane trees consisting of $n + 1$ vertices, with a marked end point.*

EXAMPLE. There exist 10 plane trees having 6 vertices with a marked end point, shown in Figure 13. They enumerate the 10 curves having $m = 2n - 1 = 9$.

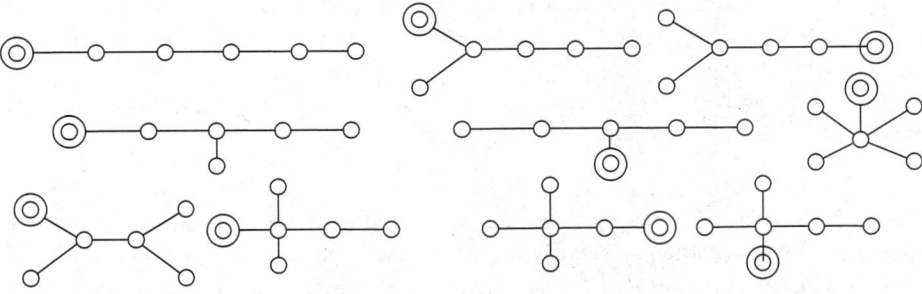

FIGURE 13. Enumeration of curves having $m = 2n - 1 = 9$ exterior edges

PROOF. Since $k = 1$, a curve having $m = 2n - 1$ is (up to equivalence) a circular cactus to which a small circle is attached, tangent interiorly to the boundary of a disk at a distinguished point. Reflect this small circle with respect to the tangent at the distinguished point. The result is a bigger cactus with a distinguished disk (the small one). It is tangent to the boundary of only one disk of the initial cactus (at the only distinguished point). Hence it corresponds to the end point of the tree of the bigger cactus.

THEOREM. *The curves with $m = 2n - 2$ are enumerated by the plane trees consisting of $n + 1$ vertices with two marked end points.*

EXAMPLE. There exist 14 plane trees having 6 vertices with 2 marked end points, shown in Figure 14. They enumerate the 14 curves having $m = 2n - 2 = 8$.

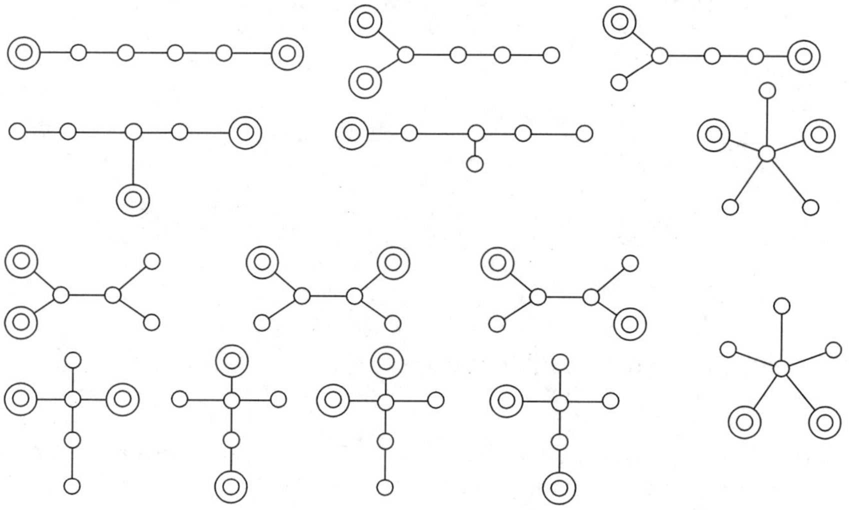

FIGURE 14. Enumeration of curves having $m = 2n - 2 = 8$ exterior edges

PROOF. Since $k = 2$, a curve having $m = 2n - 2$ is (up to equivalence) a circular cactus to which two small circles are attached, tangent interiorly to the boundaries of the disk(s) of the cactus at the two distinguished points.

Indeed, otherwise the pendant would be a circle with 2 distinguished points, which is not hangable (see §1).

Reflecting the two small circles, we construct a tree with two marked end points as in the proof of the preceding theorem.

The classification of curves having $m = 2n - k$, $k \geq 3$, is more complicated. Indeed, the pendant consisting of k segments might be more complicated than the union of k circles attached at one distinguished point each.

EXAMPLE. For $k = 3$, the possible pendants consisting of 3 segments are:
1) three circles (each attached at one distinguished point);
2) the "8" cactus, consisting of two disks, attached at one distinguished point;
3) a curve consisting of two circles tangent interiorly, attached at a distinguished point on the exterior circle;
4) one circle attached at three distinguished points.

The corresponding curves can still be enumerated by plane trees with additional structures, but when k grows such a description becomes complicated.

Moreover, some of the curves that we obtain are not new: they are conformally equivalent to some of the curves with smaller k. Namely, this happens when one of the interior domains bounded by the curve has more sides then the exterior contour.

EXAMPLE. The cactus A_3 with three pendant circles inside the middle disk ($k = 3$, $n = 5$) is conformally equivalent to a curve with $m = 8$, $k = 2$, since the remaining part of the middle disk is bounded by 8 edges.

In the case of $n = 5$ double points, one can enumerate the remaining curves (with $m = 8, 7, 6, 5$, and 4 exterior edges), using the classification of hangable A_p-pendants from §4, and Lemma 1 and Theorem 1 from §2 for the tree-like pendants (the non-tree-like pendants have at least 7 edges). The calculations are elementary, but long.

The resulting list of spherical curves having $n \leqslant 5$ double points are shown in Figure 15 and Figure 16. Each curve is accompanied by its *signature*, i.e., the numbers of sides of the complementary domains. The signature is an invariant of the spherical curve, and in most cases the signatures distinguish the curves.

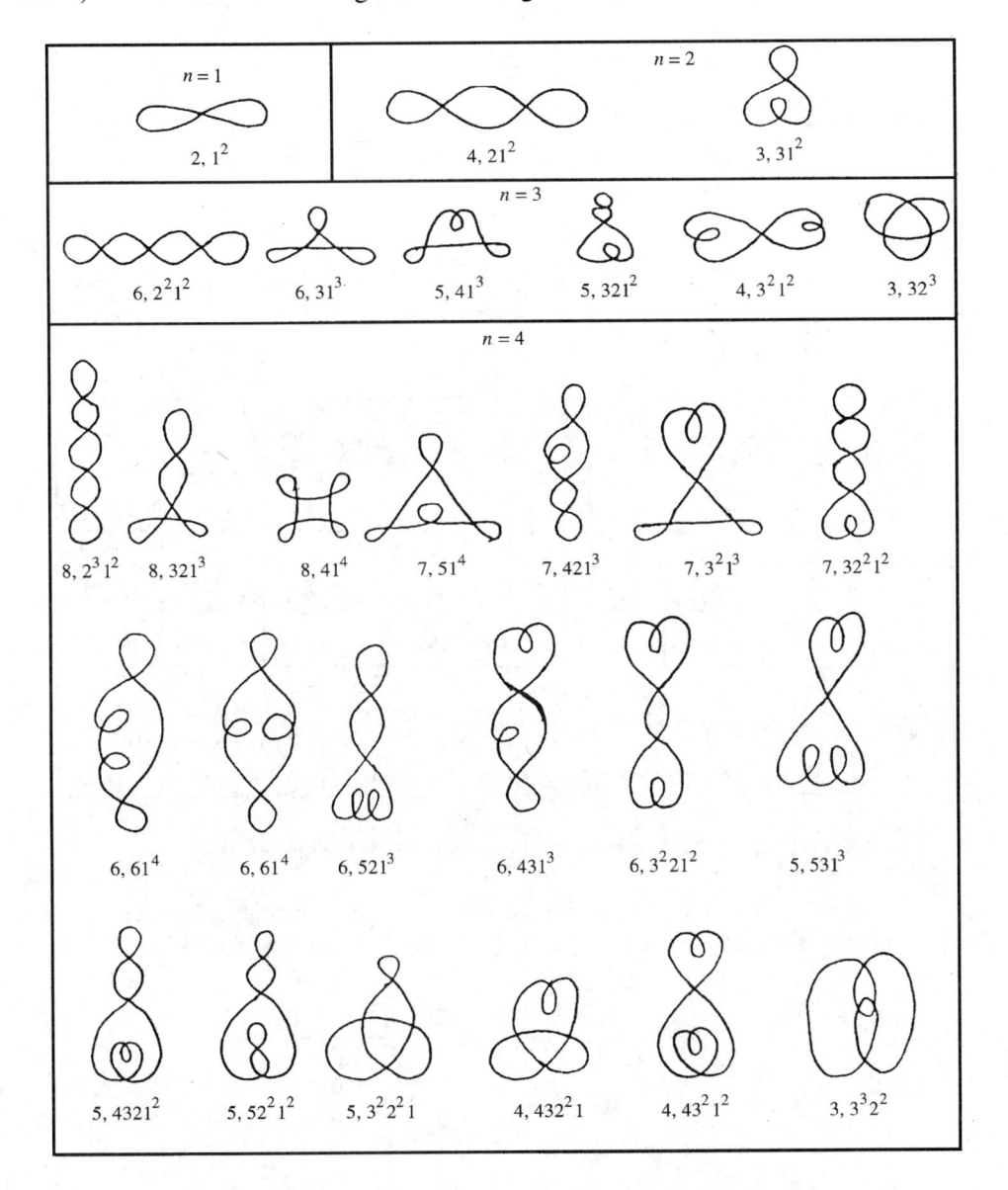

FIGURE 15. Spherical curves having $n \leqslant 4$ double points

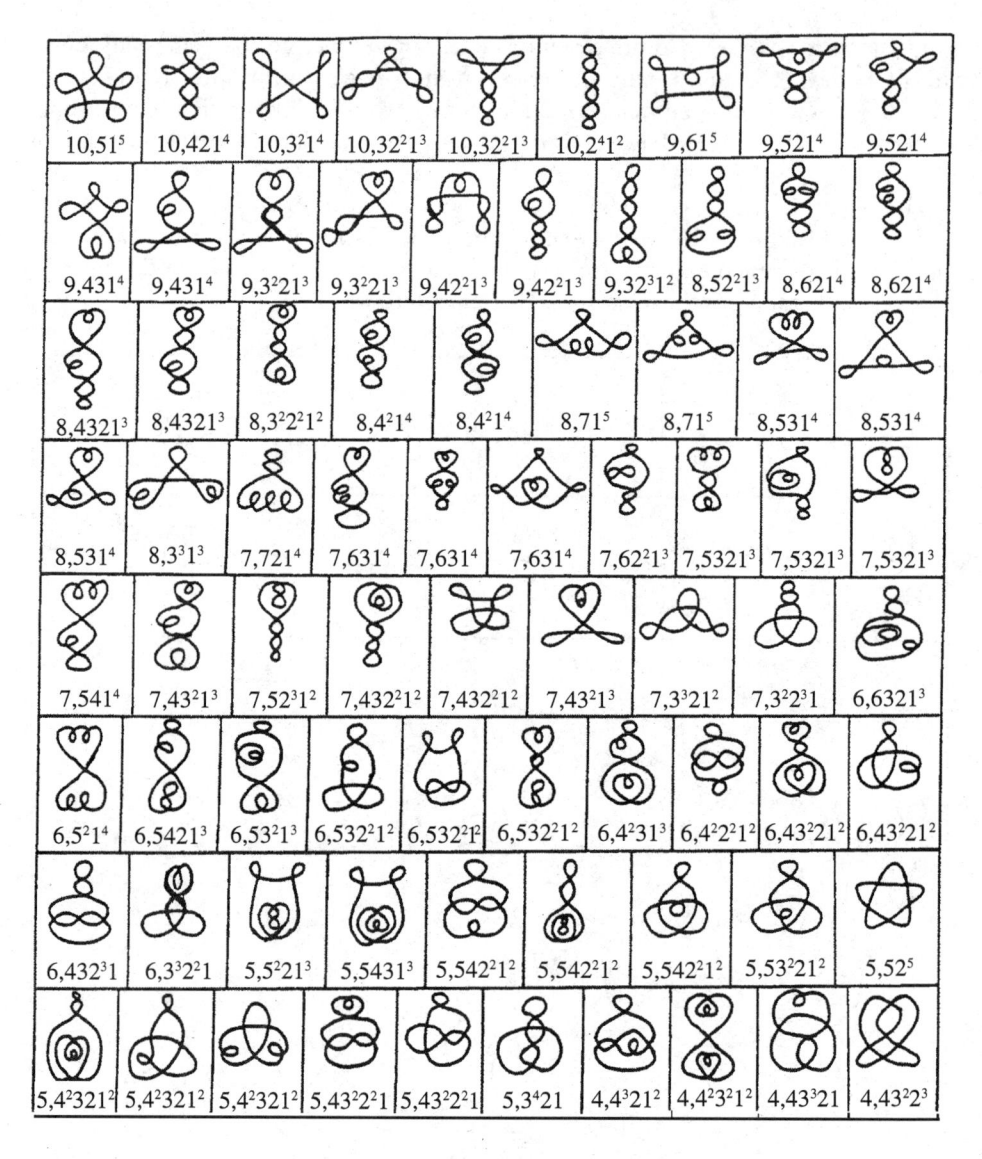

FIGURE 16. Spherical curves having 5 double points

The numbers of types of nonoriented curves on the nonoriented sphere having $n \leq 5$ double points and m exterior edges are shown in the following table:

n	$m=2$	3	4	5	6	7	8	9	10	Σ
1	1	0	0	0	0	0	0	0	0	1
2	0	1	1	0	0	0	0	0	0	2
3	0	1	1	2	2	0	0	0	0	6
4	0	1	2	4	5	4	3	0	0	19
5	0	0	4	13	13	16	14	10	6	76

DEFINITION. A curve is *reducible* if it contains a separating double point.

| $n=0$ | 1 | 2 | 3 | 4 | 5 | 6 | 7 |

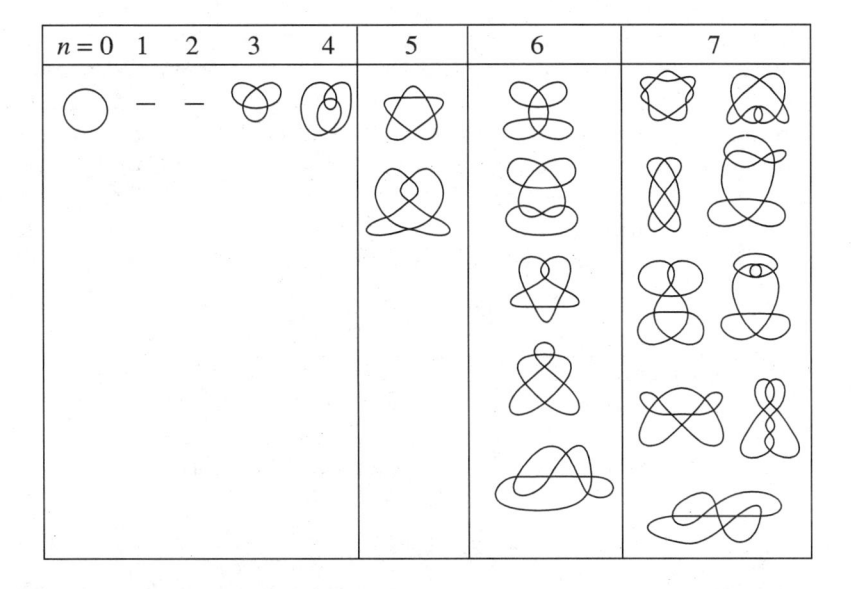

FIGURE 17. Irreducible spherical curves having $n \leq 7$ double points

The classification of the irreducible spherical curves with a given number of double points is much simpler than that of all generic curves having n double points.

LEMMA. *The number of exterior edges of an irreducible curve is at most equal to the number of double points*: $m \leq n$.

PROOF. The cactus consists of one disk, otherwise there would be a separating double point. The m distinguished points on the boundary are adjacent to $4m$ ends of the segments of the curve. The total number of segments being $2n$, one sees that $4m \leq 4n$.

The enumeration of irreducible curves having n double points starts from the case $m = n$. In this case the pendants are circles with odd numbers of distinguished points, at least 3 of them on each pendant. The classification leads to the enumeration of plane trees at whose interior vertices odd numbers of branches meet if the distances to the ends are odd (the distances between the ends always being even).

The numbers N of irreducible curves with small values of $m = n$ form a strange sequence:

$m = n$	3	4	5	6	7	8	9	10	11	12	13	14
N	1	0	1	1	1	1	3	2	5	7	9	16

The total numbers T of irreducible spherical curves with n double points are known only for $n \leq 7$:

n	0	1	2	3	4	5	6	7
T	1	0	0	1	1	2	5	9

The curves are shown in Figure 17.

To obtain the list of plane curves from the lists of spherical curves one puts the ∞ point into each of the complementary domains. Since the spherical curve type is an invariant of the plane curve type, one must only check whether some of the plane curves obtained from the given spherical one are equivalent as plane curves. In most cases the signature shows that the initial spherical curve has no such conformal symmetries. The resulting lists of plane curves having $n = 5$ double points have been compiled by F. Aicardi. Her calculations give the following table of numbers of classes of plane curves, oriented or not, on the oriented and nonoriented plane:

oriented	$n=0$	1	2	3	4	5
plane, curve	2	3	10	39	204	1262
plane	1	2	5	21	102	639
curve	1	2	5	21	102	640
—	1	2	5	20	82	435

References

1. V. I. Arnold, *Plane curves, their invariants, perestroikas and classifications*, Singularities and Bifurcations (V. I. Arnold, ed.), Adv. Sov. Math., vol. 21, 1994, pp. 33–91.

Translated by THE AUTHOR

On Diagrams of Configurations of 7 Skew Lines of $\mathbb{R}^3$

Alberto Borobia and Vladimir F. Mazurovskiĭ

ABSTRACT. A configuration of skew lines of $\mathbb{R}^3$ is defined as an (unordered) collection of nonoriented pairwise skew lines in $\mathbb{R}^3$. An isotopy of such a collection in the process of which the lines remain pairwise skew lines is called a rigid isotopy. Configurations of skew lines of $\mathbb{R}^3$ are usually specified by arrangements of lines in the plane with additional crossing information, which are called diagrams. The main result of the paper is the reduction of the classification problem of configurations of at most 7 skew lines in $\mathbb{R}^3$ up to rigid isotopy to that of the classification of configurations with a diagram of a certain very special kind. This allows us to reduce the classification problem of configurations of 7 skew lines in $\mathbb{R}^3$ up to rigid isotopy to a computational one.

§1. Introduction

Following O. Ya. Viro [V], by a *configuration of skew lines* in $\mathbb{R}^3$ we mean an (unordered) collection of nonoriented pairwise skew lines in $\mathbb{R}^3$. An isotopy of such a collection in the process of which the lines remain pairwise skew lines is called a *rigid isotopy*. It is obvious that the property of being rigidly isotopic is an equivalence relation. The corresponding equivalence class of a configuration is called its *rigid isotopy type*.

It is natural to pass from the problem concerning lines in $\mathbb{R}^3$ to a similar problem concerning lines in projective space $\mathbb{R}P^3$. A *nonsingular configuration of lines* of $\mathbb{R}P^3$ is defined as a collection of nonoriented pairwise disjoint lines in $\mathbb{R}P^3$. Two nonsingular configurations of lines of $\mathbb{R}P^3$ are called *rigidly isotopic* if they can be joined by an isotopy which consists of nonsingular configurations. It is easy to see that nonsingular configurations of lines of $\mathbb{R}P^3$ are a particular case of links in $\mathbb{R}P^3$, and the rigid isotopies of nonsingular configurations are a particular case of ordinary link isotopies in the process of which the lines of the configuration remain pairwise disjoint.

1991 *Mathematics Subject Classification.* Primary 51A20.
Key words and phrases. Configuration of skew lines, rigid isotopy, diagram, weaving.

The work of the first author is partially supported by DGICYT PB92-0716. The research described in this publication was made possible in part by grant No. NNE000 from the International Science Foundation. The work of the second author was made possible in part by Grant No. NNE 000 from the International Science Foundation and Grant No. 94-1.1-144 from the State Committee of the Russian Federation for Higher Education.

©1996, American Mathematical Society

In [V] it was shown that the classification of the rigid isotopy types of affine configurations of m skew lines is equivalent to the classification of nonsingular configurations of m projective lines. Indeed, the canonical inclusion $\mathbb{R}^3 \to \mathbb{R}P^3$ induces an embedding of the space of configurations of m skew lines of $\mathbb{R}^3$ into the space of nonsingular configurations of m lines of $\mathbb{R}P^3$. Since the image of this imbedding is the complement of a subspace of codimension 2, rigid isotopy of configurations of skew lines of $\mathbb{R}^3$ is equivalent to rigid isotopy of the corresponding nonsingular configurations of $\mathbb{R}P^3$.

Viro in [V] and Mazurovskiĭ in [M1, M2] obtained the classification of nonsingular configurations of m projective lines up to rigid isotopy for $m \leqslant 5$ and $m = 6$ respectively. In order to study the case $m = 7$, we must investigate the properties of configuration diagrams for 7 skew lines of $\mathbb{R}^3$.

The third section of this paper is due to Borobia, the rest was written by both authors.

§2. Diagrams of configurations of skew lines of $\mathbb{R}^3$

Configurations of skew lines of $\mathbb{R}^3$ are specified by pictures in the plane called *diagrams*. Let $Oxyz$ be the canonical Cartesian coordinate system in $\mathbb{R}^3$ and $L = \{l_1, \ldots, l_m\}$ a configuration of skew lines of $\mathbb{R}^3$. Denote by π the orthogonal projection of $\mathbb{R}^3$ onto the xy-plane. We assume that L satisfies the following conditions of general position: 1) $\pi(l_i)$ is a line of the xy-plane for any $i = 1, \ldots, m$; 2) $\pi(L)$ has no triple points and 3) no parallel lines.

Any configuration of skew lines of $\mathbb{R}^3$ can be modified to satisfy the conditions 1)–3) by an arbitrary small rigid isotopy.

Below we use the following notation:
- r_i denotes the line $\pi(l_i)$ for any $i = 1, \ldots, m$;
- the point $r_i \cap r_j$ is denoted by R_{ij} or R_{ji} for any $i, j, i \neq j$;
- L_{ij} denotes the point of l_i such that $\pi(L_{ij}) = R_{ij}$;
- α_i denotes the plane parallel to the z-axis that contains l_i.

We say that l_i *lies over* l_j if the z-coordinate of $L_{ij} \in l_i$ is greater than the z-coordinate of $L_{ji} \in l_j$. The upper point L_{ij} is called the *overpass point*; the lower point L_{ji} is called the *underpass point*. For each underpass point, we choose a sufficiently small connected neighborhood in L, and denote by U the union of these neighborhoods for all the underpass points. The image $\pi(L \setminus U)$ is then called the *diagram* of L. The collection $\{r_1, \ldots, r_m\}$ of m lines of the xy-plane is called an *arrangement of lines associated with the diagram of L*.

Using the above notation, we say that the line l_i *admits a rotation* at $Q_i \in l_i$ if the point $P_i = \pi(Q_i)$ does not belong to another line of $\pi(L)$ and l_i lies over all lines of L whose images under the projection π intersect r_i on one of the halflines of $r_i \setminus \{P_i\}$, while the opposite happens with the other halfline.

If the configuration L has a line l_i that admits a rotation at some point Q_i, we use the rigid isotopy which is the composition of the following three rigid isotopies of L and affects only the line l_i:

(1st) Let d_i be the line joining P_i and Q_i. Then $l_i \cup d_i$ divides α_i into four regions, two of which do not intersect any line of L. Moreover, for $j \neq i$ the plane α_j is not parallel to α_i, since r_j is not parallel to r_i. Therefore we can consider the rigid isotopy consisting of the rotation of l_i on the plane α_i with center of rotation at Q_i until the line reaches the position of the line d_i. Such a rigid isotopy is called a *rotation* of l_i.

(2nd) Let $L^i = \{l_1, \ldots, l_{i-1}, l_{i+1}, \ldots, l_m\}$ be the subconfiguration of L and $C(P_i)$ be the connected component of $Oxy \setminus \pi(L^i)$ that contains the point P_i. Let t_i be a line parallel to d_i such that $t_i \cap Oxy \in C(P_i)$. Then d_i can be moved by means of a translation to the position of t_i without intersecting any line of L^i. This kind of rigid isotopy is called a *translation* of d_i.

(3rd) Consider a cone with vertex $P'_i = t_i \cap Oxy$ and axis t_i such that the cone does not intersect any line of L^i. Then t_i can be moved by pivoting over P'_i until it reaches the position of any line lying inside the cone. Such a rigid isotopy is called a *pivotation* of t_i.

§3. m-cornered diagrams of configurations of m skew lines

In what follows we suppose that $\mathbb{R}^3$ is canonically embedded into $\mathbb{R}P^3$ and the canonical Cartesian coordinate system $Oxyz$ is fixed in $\mathbb{R}^3$.

If X is a set in $\mathbb{R}^3$, then the projective closure of X in $\mathbb{R}P^3$ is denoted by $\text{pr}(X)$. If Y is a set in $\mathbb{R}P^3$, then we denote by $\text{aff}(Y)$ the set $Y \cap \mathbb{R}^3$.

Let D be the diagram of a configuration of m skew lines of $\mathbb{R}^3$. The projective closure of the arrangement of lines associated with the diagram D divides the projective closure of the xy-plane into connected components. We say that the diagram D is *m-cornered* if one of these connected components is a polygon of m edges.

THEOREM 3.1. *For any configuration L of m skew lines of $\mathbb{R}^3$, where $m \leqslant 7$, there exists a configuration of skew lines of $\mathbb{R}^3$ rigidly isotopic to L such that its diagram is m-cornered.*

PROOF. It is convenient to divide the proof into the following three cases.

Case 1: $m \leqslant 5$.

Finashin [F] showed that for $m \leqslant 5$ any collection of m lines of $\mathbb{R}P^2$ in general position divides $\mathbb{R}P^2$ into connected components, one of which is a polygon of m edges. This implies the theorem for $m \leqslant 5$.

In order to solve the cases $m = 6$ and 7, we use the rigid isotopies described in §2.

Case 2: $m = 6$.

Let $L = \{l_1, l_2, l_3, l_4, l_5, l_6\}$ be a configuration of 6 skew lines of $\mathbb{R}^3$. Up to rigid isotopy we can assume that the line l_1 is perpendicular to the xy-plane. Up to rigid isotopy we can also assume (see Case 1) that the diagram of the subconfiguration $L^1 = \{l_2, l_3, l_4, l_5, l_6\}$ of L is 5-cornered. Let S be the pentagon determined by the arrangement of lines associated with the diagram of L^1. There are two possibilities for the location of the point $P_1 = l_1 \cap Oxy$.

Subcase 2.1: the point P_1 lies inside $\text{aff}(S)$.

Carry out a translation of l_1 until l_1 reaches a line t_1 such that the point $t_1 \cap Oxy$ is near a vertex of S and then effect a pivotation of t_1 until t_1 reaches a line l'_1 such that the orthogonal projection of l'_1 onto the xy-plane cuts the vertex from $\text{aff}(S)$. It is clear that the diagram of the configuration $L' = \{l'_1, l_2, l_3, l_4, l_5, l_6\}$ is 6-cornered.

Subcase 2.2: the point P_1 lies outside $\text{aff}(S)$.

Carry out a pivotation of l_1 until it reaches a line l'_1 such that $\pi(l'_1)$ cuts a vertex from $\text{aff}(S)$. The diagram of the configuration $L' = \{l'_1, l_2, l_3, l_4, l_5, l_6\}$ is 6-cornered again.

Case 3: $m = 7$.

Let $L = \{l_1, l_2, l_3, l_4, l_5, l_6, l_7\}$ be a configuration of 7 skew lines of $\mathbb{R}^3$. Carry out a translation of the line l_1 until the line reaches the position of a line d_1 that intersects some line l_j of L ($j \neq 1$). Without loss of generality we can assume that $j = 2$. Let $P_{12} = d_1 \cap l_2$ and F_1 be the strip between l_1 and d_1. Note that $F_1 \cap L = l_1 \cup \{P_{12}\}$. Consider a projective transformation φ isotopic to the identity such that $\varphi(P_{12}) \in \mathbb{R}P_\infty^2$ and $\varphi(P_1) \in Oxy$, where $P_1 = \text{pr}(l_1) \cap \text{pr}(d_1) \in \mathbb{R}P_\infty^2$. Since $\text{pr}(L)$ is rigidly isotopic to $\varphi(\text{pr}(L))$, it follows that aff$(\varphi(\text{pr}(L)))$ is rigidly isotopic to L. In order to simplify the notation, we denote the elements of aff$(\varphi(\text{pr}(L)))$ in the same way as for L before the action of φ. Then we have the following situation in $\mathbb{R}^3$: F_1 is a planar cone, its vertex is P_1 and its sides are l_1 and d_1; d_1 is parallel to l_2; F_1 intersects L only on l_1. Up to rigid isotopy, we can assume that l_2 is perpendicular to Oxy. Let $P_2 = Oxy \cap l_2$.

Now we consider a rigid isotopy that affects only l_1 and l_2. We can assume that the diagram of the subconfiguration $L^{12} = \{l_3, l_4, l_5, l_6, l_7\}$ of L is 5-cornered. Let S be the pentagon determined by the arrangement of lines associated with the diagram of L^{12}. There are two possibilities for the location of the point P_2.

Subcase 3.1: the point P_2 lies outside aff(S).

Consider the half-cone C such that its axis is l_2, its vertex is P_2, its planar face lies on the plane containing l_2 and P_1, and C does not intersect $F_1 \cup L^{12}$ (see Figure 1). We can pivot l_2 until it reaches a line $l_2' \subseteq \text{int } C \cup \{P_2\}$ such that $\pi(l_2')$ cuts off a vertex from aff(S).

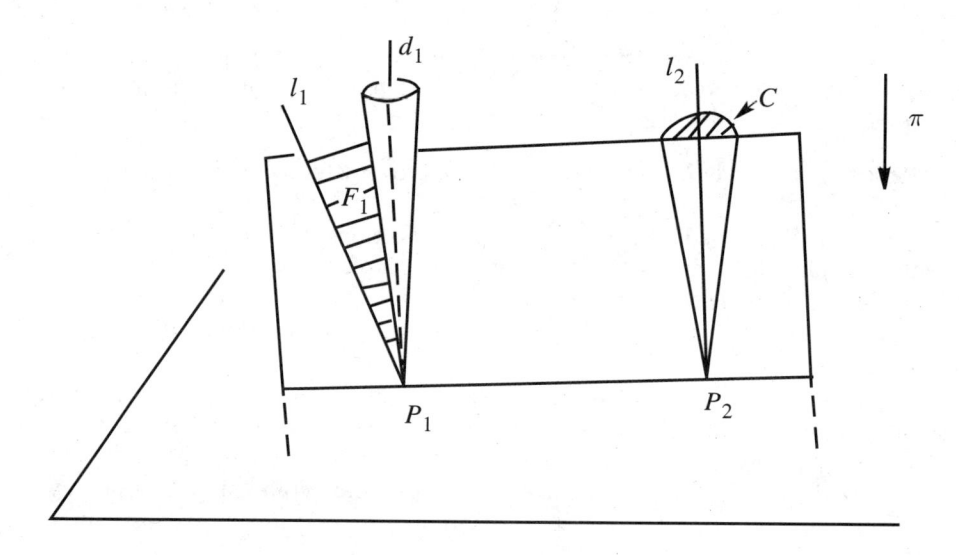

FIGURE 1

Since no line in int $C \cup \{P_2\}$ intersects F_1, after the movement of l_2 to l_2', we can carry out the rigid isotopy (of the configuration $\{l_1, l_2', l_3, l_4, l_5, l_6, l_7\}$) consisting in the rotation of l_1 in the planar cone F_1 until it reaches d_1. The line d_1 is perpendicular to the xy-plane and the diagram of the subconfiguration $\{l_2', l_3, l_4, l_5, l_6, l_7\}$ is 6-cornered. Therefore we can argue with d_1 as we did in Case 2. Finally we obtain a configuration $L' = \{l_1', l_2', l_3, l_4, l_5, l_6, l_7\}$ such that L' is rigidly isotopic to L and the diagram of L' is 7-cornered.

Subcase 3.2: the point P_2 lies inside aff(S).

First we translate l_2 until it reaches some line d_2 such that the point $d_2 \cap Oxy$ is close to a vertex of aff(S). After that we can act exactly as we did in Subcase 3.1. In

the process of the above translation, the line l_1 must not be intersected. There are two possibilities:
- the line $r_1 = \pi(l_1)$ does not intersect aff (S); then we have no problem translating the line l_2;
- $r_1 \cap \text{aff}(S) \neq \varnothing$; then $\text{pr}(r_1)$ divides S into two polygons, one of which contains P_2; it is easy to see that we can move l_2 near any vertex of this polygon without any problems.

We say that the diagram of a configuration of m skew lines of $\mathbb{R}^3$ is *unbounded m-cornered* if one of the bounded faces of the arrangement of lines associated with this diagram is an unbounded polygon of m edges (see Figure 2).

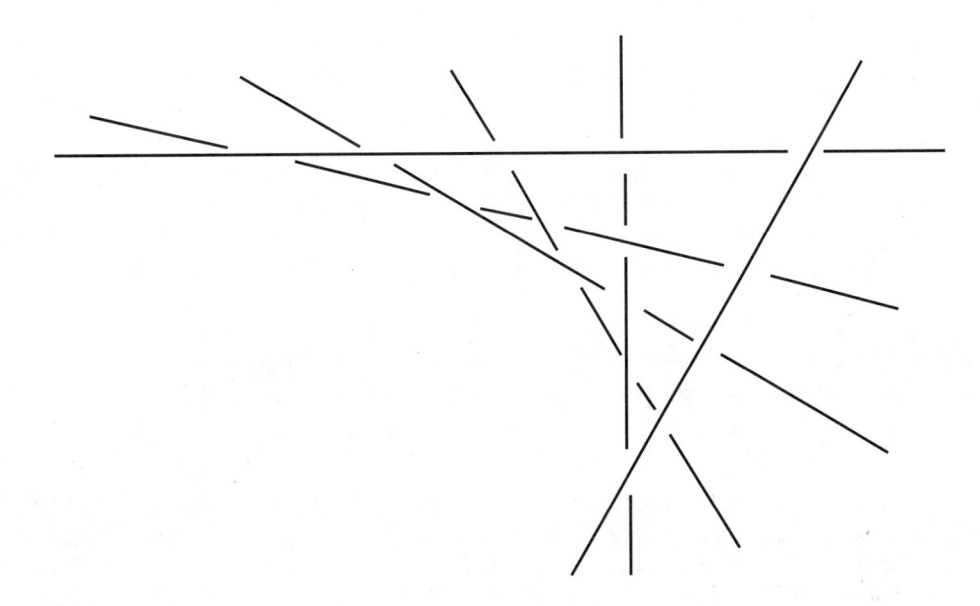

FIGURE 2. An unbounded 6-cornered diagram

COROLLARY 3.2. *For any configuration L of m skew lines of $\mathbb{R}^3$, where $m \leq 7$, there exists a configuration of skew lines of $\mathbb{R}^3$ rigidly isotopic to L such that its diagram is unbounded m-cornered.*

§4. Completely shellable unbounded m-cornered weavings of m lines

A *weaving* is defined as a set of lines of $\mathbb{R}^2$ in general position together with a choice, at every intersection point, of a gap in one of the two intersecting lines. It is easy to see that the diagrams of configurations of skew lines of $\mathbb{R}^3$ can be regarded as a particular case of weavings. But not all weavings can be obtained as diagrams of configurations of skew lines of $\mathbb{R}^3$ (see, for example, [W]). If nevertheless this is possible for a weaving, the latter is said to be *liftable*.

Let $H = \{h_1, \ldots, h_m\}$ be a weaving. We say that a line h_i of H *crosses over* a line h_j of H if h_j has a gap at the intersection point $h_i \cap h_j$, otherwise we say that h_i *crosses under* h_j. A line h_i of H is called *free* if either h_i crosses over all the other lines of H, or h_i crosses under all these lines, or there is a point X on h_i that separates the overcrossings from the undercrossings. We say that the given weaving H is *1-shellable* if it has a free line. The weaving H is called *s-shellable* if, by recursion, it has a free line h_i and the weaving $H^i = \{h_1, \ldots, h_{i-1}, h_{i+1}, \ldots, h_m\}$ of $m-1$ lines is $(s-1)$-shellable. A weaving of m lines is called *completely shellable* if it is $(m-1)$-shellable. It is easy to see that any completely shellable weaving is liftable.

OBSERVATION 4.1. It follows from Theorem 3.1 that for any configuration L of m skew lines of $\mathbb{R}^3$, where $m \leqslant 7$, there exists a configuration of skew lines of $\mathbb{R}^3$ rigidly isotopic to L such that its diagram is unbounded m-cornered and 2-shellable. R. Penne [P2] proved that the diagram of any configuration of at most 5 skew lines of $\mathbb{R}^3$ is completely shellable. Therefore any configuration of $\leqslant 7$ skew lines is rigidly isotopic to a configuration whose diagram is unbounded m-cornered and completely shellable.

An arrangement of m lines of $\mathbb{R}^2$ is called *unbounded m-cornered* if one of the bounded faces of the arrangement is an unbounded polygon of m edges. A weaving of m lines is called *unbounded m-cornered* if the arrangement of lines associated with the weaving is unbounded m-cornered.

Let W_m denote the set of all the completely shellable unbounded m-cornered weavings. Fix an orientation of $\mathbb{R}^2$. This orientation induces a linear ordering of lines of an unbounded m-cornered weaving. We call such a linear ordering of lines of an unbounded m-cornered weaving *concordant with the orientation of* $\mathbb{R}^2$. Let $H = \{h_1, \ldots, h_m\}$ and $H' = \{h'_1, \ldots, h'_m\}$ be ordered weavings. We say that H and H' *have the same crossing function* if for any $i \neq j$ the line h_i crosses over h_j if and only if h'_i crosses over h'_j. We define the following equivalence relation on W_m: two weavings from W_m are *equivalent* if they can be ordered concordantly with a fixed orientation of $\mathbb{R}^2$ in such a way that the obtained ordered weavings have the same crossing function.

REMARK 4.2. In subsequent lemmas we show that any two configurations of m skew lines of $\mathbb{R}^3$ that have equivalent diagrams from W_m are rigidly isotopic. Therefore it follows from Observation 4.1 that in order to find all the rigid isotopy types of configurations of $\leqslant 7$ skew lines of $\mathbb{R}^3$ we only need to find all the equivalence classes of W_m for $m \leqslant 7$ and then to detect what equivalence classes of the weavings determine rigidly isotopic configurations. This reduces the classification of configurations of 7 skew lines of $\mathbb{R}^3$ to a computational problem. A description of the classification will be published separately.

Let B_m be an equivalence class of W_m. Denote by E_m the set of all configurations of m skew lines of $\mathbb{R}^3$ whose diagrams belong to B_m. We consider E_m and B_m as topological spaces with the natural topologies. Let $p: E_m \to B_m$ be the map taking each configuration of E_m to its diagram.

LEMMA 4.3. *The space B_m is arcwise connected.*

PROOF. It is not difficult to prove that any two unbounded m-cornered arrangements of m lines of $\mathbb{R}^2$ can be connected by an isotopy that consists of unbounded m-cornered arrangements of lines. It immediately follows that any two points of B_m can be connected by a path.

Fix some point $b \in B_m$. Let F_b be the set of all configurations of E_m that have the diagram b.

LEMMA 4.4. *The space F_b is arcwise connected.*

PROOF. Let us coordinatize the problem, following [**P1**]. Without loss of generality, we can assume that the origin of the xy-plane does not belong to any line of the weaving b. If l_i is a line in $\mathbb{R}^3 \setminus \{O\}$, then we can define the plane $\alpha(l_i)$ containing l_i and the origin. Furthermore, l_i is uniquely determined by $r_i = \pi(l_i)$ and $\alpha(l_i)$. So we can say that the preimages of r_i under π are in one-to-one correspondence with the planes $\alpha(l_i)$, which can be represented by the equations $z = a_i x + b_i y$, and hence by the pairs $(a_i, b_i) \in \mathbb{R}^2$. Further, if $R_{ij} = (x_{ij}, y_{ij})$ is the intersection of r_i and r_j, then the lifting l_i lies over l_j if and only if $a_i x_{ij} + b_i y_{ij} > a_j x_{ij} + b_j y_{ij}$.

Consequently, each lifting of a given weaving with the associated arrangement $\{r_1, \ldots, r_m\}$ is the solution of a system of homogeneous linear inequalities in the unknowns $L = \{a_1, b_1, \ldots, a_m, b_m\}$.

Since the weaving b is completely shellable, hence liftable, it follows that F_b is homeomorphic to a nonempty open convex subset of $\mathbb{R}^{2m}$. Therefore F_b is arcwise connected.

LEMMA 4.5. *The bundle $p: E_m \to B_m$ is topologically trivial.*

PROOF. Let $L, L' \in E_m$, and let b and b' be the diagrams of L and L' respectively. Denote by U and U' the open convex subsets of $\mathbb{R}^{2m}$ corresponding to the fibers F_b and $F_{b'}$ respectively, in accordance with the proof of Lemma 4.4. Using the notation of Lemma 4.4, we can say that the collections

$$(b, a_1, b_1, \ldots, a_m, b_m) \in B_m \times U \quad \text{and} \quad (b', a'_1, b'_1, \ldots, a'_m, b'_m) \in B_m \times U'$$

uniquely determine L and L' respectively.

In $\mathbb{R}^{2m+1}$ fix the canonical Cartesian coordinate system $Ox_1 \cdots x_{2m+1}$ and consider the stereographic projection pr_1 of the hemisphere

$$\{(x_1, \ldots, x_{2m}, x_{2m+1}) \in \mathbb{R}^{2m+1} \mid x_{2m+1} = 1 - \sqrt{1 - x_1^2 - \cdots - x_{2m}^2},$$
$$x_1^2 + \cdots + x_{2m}^2 < 1\}$$

to the plane $x_{2m+1} = 0$. We assume that the space $\mathbb{R}^{2m}$ is embedded into $\mathbb{R}^{2m+1}$ as the plane $x_{2m+1} = 0$. Let $\bar{U} = \mathrm{pr}_1^{-1}(U)$, $\bar{U}' = \mathrm{pr}_1^{-1}(U')$, and $\bar{u}$ and $\bar{u}'$ be the centers of mass of the closed subsets $\mathrm{cl}\,\bar{U}$ and $\mathrm{cl}\,\bar{U}'$ respectively. Denote by $\tilde{U}$ the image of $\bar{U}$ under the translation along the geodesic joining the point $\bar{u}$ with the origin, and denote by $\tilde{U}'$ the image of $\bar{U}'$ under the translation along the geodesic joining the point $\bar{u}'$ with the origin. Let $\tilde{\varphi}: \tilde{U}' \to \tilde{U}$ be the homeomorphism of the subsets $\tilde{U}'$ and $\tilde{U}$ that assigns to each point $\tilde{y}' \in \tilde{U}'$ the point $\tilde{y} \in \tilde{U}$ such that:

1) $\tilde{y}'$ and $\tilde{y}$ belong to the common geodesic Γ with the beginning in the origin;

2) $\dfrac{\mathrm{dist}(O, \tilde{y}')}{\mathrm{dist}(O, \partial \tilde{U}' \cap \Gamma)} = \dfrac{\mathrm{dist}(O, \tilde{y})}{\mathrm{dist}(O, \partial \tilde{U} \cap \Gamma)}$ (we regard the hemisphere as a metric space with the Riemannian metric).

The homeomorphism $\tilde{\varphi}$ induces the canonical homeomorphism $\varphi: U' \to U$. Then a trivialization $f: E_m \to B_m \times F_b$ is determined by the formula

$$(b', a'_1, b'_1, \ldots, a'_m, b'_m) \xrightarrow{f} (b', \varphi(a'_1, b'_1, \ldots, a'_m, b'_m)).$$

COROLLARY 4.6. *The space E_m is arcwise connected.*

PROOF. The fiber and the base of the topologically trivial bundle $p\colon E_m \to B_m$ are arcwise connected. Therefore, the total space E_m is also arcwise connected.

References

[F] S. M. Finashin, *Projective configurations and real algebraic curves*, Ph. D. Thesis, Leningrad Univ., 1985. (Russian)

[M1] V. F. Mazurovskiĭ, *Configurations of six skew lines*, Zap. Nauchn. Sem. Leningrad. Otdel. Mat. Inst. Steklov. (LOMI) **167** (1988), 121–134; English transl., J. Soviet Math. **52** (1990), no. 1, 2825–2832.

[M2] _____, *Configurations of at most six lines of* $\mathbb{R}P^3$, Real Algebraic Geometry, Proceedings of the conference held in Rennes, France, June 24–28, 1991 (M. Coste, L. Mahe, and M.-F. Roy, eds.), Lecture Notes in Math., vol. 1524, Springer-Verlag, Berlin, Heidelberg, and New York, 1992, pp. 354–371.

[P1] R. Penne, *Lines in 3-space: isotopy, chirality and weavings*, Ph. D. Thesis, Antwerp Univ., 1992.

[P2] _____, *Configurations of few lines in 3-space: isotopy, chirality, and planar layouts*, Geom. Dedicata **45** (1993), 49–82.

[V] O. Ya. Viro, *Topological problems concerning lines and points of three-dimensional space*, Dokl. Akad. Nauk SSSR **284** (1985), no. 5, 1049–1052; English transl., Soviet Math. Dokl. **32** (1985), no. 2, 528–531.

[W] W. Whiteley, *Rigidity and polarity* 2: *Weaving lines and tensegrity frameworks*, Geom. Dedicata **30** (1989), 255–279.

Translated by V. F. MAZUROVSKIĬ

DEPARTAMENTO DE MATEMATICAS, FACULTAD DE CIENCIAS, U.N.E.D. 28040 MADRID, SPAIN

DEPARTMENT OF MATHEMATICS, IVANOVO STATE UNIVERSITY, UL. ERMAKA, 39, IVANOVO, 153025, RUSSIA

Smoothing Isolated Singularities on Real Algebraic Surfaces

Benoît Chevallier

Dedicated to the memory of Professor D. A. Gudkov

ABSTRACT. We investigate M-surfaces using the Viro gluing theorem together with choices in the combinatorics of Newton polytopes. They lead to some "central" types of singularities and explicit connections between curves and surfaces.

§I. Introduction

Let A be a smooth real algebraic hypersurface of degree d in the projective space $P(\mathbb{R}^q)$, and let $\mathbb{C}A$ be the complex locus. Recall the following inequality between the sum of $\mathbb{Z}_2$-Betti numbers when $\mathbb{C}A$ is smooth (Harnack–Thom–Oleĭnik theorem [T, G, O]):

$$\dim H_*(A, \mathbb{Z}_2) \leqslant \dim H_*(\mathbb{C}A, \mathbb{Z}_2)$$

$$\left(\dim H_*(\mathbb{C}A, \mathbb{Z}_2) = \frac{(d-1)^{q+1} + (-1)^q}{d} + q - (-1)^q \right).$$

A is said to be a *maximal* or an *M-hypersurface* if

$$\dim H_*(A, \mathbb{Z}_2) = \dim H_*(\mathbb{C}A, \mathbb{Z}_2).$$

This paper introduces a class of surfaces with isolated singular points, which turn out to be rather central in the sense that their smoothings give many topological types of pairs $(P(\mathbb{R}^3), A)$, where A is maximal. The deformations are based on Viro's method for constructing real algebraic varieties with prescribed topology. Some introductory material is given in §§I and III. We begin with surfaces of degree four. A new simple constructive approach to their topological classification is shown in §II (Theorem II.1). This is generalized in §III towards a type of result represented by Theorem III.3. Some concrete topological types are listed in Theorem III.4 (mostly not new at this stage). Their purpose is to show that Viro's method, widely used for curves, is powerful for surfaces. Note also Proposition IV.1. Sections IV, V, VI, VII, and VIII are devoted to proofs.

1991 *Mathematics Subject Classification.* Primary 14P25, 14J17.

Besides plane algebraic curves, the technical tools are of three kinds:
1. Real polynomials of degree two in one variable.
2. Lemma III.1.
3. Viro gluing theorem ([V1, V3, R]).

Notation and definitions. Let f be a real polynomial in q variables or homogeneous in $q+1$ variables. The same letter or the same symbol $\{f = 0\}$ will denote an algebraic hypersurface in $P(\mathbb{R}^q)$ as well as its restrictions to affine charts.

If $S = \{f = 0\} \subset \mathbb{R}^3$ is an algebraic surface of degree d with Δ as its Newton polyhedron, it is often convenient to consider the set

$$T = \{(\alpha, \beta, \gamma, \delta) \in \mathbb{R}_+^4 : \alpha + \beta + \gamma + \delta = d, (\alpha, \beta, \gamma) \in \Delta\}$$

instead of Δ. The set T will also be called the *Newton polyhedron* of S. Each point $O = (0,0,0,1), A = (0,1,0,0), B = (1,0,0,0), C = (0,0,1,0) \in P(\mathbb{R}^3)$ is associated in a natural way to a vertex of the tetrahedron:

$$\mathrm{Tr} = \{(\alpha, \beta, \gamma, \delta) \in \mathbb{R}_+^4 : \alpha + \beta + \gamma + \delta = d\}.$$

The connected part of $\mathrm{Tr} \setminus T$ containing the vertex associated to the point K ($K \in \{O, A, B, C\}$) is called the *covolume* of T (or f or S) at K and is denoted by $\mathrm{Cv}_f(K)$. The sets Tr and T will be considered in the linear 3-space containing them. Tr will be represented as a tetrahedron in $\mathbb{R}^3$. The points $(\alpha, \beta, \gamma, \delta)$ and (α, β, γ) correspond to each other in the use of T and Δ. Thus $\mathrm{Tr} \cap \mathbb{N}^3$ has a meaning.

If $\Delta' \subset \Delta$, the sum $f^{\Delta'}$ of all the monomials $kx^\alpha y^\beta z^\gamma$ of $f \in \mathbb{R}[x, y, z]$, such that $(\alpha, \beta, \gamma) \in \Delta' \cap \mathbb{N}^3$, is called the *restriction* of f to Δ'. If f is involved in a gluing process, the Viro theorem requires a standard nondegeneracy condition, called NDC here, for all $f^{\Delta'}$, where Δ' is a face of Δ ([V3, R]).

We shall consider surfaces S with isolated singularities at some points K and deformations of S at K of the form

$$S_\varepsilon = \{f_\varepsilon = 0\} \quad \text{with} \quad f_\varepsilon = f + \sum_{\substack{i=1 \\ (\alpha_i, \beta_i, \gamma_i) \in \mathrm{Cv}_f(K)}}^{w} \varepsilon_i x^{\alpha_i} y^{\beta_i} z^{\gamma_i}.$$

Let $K = O$, $\eta > 0$, be the center and the radius of the ball $B(O, \eta)$. Assume that $\partial B(O, \eta')$ is transversal to S for all $\eta' \in \,]0, \eta]$ and that $(S \cap B(O, \eta') \setminus \{O\})$ is smooth or empty. The surface S_ε is said to be a *local deformation* of S at O if there exists an open set $U \subset \mathbb{R}^w$ with the origin in its boundary $\overline{U} \setminus U$ and the following property:

$\forall \varepsilon' \in U, \exists h_{\varepsilon'}$ homeomorphism of pairs:

$$(B(O, \eta), S_{\varepsilon'} \cap B(O, \eta)) \to (B(O, \eta), S_\varepsilon \cap B(O, \eta))$$

whose restriction outside O is C^∞ smooth.

Such a deformation is said to be *smooth*, or a *smoothing*, if $S_\varepsilon \cap B(O, \eta)$ does not contain any singular points. In this case the topological type of the pair $(B(O, \eta), S_\varepsilon \cap B(O, \eta))$ is called the *local topological type* of S_ε at O. Considering the sum of the Betti numbers of $S_\varepsilon \cap B(O, \eta)$, the notion of *maximal local deformation* (i.e., a local version of the Harnack–Thom–Oleĭnik theorem, see [R1]) appears.

The definitions above have straightforward analogs for plane curves or for hypersurfaces of any dimension.

In the next section the symbol τ_d^c (resp. τ_d^s, resp. $\chi_{2k}^{(m)}$) denotes the isotopy types of smooth projective curves of degree d (resp. smooth projective surfaces of degree d, resp. smooth local deformation of singular curves $\{g = 0\} \subset \mathbb{R}^2$ (O is the singular point) with $(0,0), (k,0), (0,2k) \in \mathbb{N}^2$ as vertices of the covolume and m real local distinct branches ($0 \leqslant m \leqslant k$) which all have a quadratic contact and the same sign of curvature at O). Isotopy types mean isotopy types of embeddings (i.e., topological types of pairs).

Consider the set

$$\chi_{2k} \underline{\times} \chi_{2k} = \{(\alpha, \beta) : \alpha \in \chi_{2k}^{(m)}, \beta \in \chi_{2k}^{(m)}, m \in \{0, 1, \ldots, k\}\}.$$

Using [C2] or [K] one can construct a natural map $V' : \chi_{2k} \underline{\times} \chi_{2k} \to \tau_{2k}^c$. Define the subset $\chi_{2k} * \chi_{2k}$ of $\chi_{2k} \underline{\times} \chi_{2k}$ by throwing out all but one element in each preimage $V'^{-1}(V'(\alpha, \beta))$, $(\alpha, \beta) \in \chi_{2k} \underline{\times} \chi_{2k}$. If $k = 3$, the restriction V of V' to $\chi_6 * \chi_6$ becomes a bijection when τ_6^c is replaced by $\tau_6^{c(*)} = \tau_6^c \setminus \{\langle 1\langle 9 \rangle\rangle\}$ ([V2, C2]).

It is a conjecture (referred in III.5.3 as Cjr) that $\chi_{2k}^{(m)}$ does not depend on the value of the curvatures of the k local branches at O. According to [S], the conjecture is true when $k \leqslant 5$.

The inequality $0 < a \ll b$ means that a/b can be chosen as small as it is needed. "Component" must always be understood as "connected component".

§II. A class of singular surfaces of degree 4

Let $\widetilde{\tau}_4^S$ be the set of all isotopy types of smooth real algebraic surfaces of degree 4 nonhomotopic to a point (it contains the three types of surface A such that $\dim H_*(A, \mathbb{Z}_2) = 24$ (maximal)) and $\widetilde{\tau}_4^{S(*)} = \widetilde{\tau}_4^S \setminus \{S_1 \amalg 9S_0\}$ (see III.2).

II.1. THEOREM. *There exist polynomials*

$$f(x, y, z, t) = az^2t^2 + bxyzt + cy^3z + dx^3t + ex^2y^2$$

such that all elements of $\widetilde{\tau}_4^{S(*)}$ are realized by gluing local deformations of the singular surface $S = \{f = 0\}$ at O and local deformations of S at C. They are in the image of a bijective map $L : \chi_6 * \chi_6 \to \widetilde{\tau}_4^{c(*)}$, where L is defined by the gluing.

II.2. COROLLARY. *There exists a bijective map $K : \tau_6^{c(*)} \to \widetilde{\tau}_4^{S(*)}$ ($K = L_0 V^{-1}$).*

II.3. REMARKS. 1) The "onto" property of L and $L_0 V^{-1}$ will only be deduced from [Kh] (the deep conclusive article, together with [G1], of the original version of the first part of the sixteenth Hilbert's problem ([H])) where the one-to-one correspondence between τ_6^c and $\widetilde{\tau}_4^S$ is originally proved.

2) Other elementary relationships between τ_6^c and τ_4^S have been studied by Hilbert [H1], Rohn [Rn], Gudkov, Utkin ([G, G-U]), and Viro ([V5]).

II.4. PROOF OF THEOREM II.1. Consider five real coefficients a, b, c, d, e and five polynomials with real variables:

$$f(x, y, z, t) = az^2t^2 + bxyzt + cy^3z + dx^3t + ex^2y^2,$$
$$f(x, y, z, 1) = az^2 + z(bxy + cy^3) + dx^3 + ex^2y^2,$$
$$f(x, y, 1, t) = at^2 + t(bxy + dx^3) + cy^3 + ex^2y^2,$$
$$D_0(x, y, 1) = (bxy + cy^3)^2 - 4a(dx^3 + ex^2y^2),$$
$$D_c(x, y, 1) = (bxy + dx^3)^2 - 4a(cy^3 + ex^2y^2)$$

such that the affine curves $\{D_0 = 0\}$, $\{D_c = 0\}$ are unions of three distinct parabolas with the same sign of curvature at the origin.

To meet this condition choose for instance: $b < 0$, $c > 0$, $d > 0$, $e < 0$, $-e/d \ll -c/b$, $-e/c \ll -d/b$, $0 < a \ll 1$, then look at the Newton polygons and the two intersections of the surfaces in $\mathbb{R}^3$ defined by:

$$\begin{cases} u = [y(bx + cy^2)]^2, \\ u = 4a(dx^3 + ex^2y^2), \end{cases} \qquad \begin{cases} u = [x(by + dx^2)]^2, \\ u = 4a(cy^3 + ex^2y^2). \end{cases}$$

The topology of the double coverings:

$$\{f(x, y, z, 1) = 0\} \quad \text{over} \ \{D_0 \geqslant 0\} \qquad \text{(Figure 1-a)},$$
$$\{f(x, y, 1, t) = 0\} \quad \text{over} \ \{D_c \geqslant 0\} \qquad \text{(Figure 1-b)}$$

follows and implies that $S = \{f = 0\} \subset P(\mathbb{R}^3)$ is diffeomorphic to a sphere with three handles and two degenerate singular points at O and C. See Figure 1-c. S is not homotopic to a point.

The Newton polyhedron T of the surface S divides Tr ($d = 4$) in two symmetric parts. It is easy to check that any deformation $S_{\varepsilon,0}$ of S defined by:

$$f(x, y, z, 1) + \sum_{(\alpha_i, \beta_i, \gamma_i) \in \mathrm{Cv}_f(0)} \varepsilon_i x^{\alpha_i} y^{\beta_i} z^{\gamma_i} = 0$$

corresponds to a deformation of $\{D_0 = 0\}$ defined by:

$$D_{\varepsilon,0}(x, y, 1) = D_0(x, y, 1) + \sum_{(u_j, v_j) \in \mathrm{Cv}_{D_0}(0)} \varepsilon'_j x^{u_j} y^{v_j} = 0.$$

The polygon T and the Newton polygon of D_0 (resp. the covolumes $\mathrm{Cv}_f(O)$ and $\mathrm{Cv}_{D_0}(O)$) are represented in Figure 2-a (resp. 2-b); see p. 46.

REMARK. $\{D_{\varepsilon,0} = 0\}$ can be arbitrarily chosen.

PROOF. $D_{\varepsilon,0}(x, y, 1)$ has the form $Q_3^2(x, y) - 4aP_4(x, y)$, where the Newton polygons Δ_3 of Q_3^2 and Δ_4 of P_4 are as in Figure 2-c ($\deg Q_3 = 3$, $\deg P_4 = 4$). Face restrictions $Q_3^{2\Gamma'}$ and P_4^{Γ} are defined by $D_0 = Q_3^{2\Gamma'} - 4aP_4^{\Gamma}$ (i.e., by S). $P_4^{\Delta_4 \setminus \Gamma}$ is arbitrary. If

$$Q_3(x, y) = a_3 y^3 + a_2 y^2 + a_1 y + a_0 + x(b_1 y + b_0),$$

then

$$Q_3^{2(\Delta_3 \setminus (\Delta_4 \cap \Delta_3))} = \underline{a_3^2 y^6} + 2a_3 a_2 y^5 + \underline{2b_1 a_3 x y^4}.$$

The underlined parts are monomials of $Q_3^{2\Gamma'}$, thus a_3 and b_1 are fixed. The third coefficient $2a_3 a_2$ is arbitrary (there is no condition on a_2). Finally the hypothesis on S does not impose a condition on $D_{\varepsilon,0} - D_{\varepsilon,0}^{\Gamma \cup \Gamma'} = D_{\varepsilon,0} - D_0$, thus on the deformation $\{D_{\varepsilon,0} = 0\}$ of $\{D_0 = 0\}$.

CONSEQUENCE. *There exist local deformations $S_{\varepsilon,0} = \{f_{\varepsilon,0} = 0\}$ of S at O, with $f_{\varepsilon,0}(x, y, z, 1) = az^2 + zQ_3(x, y) + P_4(x, y)$, that are double coverings over $\{D_{\varepsilon,0} = 0\}$, where $\{D_{\varepsilon,0} = 0\}$ can realize any element of $\chi_6^{(3)}$. For symmetrical reasons the same can be said in the neighborhood of the singular point C (change affine charts).*

The assumptions of the Viro theorem on gluing are verified for the given surfaces. Namely $f_{\varepsilon,0}$, $f_{\varepsilon,c}$ have NDC, the equality $f_{\varepsilon,0}^T = f_{\varepsilon,c}^T = f$ holds and there exists a

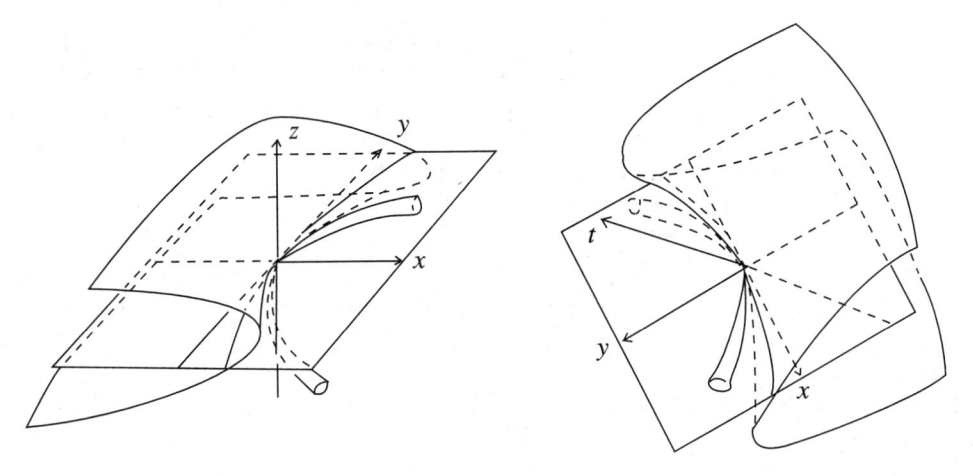

FIGURE 1-a

FIGURE 1-b

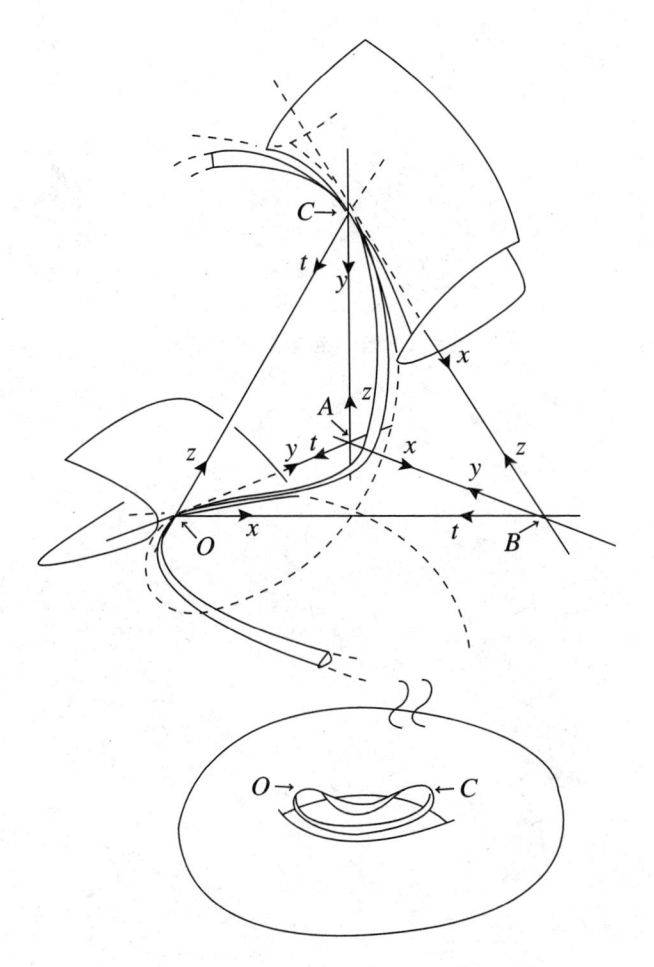

FIGURE 1-c

convex function $\nu\colon \mathrm{Tr} \to \mathbb{R}$, $\nu(\mathrm{Tr} \cap \mathbb{N}^3) \subset \mathbb{Z}$, linear on each connected subset of $\mathrm{Tr} \setminus \mathrm{T}$. The use of this theorem proves (in the shortest way) the existence of a (injective) map $L\colon \chi_6 * \chi_6 \to \widetilde{\tau}_4^{S(*)}$. In other words, gluing various deformations $S_{\varepsilon,0}$ and $S_{\varepsilon,c}$, which can be viewed as performing independent local smoothings of S at O and C, gives Theorem II.1 by considering II.3.1 and the Figures below.

Figure 3-a represents maximal deformations of $\{f = 0\}$ and $\{D_0 = 0\}$ at O. Figure 3-b represents the set $\widetilde{\tau}_4^S$ deduced from the well-known set τ_6^c ([**G, V4**]). Each element (p, q) (resp. [1,1]) is the disjoint union of a genus p orientable component and q embedded spheres (resp. two nonlinked disjoint torus) outside each other.

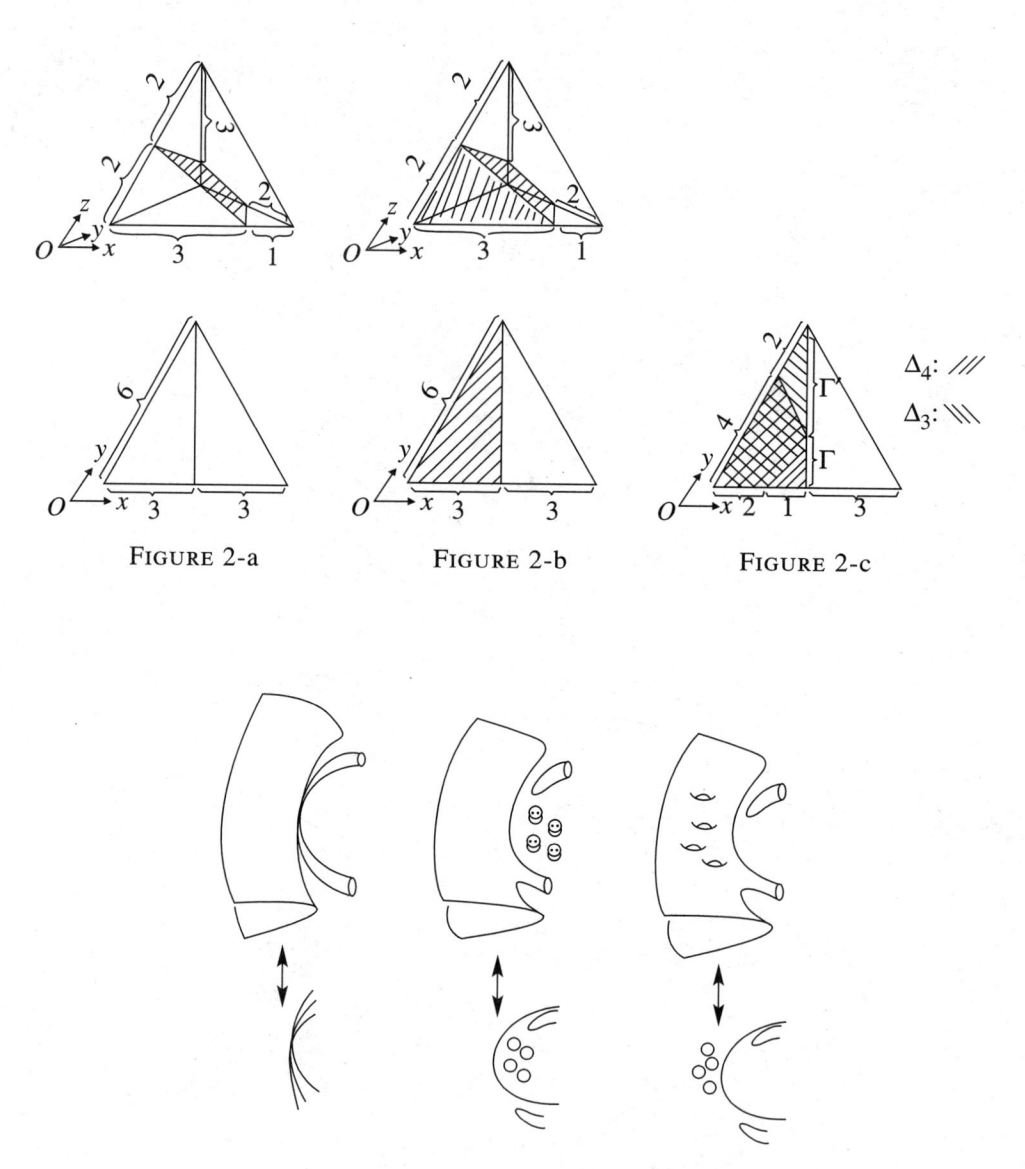

FIGURE 2-a   FIGURE 2-b   FIGURE 2-c

FIGURE 3-a

$$\begin{array}{ccc}
(10,1) & (6,5) & (2,9) \\
(10,0)\ (9,1) & (6,4)\ (5,5) & (2,8)\ (1,9)
\end{array}$$

$(9,0)\ (8,1)\ (7,2)\ (6,3)\ (5,4)\ (4,5)\ (3,6)\ (2,7)\ (1,8)$

$(8,0)\ (7,1)\ (6,2)\ (5,3)\ (4,4)\ (3,5)\ (2,6)\ (1,7)$

$(7,0)\ (6,1)\ (5,2)\ (4,3)\ (3,4)\ (2,5)\ (1,6)$

$(6,0)\ (5,1)\ (4,2)\ (3,3)\ (2,4)\ (1,5)$

$(5,0)\ (4,1)\ (3,2)\ (2,3)\ (1,4)$

$(4,0)\ (3,1)\ (2,2)\ (1,3)$

$(3,0)\ (2,1)\ (1,2)$

$(2,0)\ (1,1)\qquad [1,1]$

$(1,0)$

FIGURE 3-b

§III. Generalization

Consider a surface $S \subset P(\mathbb{R}^3)$ of degree d with isolated singularities at O and C whose Newton polyhedron is the polygon containing the point F with vertices at D_1, D_2, D_3, D_4 represented in Figures 4-a, 4-b (d odd, d even). Such surfaces define the class of central singular surfaces mentioned in §I as well as in Theorem III.3.

Let $\Gamma = [D_3, F]$, $\Gamma' = [D_2, D_1]$. The equation of S restricted to the affine chart centered at O is

$$az^2 + zK^\Gamma(x, y) + N^{\Gamma'}(x, y) = 0,$$

$$K^\Gamma(x, y) = \prod_{i=1}^{(d-1)/2} (x - k_i y^2) \quad \text{if } d \text{ is odd},$$

$$K^\Gamma(x, y) = y \prod_{i=1}^{(d-2)/2} (x - k_i y^2) \quad \text{if } d \text{ is even},$$

$$N^{\Gamma'}(x, y) = x^{d-2}(x - \ell y^2),$$

where all the real numbers k_i (and ℓ) are positive and pairwise distinct.

From now on τ_d^c (resp. τ_d^S, resp. $\chi_{2(d-1)}^{(d-1)}$) defined in §I will be restricted to the subset of the isotopy types of maximal curves (resp. maximal surfaces, resp. maximal deformations with $\chi_{2(d-1)}^{(d-1)} = \chi_{2(d-1)}$). Most difficulties appear at this level.

The surface $S \subset P(\mathbb{R}^3)$ will be deformed locally at O and C by gluing four surfaces $S^{(1)}$, $S^{(2)}$, $S^{(3)}$, $S^{(4)}$ whose Newton polyhedra coincide with the regions 1, 2, 3, 4, bounded by polygons, represented in Figure 4-c.

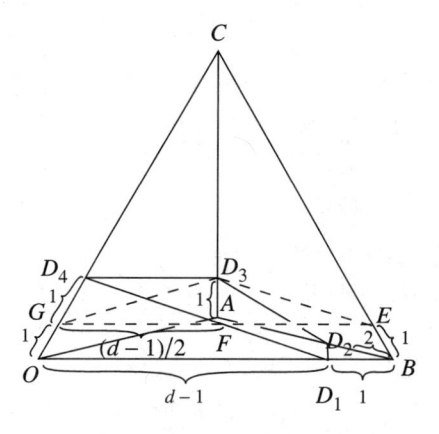

FIGURE 4-a

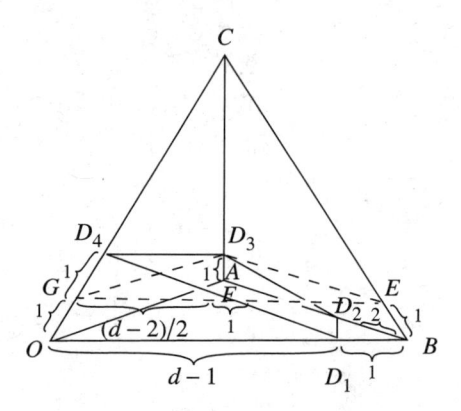

FIGURE 4-b

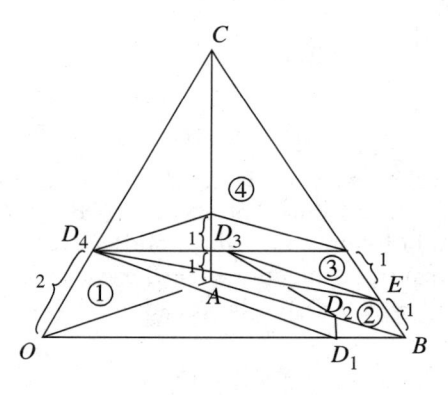

FIGURE 4-c

III.1 LEMMA.[1] *Set*

$$D(x,y,1) = (K^\Gamma(x,y))^2 - 4aN^{\Gamma'}(x,y),$$
$$D_\varepsilon(x,y,1) = (K(x,y))^2 - 4aN(x,y), \qquad 0 < a \ll 1.$$

A local deformation of S at O induces a local deformation $\{D_\varepsilon = 0\}$ of the curve $\{D = 0\}$ at O, which can be arbitrarily chosen. Thus the surface S can be deformed as a double covering over $\{D_\varepsilon \geq 0\}$ such that the set of isotopy types of $\{D_\varepsilon = 0\}$ is $\chi_{2(d-1)}$.

PROOF. It suffices to show that the coefficients of $D_\varepsilon^{\overline{\Delta}\setminus\overline{\Gamma}} = (K^2)^{\overline{\Delta}\setminus\overline{\Gamma}}$ can be controlled, where $\overline{\Delta} \subset \mathbb{R}^2$ (resp. $\overline{\Gamma} \subset \partial\overline{\Delta}$) is the convex hull of the points $(0, d+1)$, $(0, 2(d-1))$, $(d-3, 4)$ (resp. $(0, 2(d-1))$, $(d-3, 4)$). Write

$$K(x,y) = \sum_{i=0}^{u} \left(x^i \sum_{j=1}^{d-1-2i} a_{i,j} y^j \right)$$

---

[1] This collection was ready for publication when I was informed of an obvious mistake in Lemma III.1, which is actually true only for $d \leq 6$. Therefore, Theorem III.3 is also true only with this restriction. For $d > 6$, some constraints appear in the connection between surfaces of degree d and plane curves of degree $2(d-1)$. It will be looked at again in a subsequent paper.

and $u = (d-1)/2$ (resp. $d/2 - 1$) if d is odd (resp. even).

If $\alpha_k = 2(d-1) - (k-1)$, $a_{0,d-1} \neq 0$, then $(K^2)^{\overline{\Delta}}$ is equal to

$$\left[(a_{0,d-1}^2)y^{\alpha_1} + \sum_{k=2}^{d-2}(\cdots + 2a_{0,d-1}\dot{a}_{0,d-k} + \cdots)y^{\alpha_k} \right]$$
$$+ x\left[(2a_{0,d-1}a_{1,d-3})y^{\alpha_3} + \sum_{k=4}^{d-2}(\cdots + 2a_{0,d-1}\dot{a}_{1,d-k} + \cdots)y^{\alpha_k} \right]$$
$$+ x^2\left[(a_{1,d-3}^2 + 2a_{0,d-1}a_{2,d-5})y^{\alpha_5} + \sum_{k=6}^{d-2}(\cdots + 2a_{0,d-1}\dot{a}_{2,d-k} + \cdots)y^{\alpha_k} \right]$$
$$+ \cdots + x^{d-4}[\cdots + (\cdots + 2a_{0,d-1}\dot{a}_{d-4,2} + \cdots)y^5]$$
$$+ x^{d-3}[(\cdots + 2a_{0,d-1}\dot{a}_{d-3,2} + \cdots)y^4].$$

The undefined coefficients (associated with $\overline{\Gamma}$) are fixed when D is given. One can choose all the coefficients $\dot{a}_{i,j}$ marked with a spot in decreasing order of the second index and increasing order of the first index:

$$\dot{a}_{0,d-2}, \ldots, \dot{a}_{0,2}, \dot{a}_{1,d-4}, \ldots, \dot{a}_{1,2}, \dot{a}_{2,d-6}, \ldots, \dot{a}_{d-4,2},$$

therefore one can choose all the coefficients of the monomials of $(K^2)^{\overline{\Delta}\setminus\overline{\Gamma}}$. In other words, the underlined coefficients are those of monomials that define altogether the singularity but not the deformations. The deformation is controlled by the coefficients with a spot.

REMARK. In what follows (as in §II), maximal surfaces are reached with the condition $0 < a \ll 1$, which implies that $\{D = 0\}$ is locally the union of $d-1$ real branches with quadratic contacts and positive curvatures at O.

III.2. **Notation.** The following symbols are currently used to describe components of smooth algebraic surfaces in $P(\mathbb{R}^3)$:

qS_p is the isotopy type of the disjoint union of q smooth connected orientable surfaces of genus p outside each other (for the natural notion of interior and exterior). R_m is the isotopy type of a smooth nonorientable surface of genus m. Such a component exists for a smooth surface A if and only if deg A is odd.

$K_1 \amalg K_2$ denotes the isotopy type of the disjoint union of two smooth surfaces with isotopy types K_1 and K_2.

$K_1 \langle K_2 \langle \cdots \langle K_n \rangle \rangle \cdots \rangle$ stands for the isotopy type of n inclusions of connected orientable smooth surfaces with isotopy types $K_1, \ldots, K_n$. The surface associated to K_j contains the surface associated to K_{j+1}, $j \subset \{0, \ldots, n-1\}$.

If the disjoint union $C \amalg C'$ is the set of the ovals of a real plane algebraic curve, then C is said to be *free* if any oval of C does not contain, and is not contained in, any oval of C'.

Some constructions use maximal curves of degree i with a condition on oscillation for one component (see the proof of III.4). Their isotopy types form a subset $\tau_i^{c'}$. (It is not clear whether $\tau_i^c \setminus \tau_i^{c'}$ is empty or not.)

III.3. THEOREM. *For any degree $d \geqslant 3$ there exist singular surfaces S of the type introduced above, together with a set of local smoothings at O and C related to plane curves such that their gluing defines a map*

$$X_d: \chi_{2(d-1)} \times \tau_{d-2}^{c'} \times \tau_{d-3}^{c'} \times \cdots \times \tau_1^{c'} \longrightarrow \tau_d^S.$$

III.4. THEOREM. *Let $\mathcal{F}_d$ be the image of X_d. Then*
1. *$\mathcal{F}_5$ contains $R_{22} \amalg 4S_0$, $R_{18} \amalg 8S_0$, $R_{14} \amalg 12S_0$;*
2. *$\mathcal{F}_6$ contains $S_{43} \amalg 10S_0$, $S_{39} \amalg 14S_0$, $S_{35} \amalg 18S_0$, $S_{31} \amalg 22S_0$, $S_{42} \amalg S_1 \amalg 9S_0$, $S_{38} \amalg S_1 \amalg 13S_0$, $S_{34} \amalg S_1 \amalg 17S_0$, $S_{36} \amalg S_3 \amalg 13S_0$, $S_{34} \amalg S_5 \amalg 13S_0$, $S_{37} \amalg S_6 \amalg 9S_0$.*

III.5. REMARKS. 1) The steps to prove III.3 and III.4 are Lemma III.1 and §§IV, V, VI, and VII. Theorem III.3 shows that surfaces of degree d can be connected with curves of degree $2(d - 1)$. Improvements would consist of changing some sets $\tau_k^{c'}$ into sets of isotopy types of curves of degree higher than k. An example is the map L' extended from L:

$L': \chi_6 \times \chi_6 \to \widetilde{\tau}_4^S \subset \tau_4^S$ in Theorem II.1. There also exists a map $\chi_8 \times \chi_8 \to \tau_5^S$ such that its image contains the three elements of III.4.1 as well as $R_{10} \amalg 16S_0$, $R_6 \amalg 20S_0$ (author's unpublished result).

2) Most of the explicit isotopy types above are known to V. Kharlamov and O. Viro ([**Kh1, Kh2**], unpublished). Our method is different and $\mathcal{F}_d$ might not coincide with what they construct when $d > 6$.

3) Lemma III.1 and the knowledge of $\chi_{2(d-1)}$ give surfaces $S^{(1)}$ with all possible isotopy types. The next sections will describe constructions for surfaces $S^{(4)}$, $S^{(3)}$, $S^{(2)}$ and explain

$$\tau_{d-2}^{c'} \times \tau_{d-3}^{c'} \times \cdots \times \tau_1^{c'}$$

in the source set of X_d. Note that the complete description of $\chi_{2(d-1)}$ is not yet achieved when $d > 5$ ([**P, Ch**]). If conjecture Cjr is not true (it could happen if $d > 6$, see §I), then the set $\chi_{2(d-1)}$ of Theorem III.3 depends on the actual curves constructed in §VI (Theorem III.4 relies on the proved cases of Cjr).

4) Lemma VI.1 and thus Theorem III.3 can be extended if all the possible isotopy types of the curves $\{L = 0\}$ defined in §VI (i.e., χ_d when d is even) are considered.

§IV. Surfaces $S^{(4)}$

Let H_i be a homogeneous polynomial of degree i. All isotopy types of smooth hypersurfaces $\{f = 0\}$ of degree d can theoretically be realized as local deformations at O of cones $\{H_d = 0\}$. (Write $f = H_d + H_{d-1} + \cdots + H_1$ and apply $(x, y, z) \mapsto (\lambda x, \lambda y, \lambda z)$).

Proposition IV.1 gives an explicit element of τ_d^S. It is more precise than only the genus of the components calculated in [**V6**], where Harnack's technique ([**Ha**]) is adapted to surfaces. In fact the constructive methods here and in [**V6**] are connected by blowing ups. As Viro conjectured, straightforward generalizations of [**V6**] (and thus of Proposition IV.1) prove the existence of maximal hypersurfaces for any dimension and any degree ([**V3**]).

IV.1. PROPOSITION. *There exists a maximal surface σ of degree d, not homotopic to a point, with the following isotopy type*:

$$S_{(2d^3-6d^2+7d)/6} \langle \underbrace{S_0 \amalg \cdots \amalg S_0}_{(d^3-12d^2+44d)/48-1} \rangle \amalg \underbrace{S_0 \amalg \cdots \amalg S_0}_{(7d^3-36d^2+44d)/48} \quad \text{if } d \text{ is even,}$$

$$R_{(2d^3-6d^2+7d-3)/6} \amalg \underbrace{S_0 \amalg \cdots \amalg S_0}_{(d^3-6d^2+11d)/6-1} \quad \text{if } d \text{ is odd.}$$

PROOF. Consider a polynomial $f_\varepsilon(x, y, z)$ and the following hypothesis (P):

$$f_\varepsilon = H_d + \sum_{i=1}^{d} \varepsilon_{d-i} H_{d-i}, \quad |\varepsilon_0| \ll |\varepsilon_1| \ll \cdots \ll |\varepsilon_{d-1}| \ll 1,$$

each curve $C_j = \{H_j(x, y, 1) = 0\}$ is maximal with the isotopy type described in Figure 5-a and has a component intersecting transversally a component of C_{j+1} (resp. C_{j-1}) in $j(j+1)$ (resp. $j(j-1)$) points in the fashion of Figure 5-b, $2 \leqslant j \leqslant d-1$. (There are no other intersection points according to the Bezout theorem.)

Those requirements (P) are met by affine transformations on Harnack's constructions (or others: [**C2, K, V4**]) which give "oscillating" conditions along a straight line.

To obtain the theorem it suffices to see $\{f_\varepsilon = 0\} = \sigma \cap \mathbb{R}^3$ as a gluing (in the sense of Viro) of all the surfaces $(F_n)_{0 \leqslant n \leqslant d-1}$, $F_n = \{\varepsilon_{n+1} H_{n+1} + \varepsilon_n H_n = 0\} \subset \mathbb{R}^3$, $\varepsilon_d = 1$.

Indeed, the topology of F_n is well understood in the affine chart $(x, y, 1, t)$. It is the graph of the rational function:

$$t = -\frac{\varepsilon_{n+1} H_{n+1}(x, y, 1)}{\varepsilon_n H_n(x, y, 1)}.$$

Take x or y as a parameter. The saddle points (all with $t \neq 0$ after small perturbations if necessary) are detected through a finite set of graphs of rational functions in one variable. See Figure 5-c, where the signs of $\varepsilon_n \varepsilon_{n+1} H_n H_n(x, y, 1)$ are indicated. If $\{H(x, y, z)_{n+1} = 0\} \cap \{H(x, y, z)_n = 0\}$ is one-dimensional, the surface F_n has handles (i.e., connected sums with orientable surfaces of genus 1). One of them carries the unique singular point of multiplicity n at O.

According to the easy lemma below, the ovals of C_n not intersecting C_{n+1} or C_{n-1} increase the genus and the number of local components with zero or nonzero genus in the gluing process of F_{n-1} and F_n. See Figure 5-d, where the sphere and the hole are due to the oval of C_3.

LEMMA. *If $\{H(x, y, z)_{n+1} = 0\} \cap \{H(x, y, z)_n = 0\} = \{O\}$ and V is an open ball centered at O with a small radius, then the affine surface F_n is isotopic to the union of the three sets*:

$$\{H_{n+1} = 0\} \cap \mathbb{R}^3 \setminus \overline{V}, \quad \{H_n = 0\} \cap V, \quad \{\varepsilon_{n+1} \varepsilon_n H_{n+1} H_n \leqslant 0\} \cap \partial V.$$

Now use Figures 5-a and 5-b to count the genus and the number of components after gluing all the F_n, $1 \leqslant n \leqslant d-1$. This leads to a map: $\tau_{d-1}^{c'} \times \tau_{d-2}^{c'} \times \cdots \times \tau_1^{c'} \to \tau_d^{S}$, from the isotopy types of C_j, $1 \leqslant j \leqslant d-1$, to the isotopy types of maximal projective surfaces. (The curve C_d (not empty) does not influence the isotopy type of the final surface when d is fixed.)

j even $\frac{j(j-6)}{8}$ ovals $\frac{3j(j-2)}{8}$ ovals

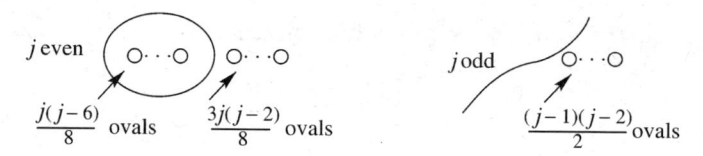

j odd $\frac{(j-1)(j-2)}{2}$ ovals

FIGURE 5-a

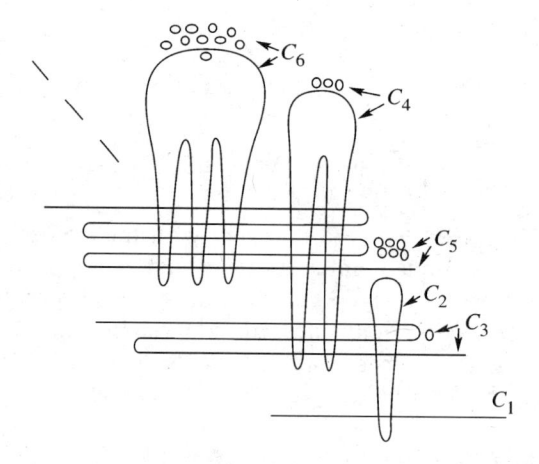

FIGURE 5-b

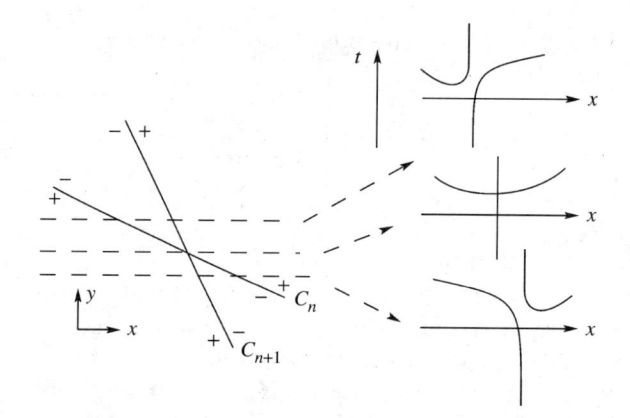

FIGURE 5-c

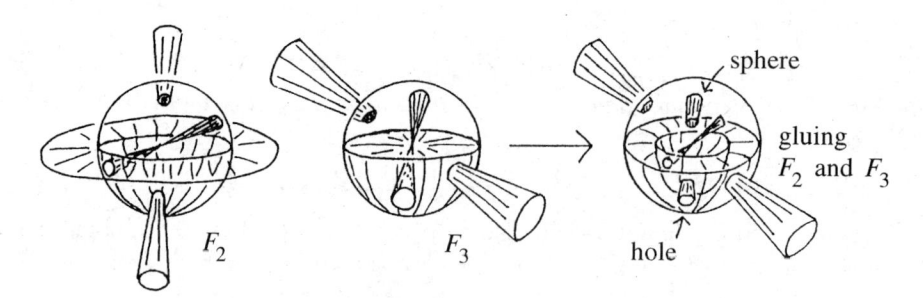

FIGURE 5-d

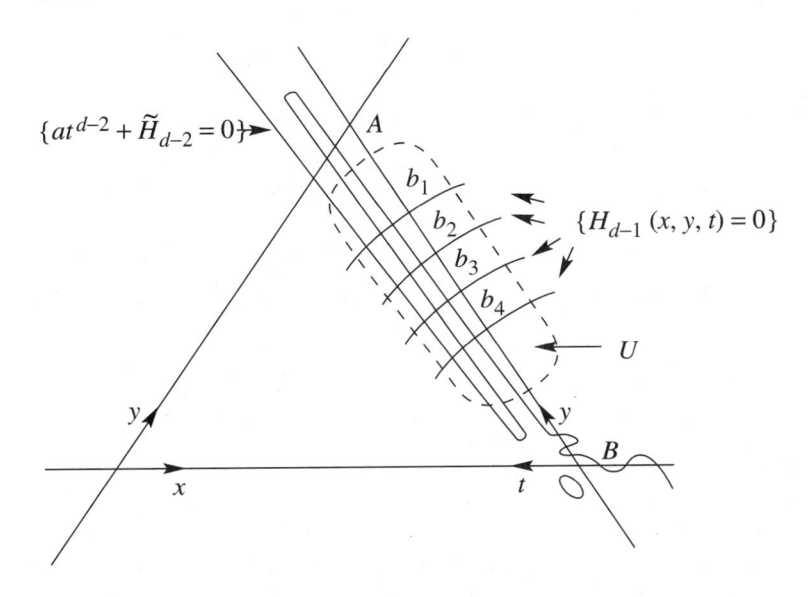

FIGURE 6

§V. Surfaces $S^{(3)}$

Let H_{d-2} (resp. H_{d-1}, resp. $\widetilde{H}_{d-2}$) be a homogeneous polynomial of degree $d-2$ (resp. $d-1$, resp. $d-2$) in x, y, t (resp. such that $\{H_{d-1}(x, y, 0) = 0\}$ is the union of $d-1$ distinct straight lines, resp. without monomials in t^{d-2}), and let h be the polynomial $h(x, y, t) = at^{d-2} + H_{d-1}(x, y, 0)$. If $S^{(3)} = \{h + \widetilde{H}_{d-2} = 0\} \subset \mathbb{R}^3 = (x, y, 1, t)$, then we have the following

V.1. LEMMA. *Fix a small neighborhood V of $C = (0, 0, 1, 0) \in P(\mathbb{R}^3)$. There exist polynomials $\widetilde{H}_{d-2}$ and $\varepsilon_{d-2} H_{d-2}$, $|\varepsilon_{d-2}| \ll 1$, such that*:
  (i) $S^{(3)} \cap (\mathbb{R}^3 \setminus V)$ *is isotopic to* $\{h = 0\} \cap (\mathbb{R}^3 \setminus V)$;
  (ii) $S^{(3)} \cap V$ *is isotopic to the local deformation* $F_{d-2} = \{H_{d-1} + \varepsilon_{d-2} H_{d-2} = 0\}$ *of* $\{H_{d-1} = 0\}$ *at C, where F_{d-2} is the surface of* §IV *verifying* (P).

PROOF. The Viro gluing theorem implies the existence of a local deformation $S^{(3)}$ of $\{h = 0\}$ which can be interpreted as a gluing of $\{h = 0\}$ with the surface

$$\{H_{d-1}(x, y, 0) + \varepsilon_{d-2} H_{d-2}(x, y, t) = 0\}.$$

It remains to prove how the two local parts $S^{(3)} \cap V$, $F_{d-2} \cap V$ can be isotopic.

Let U be a tubular neighborhood in the $(1, y, 0, t)$-affine plane of the 1-simplex $[b_1, b_{d-1}]$ included in the $(1, y, 0, 0)$-axis, where $\{b_1, b_2, \ldots, b_{d-1}\}$ is the zero set of $H_{d-1}(1, y, 0)$. Let L be the cone lying on ∂U with vertex at $C = (0, 0, 1, 0)$. Usual methods ([A, C1, K, V4] and affine transformations) allow us to construct a maximal curve $\{at^{d-2} + \widetilde{H}_{d-2}(1, y, t) = 0\}$, where the real a is given, containing a component C which verifies:
  - $C \cap U$ is in one connected part of $U \setminus (U \cap \{z = t = 0\})$;
  - $C \cap U$ is isotopic to $d-2$ straight lines parallel to $[b_1, b_{d-1}]$ (see Figure 6 for $d = 5$);
  - $C \cap U \cap \{H_{d-1} = 0\}$ is a set of $(d-1)(d-2)$ transversal intersection points.

If $\varepsilon_{d-2}H_{d-2}(x,y,t) = at^{d-2} + \widetilde{H}_{d-2}(x,y,t)$, where a is taken small enough, the surface

$$S^{(3)} \cap \Lambda = \{h + \widetilde{H}_{d-2} = 0\} \cap \Lambda = \{H_{d-1}(x,y,0) + \varepsilon_{d-2}H_{d-2}(x,y,t) = 0\} \cap \Lambda$$

is isotopic to the surface

$$F_{d-2} \cap \Lambda = \{H_{d-1}(x,y,t) + \varepsilon_{d-2}H_{d-2}(x,y,t) = 0\} \cap \Lambda,$$

since their restrictions to the $(1,y,z,t)$-affine chart are respectively the graphs of the two functions:

$$z = \frac{H_{d-1}(1,y,0)}{\varepsilon_{d-2}H_{d-2}(1,y,t)}, \qquad z = \frac{H_{d-1}(1,y,t)}{\varepsilon_{d-2}H_{d-2}(1,y,t)}$$

and $\{H_{d-1}(1,y,t) = 0\} \cap U$ is isotopic to $\{H_{d-1}(1,y,0) = 0\} \cap U$.

Stretch U along the $(1,y,0,0)$-axis so that the ovals of $\{H_{d-2} = 0\}$ are inside U. Section IV tells us that all the "homology" of $S^{(3)}$ is inside Λ. Since the surfaces $S^{(3)}$ and F_{d-2} are constructed as local deformations, Λ can be replaced by $\Lambda \cap V$. Thus $S^{(3)} \cap V$ is isotopic to $F_{d-2} \cap V$.

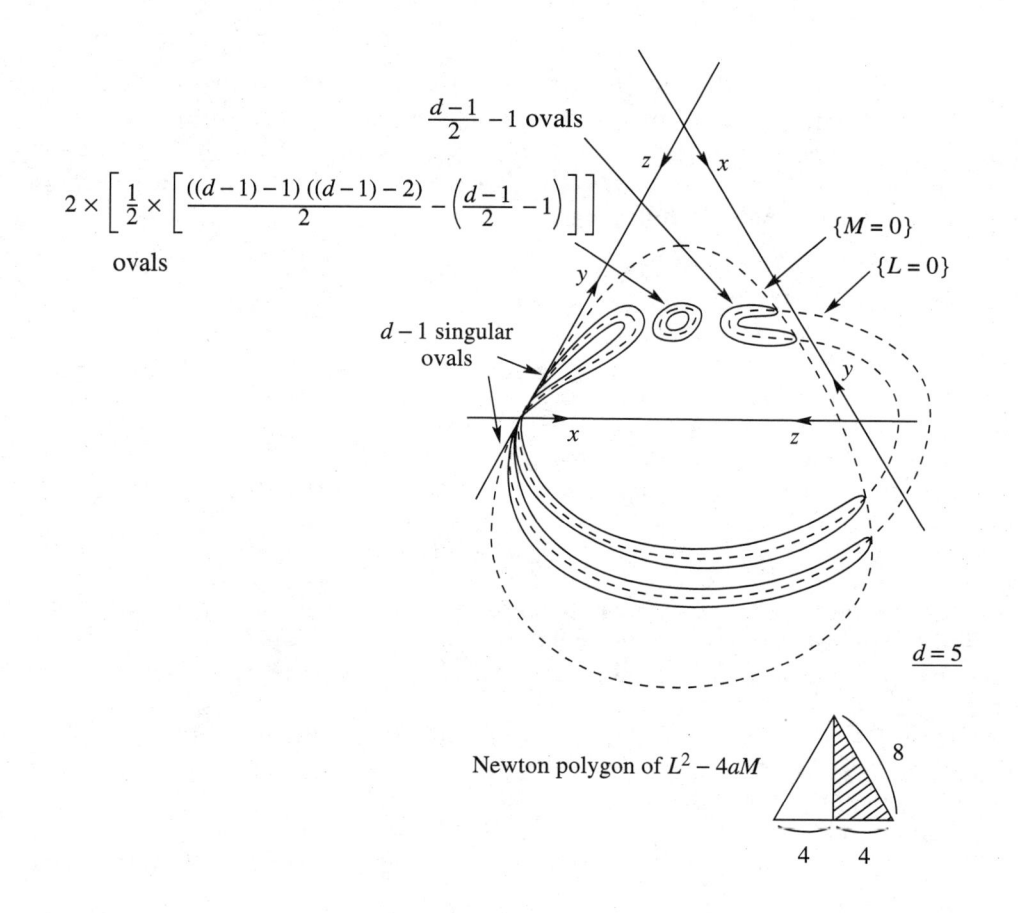

FIGURE 7-a

{L = 0}

{M = 0}

$\frac{d}{2} - 1$ ovals

$d - 1$ singular ovals

$d = 6$

$2 \times \left[\frac{1}{2} \times \left[\frac{((d-1)-1)((d-1)-2)}{2} - \left(\frac{d}{2} - 1\right) \right] \right]$ ovals

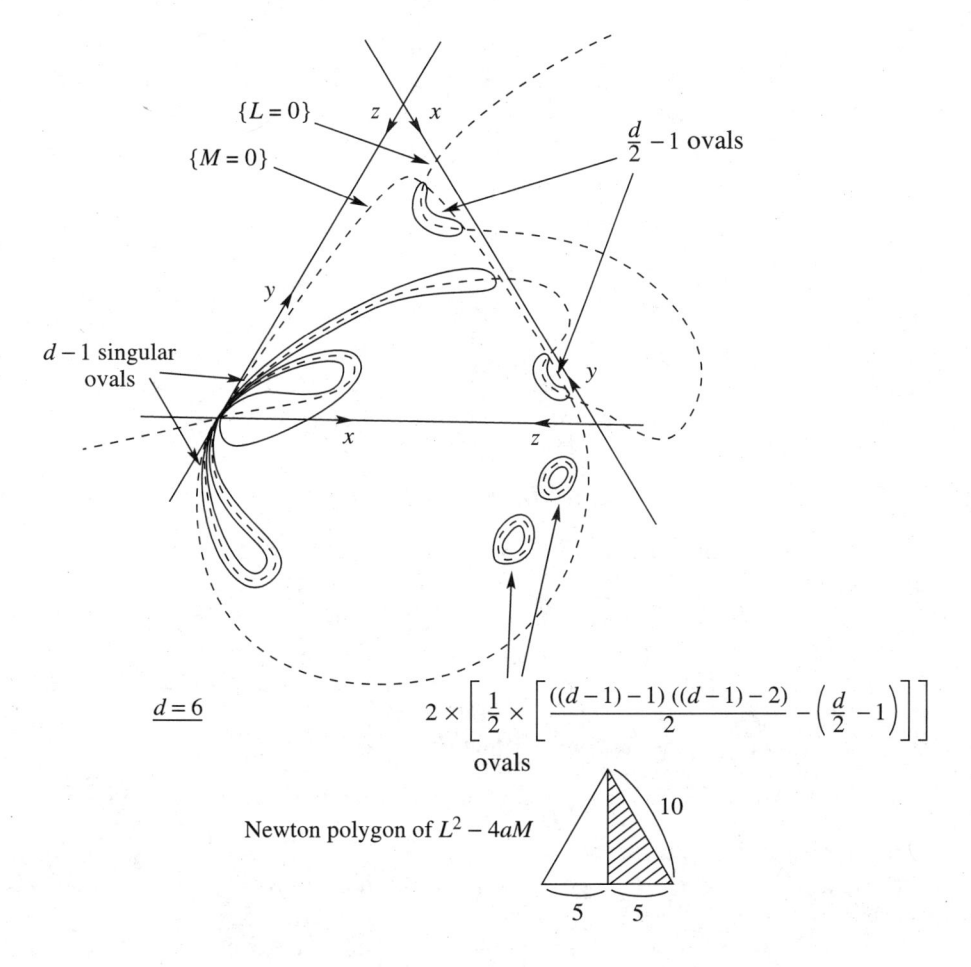

Newton polygon of $L^2 - 4aM$

FIGURE 7-b

§VI. The surfaces $S^{(2)}$

The equation of $S^{(2)}$ in the affine chart centered at O is $az^2 + zL(x, y) + M(x, y) = 0$. Thus $S^{(2)}$ restricted to this chart is a double covering over $\{L^2 - 4aM \geq 0\}$.

VI.1. LEMMA. *If $\deg S = d$, then there exists a curve $\{L^2 - 4aM = 0\}$, the disjoint union of three sets O_1, O_e, O_s represented in Figure 7-a (resp. 7-b) when $d = 5$ (resp. $d = 6$), which also suggests the following general case when d is odd (resp. even):*
- O_1 *is the union of* $(d-1)/2 - 1$ *(resp.* $d/2 - 1$*) free ovals.*
- O_ε *is the union of*

$$\frac{1}{2}\left[\frac{((d-1)-1)((d-1)-2)}{2} - \left(\frac{d-1}{2} - 1\right)\right];$$

(*resp.*

$$\frac{1}{2}\left[\frac{((d-1)-1)((d-1)-2)}{2} - \left(\frac{d}{2} - 1\right)\right])$$

free nests of two ovals: ⊚.

- O_s is a singular connected curve (*the union of* $d - 1$ *loops*) *with a singular point at* O *and locally the union of* $2(d - 1)$ *distinct components having a quadratic contact with the* Oy *axis at* O.

PROOF. Such a curve $\{L^2 - 4aM = 0\}$ is deduced from classical curve constructions and from the hypotheses (i), (ii) below:
  (i) $L^\Gamma = K^\Gamma$, $M^{\Gamma'} = N^{\Gamma'}$ (where K, N, Γ, Γ' are defined in §III) and $\{L = 0\}$ is a maximal local deformation of $\{L^\Gamma = 0\}$ at $B = (1, 0, 0)$ (see [C2, K]).
  (ii) $M(x, y) = \tau x^{d-2}(x - \ell y^2 - \tau_1 x^2)$ with $0 < \tau \ll 1$, $0 < \ell \ll 1$, $1 \ll \tau_1$.

It follows from (i), (ii) that $\{L^2 - 4aM = 0\}$ is a small perturbation of $\{L^2 = 0\}$ in the sense of Harnack or Hilbert (see [A, G]). Note that $0 < a \ll 1$ is compatible with the condition on a in V. The curve $\{(L^\Gamma)^2 - 4aM^{\Gamma'} = 0\}$ is reduced and thus is isotopic to $\{L^2 - 4aM = 0\}$ in a small neighborhood of O.

§VII. Proof of Theorem III.3

Polynomials defining surfaces $S^{(1)}$, $S^{(2)}$, $S^{(3)}$, $S^{(4)}$ can be chosen with NDC and common restrictions to the faces of the Newton polyhedra (1), (2), (3), (4). In particular, take the polynomial $H_{d-1}(x, y, t)$ of §V such that $zH_{d-1}(x, y, 1)$ is the polynomial of type $zg(x, y)$ in the equation of $S^{(1,2)}$, where $S^{(1,2)}$ is defined as a gluing of $S^{(1)}$ and $S^{(2)}$ or equivalently as a double covering over $\{G \geqslant 0\}$, where $\{G = 0\}$ is a gluing of $\{L^2 - 4aM = 0\}$ and $\{K^2 - 4aN = 0\}$.

It is clear that there exists a convex function:

$v\colon \mathrm{Tr} \to \mathbb{R}$ with $v(\mathrm{Tr} \cap \mathbb{N}^3) \subset \mathbb{Z}$, v linear on each part (1), (2), (3), (4).

Thus the Viro theorem can be used to glue $S^{(1)}$, $S^{(2)}$, $S^{(3)}$, $S^{(4)}$ ([V3, R]), which leads to surfaces A whose isotopy types depend on the isotopy types of $S^{(1)}$ and $S^{(4)}$ (III.1, IV).

Theorem III.3 now follows from the following statement proved in 1), 2), 3) below:
the surfaces A *can be maximal, i.e.,*

$$\dim H_*(A, \mathbb{Z}_2) = \frac{(d-1)^4 - 1}{d} + 4 = d^3 - 4d^2 + 6d,$$

when the deformation $\{K^2 - 4aN = 0\}$ of $\{(K^\Gamma)^2 - 4aN^{\Gamma'} = 0\}$ has its isotopy type in $\chi_{2(d-1)}$ and both curves $\{H_{d-2} = 0\}$, $\{L = 0\}$ in §§V, VI, and the surface $S^{(4)}$ are maximal.

1) Write $\dim H_*(A, \mathbb{Z}_2) = P + 2p$, $\dim H_*(A, \mathbb{Z}_2) = P' + 2(p - 1)$, where d is respectively even and odd, p is the number of ovals of the curve $\{G = 0\}$ and P, P' are some positive integers not depending on G.

Why does the summand $2(p - 1)$ appear when d is odd? $S^{(1,2)}$ becomes a non-ramified double covering over $\mathbb{R}^2$ after a topological contraction of all the ovals of $\{G = 0\}$. It is homeomorphic to the union of a sphere (containing $C = (0, 0, 1, 0)$) and a plane. Therefore two ovals outside each other are needed to create the first handle of $S^{(1,2)}$.

From constructive methods in [C2] or [K] and the upper bound given in [R1], it follows that the maximal number of ovals for $\{K^2 - 4aN = 0\}$ is

$$r = \frac{(2(d-1) - 1)((d-1) - 1)}{2} - \frac{(d-1) - 1}{2}.$$

Note that the gluing of $\{K^2 - 4aN = 0\}$ and $\{L^2 - 4aM = 0\}$ "breaks" the singular component of $\{L^2 - 4aM = 0\}$ apart into $d - 1$ ovals.

Then, if $u = \text{card } O_e + \text{card}(O_l)$ (see §VI) one has:

$$2p = 2u + 2r + 2(d-1) = 3(d-1)^2 - 5(d-1) + 4.$$

2) Consider a deformation of $S^{(3)}$ at C by gluing $S^{(3)}$ and $S^{(4)}$. It is a surface $S^{(3,4)}$ of degree $d - 1$ whose local isotopy type in the neighborhood of C is the isotopy type of the affine part of the maximal surface constructed in §IV. Define what could be called the nonaffine part of such a maximal surface as a sphere Σ if $d - 1$ is even (resp. a plane Σ if $d - 1$ is odd).

Clearly $c = \dim H_*(\Sigma, \mathbb{Z}_2) = 2$ (resp. 3). If

$$m = \frac{((d-1)-1)^4 - 1}{d-1} + 4 = (d-1)^3 - 4(d-1)^2 + 6(d-1),$$

the contribution of $S^{(3,4)}$ to $\dim H_*(A, \mathbb{Z}_2)$ is $m - c$.

3) Finally the gluing of $S^{(1,2)}$ and $S^{(3,4)}$ creates a surface A with a "global" component (the deformation of the component of $S^{(1)}$ singular at C) whose nonaffine part is a plane Σ if d is odd (resp. a sphere Σ if d is even). Since $c' = \dim H_*(\Sigma, \mathbb{Z}_2) = 3$ (resp. 2), the total homological dimension of A is:

$$\dim H_*(A, \mathbb{Z}_2) = 2(p-1) + m - c + c' = 2(p-1) + m - 2 + 3 = d^3 - 4d^2 + 6d$$

(resp. $\dim H_*(A, \mathbb{Z}_2) = 2p + m - c + c' = 2p + m - 3 + 2 = d^3 - 4d^2 + 6d$).

The Harnack–Thom–Oleĭnik bound is reached: A is maximal.

§VIII. Proof of Theorem III.4

If $d - 1 = 4$ (resp. 5), then the surface constructed in §IV has the topological type $S_{10} \amalg S_0$ (resp. $R_{22} \amalg 4S_0$). Consequently the maximal surface A obtained in §VII has the topological type:

$$R_{10+n-1} \amalg \underbrace{S_0 \amalg \cdots \amalg S_0}_{1+n'},$$

where n, n' are related to χ_8 (resp.

$$S_{22+n} \amalg S_{n_1} \amalg \cdots \amalg S_{n_k} \amalg \underbrace{S_0 \amalg \cdots \amalg S_0}_{4+n'},$$

where $n, n_1, \ldots, n_k, n'$ are related to χ_{10}), as in Figure 8-a (resp. 8-b). The dashed regions describe the construction of $S^{(1,2)}$ (double covering over $G \geqslant 0$) in the affine chart $(x, y, z, 1)$. The eye of the reader is the point $C = (0, 0, 1, 0)$. In this example, $n = 5, n' = 11$ (resp. $n = 12, n' = 9, k = 1, n_1 = 5$).

Theorem III.4 is easily deduced in this way from χ_8 (reviewed in [P]) as well as old and new realizations of some elements in the partially known set χ_{10} ([Ch], author's unpublished results).

REMARK. The congruence modulo 8 is naturally propagated from maximal curves to maximal surfaces. It is coherent with the results of V. Arnold, V. Rokhlin, and V. Kharlamov on the Euler characteristic of real algebraic varieties and the signature mod 16 of their complex locus (reviewed in [G] or [W]; see also [G-M]) or the new ideas of V. Kharlamov and O. Viro on the origin of prohibitions for surfaces. Let s (resp. a) be the number of components (resp. handles) for the surfaces of Theorem III.4. Then

$\{K^2 - 4aN \geq 0\}$ $\{L^2 - 4aM \geq 0\}$ $\{G \geq 0\}$

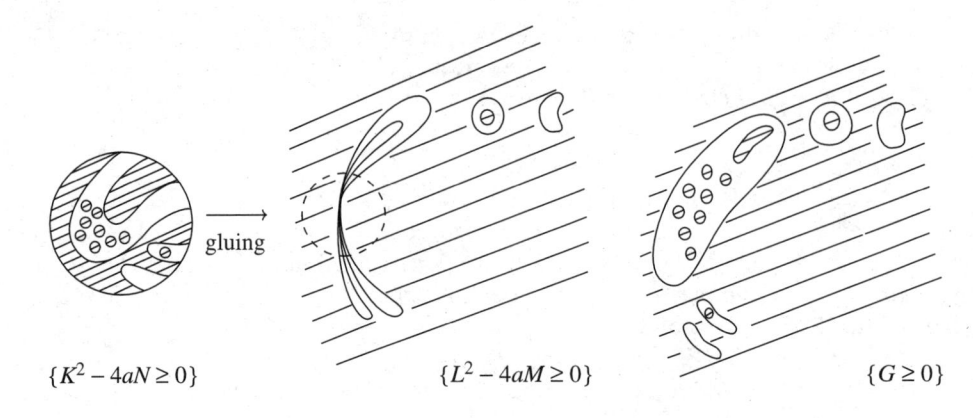

FIGURE 8-a

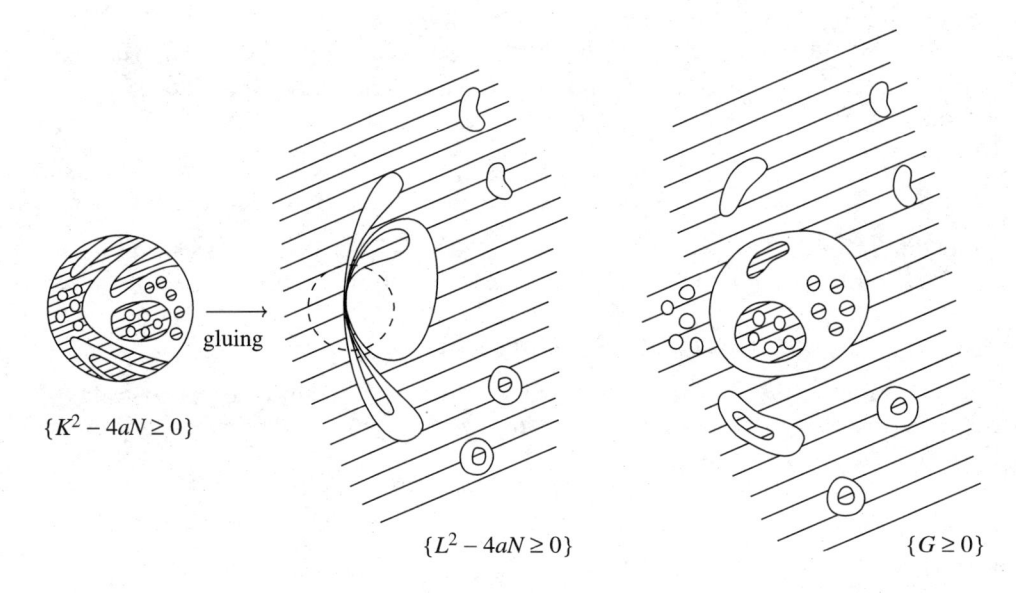

$\{K^2 - 4aN \geq 0\}$

$\{L^2 - 4aN \geq 0\}$ $\{G \geq 0\}$

FIGURE 8-b

$s - a \equiv -1$ (8) when the degree is five, $s - a \equiv 0$ (8) when the degree is six, compatible with the Rokhlin congruence for surface: $\chi(A) \equiv \sigma(\mathbb{C}A)$ (16), $\sigma(\mathbb{C}A) = \frac{1}{3}d(4 - d^2)$, $d = \deg A$ ([**Rk**]).

QUESTION. Let $\widetilde{\tau}_4^{q-1}$ be the set of topological types of smooth hypersurfaces of degree four not homotopic to a point in $P(\mathbb{R}^q)$. Is the gluing of double coverings effective to construct representatives of all elements of $\widetilde{\tau}_4^3$?

References

[A] N. A'Campo, *Sur la première parte du seizième problème de Hilbert*, Séminaire Bourbaki **31** (1978–1979), no. 537, 1–20; Lecture Notes in Math., vol. 770, Springer-Verlag, Berlin and Heidelberg, 1980, pp. 208–227.

[C1] B. Chevallier, *Sur les courbes maximales de Harnack*, C. R. Acad. Sci. Paris Sér. I Math. **300** (1985), no. 4, 109–114.

[C2] _____, *A propos des courbes de degré* 6, C. R. Acad. Sci. Paris Sér. I Math. **302** (1986), no. 1.

[Ch] Y. S. Chislenko, *M-curves of degree ten*, Zap. Nauchn. Sem. Leningrad. Otdel. Mat. Inst. Steklov. (LOMI) **122** (1982), 146–161; English transl., J. Soviet Math. **26** (1984), no. 1, 1689–1699.

[G] D. A. Gudkov, *The topology of real projective algebraic varieties*, Uspekhi Mat. Nauk **29** (1974), no. 4, 3–79; English transl., Russian Math. Surveys **29** (1974), no. 4, 1–79.

[G1] _____, *The construction of 6th order curve of the type* $\frac{5}{1}5$, Izv. Vyssh. Uchebn. Zaved. Mat. **130** (1973), no. 3, 28–36.

[G-U] D. A. Gudkov and G. A. Utkin, *Nine papers on Hilbert's 16th problem*, Amer. Math. Soc. Transl. Ser. 2 **112** (1978).

[G-M] L. Guillou and A. Marin, *Une extension d'un theoreme de Rohlin sur la signature*, A la recherche de la topologie perdue, Birkhäuser, Boston, 1986, pp. 97–118.

[H] F. Browder (ed.), *Mathematical developments arising from Hilbert problems*, Proc. Symp. Pure Math., vol. 38, Amer. Math. Soc., Providence, RI, 1976.

[H1] D. Hilbert, *Uber die Gestalt einer Flasche vierter Ordnung*, Gött. Nachr. (1909), 308–313; Ges. Abh., vol. 2, Springer-Verlag, Berlin, 1933, pp. 449–453.

[Ha] A. Harnack, *Über die Vieltheiligkeit der ebenen algebraischen Kurven*, Math. Ann. **10** (1876), 189–199.

[K] A. B. Korchagin, *On the reduction of singularities and the classification of nonsingular affine curves of degree* 6, Manuscript No. 1107-B86, deposited at VINITI, 1986, pp. 1–16. (Russian)

[Kh] V. M. Kharlamov, *Topological types of nonsingular surfaces of degree* 4 *in* $\mathbb{R}P^3$, Funktsional. Anal. i Prilozhen. **10** (1976), no. 4, 55–68; English transl., Functional Anal. Appl. **10** (1976), no. 4, 295–305.

[Kh1] _____, *On a number of components of a* M*-surface of degree* 5 *in* $\mathbb{R}P^3$, Proc. of the XVI Soviet Algebraic Conference, Leningrad, 1981, pp. 353–354. (Russian)

[Kh2] _____, *Real algebraic surfaces*, Proc. Intern. Congress Math., (Helsinki, 1978), vol. 1, Acad. Sci. Fennica, Helsinki, 1980, pp. 421–428. (Russian)

[O] O. A. Oleĭnik, *Bounds for Betti numbers of real algebraic hypersurfaces*, Mat. Sb. **28 (70)** (1951), 635–640. (Russian)

[P] G. M. Polotovskiĭ, *On the classification of nonsingular curves of degree* 8, Topology and Geometry. Rokhlin seminar. Lecture Notes in Math., vol. 1346, 1988, pp. 455–485.

[R] J. J. Risler, *Construction d'hypersurfaces réelles (d'après Viro)*, Séminaire Bourbaki **763** (Novembre 1992).

[R1] _____, *Un analogue local du théorème de Harnack*, Invent. Math. **89** (1987), 119–137.

[Rk] V. A. Rokhlin, *Congruences modulo* 16 *in sixteenth Hilbert's problem*, I *and* II, Funktsional. Anal. i Prilozhen. **6** (1972), no. 4, 58–64; **7** (1973), no. 2, 91–92; English transl., Functional Anal. Appl. **6** (1972), no. 4, 301–306; **7** (1973), no. 2, 163–164.

[Rn] K. Rohn, *Die maximalzahl und anordnung der ovale bei der ebenen kurve* 6, Ordnung und bei der flä che 4. Ordnung, Math. Ann., vol. 73, 1913, pp. 177–229.

[S] E. I. Shustin, *Versal deformations in a space of planar curves of fixed degree*, Funktsional. Anal. i Prilozhen. **21** (1987), no. 1, 90–91; English transl., Functional Anal. Appl. **21** (1987), no. 1, 82–84.

[T] R. Thom, *Sur l'homologie des variétés algébriques réelles*, Differential and Combinatorial Topology (a symposium in honor of Marston Morse), Princeton Univ. Press, Princeton, 1965, pp. 255–265.

[V1] O. Ya. Viro, *Curves of degree* 7, *curves of degree* 8, *and the Ragsdale conjecture*, Dokl. Akad. Nauk SSSR **254** (1980), no. 9, 1306–1310; English transl., Soviet Math. Dokl. **22** (1980), no. 2, 566–570.

[V2] _____, *Gluing of plane real algebraic curves and construction of curves of degree* 6 *and* 7, Lecture Notes in Math., vol. 1060, Springer-Verlag, Berlin, 1984, pp. 187–200.

[V3] _____, *Constructing real algebraic varieties with prescribed topology*, Thesis and book (to appear).

[V4] _____, *Real algebraic plane curves: constructions with controlled topology*, Algebra i Analiz **1** (1989), no. 5, 1–73; English transl., Leningrad Math. J. **1** (1990), no. 5, 1059–1134.

[V5] _____, *Construction of multicomponent real algebraic surfaces*, Dokl. Akad. Nauk SSSR **248** (1979), no. 2, 279–282; English transl., Soviet Math. Dokl. **20** (1979), no. 5, 991–995.

[V6] _____, *Construction of* M*-surfaces*, Funktsional. Anal. i Prilozhen. **13** (1980), no. 3, 71–73; English transl. in Functional Anal. Appl. **13** (1980), no. 3.

[W] G. Wilson, *Hilbert's sixteenth problem*, Topology **17** (1978), no. 1, 53–74.

Université Toulouse II, U.F.R. de Mathématiques 5 allées Antonio Machado, 31058 Toulouse Cédex, France

Université Toulouse III, Laboratoire de Topologie et Géométrie, URA CNRS 1408, 118 route de Narbonne, 31062 Toulouse Cédex, France

Quadratic Transformations $\mathbb{R}P^2 \to \mathbb{R}P^2$

Alexander Degtyarev

This paper is dedicated to the memory of D. A. Gudkov

ABSTRACT. The complete projective classification of 1- and 2-dimensional linear systems of real plane quadrics is given. In other words, this is a classification of quadratic transformations $\mathbb{R}P^2 \to \mathbb{R}P^1$ and $\mathbb{R}P^2 \to \mathbb{R}P^2$, respectively.

Introduction

The purpose of this paper is to give the projective classification of quadratic transformations $\mathbb{R}P^2 \to \mathbb{R}P^2$, i.e., transformations given by quadratic polynomials in some homogeneous coordinates. Originally, this problem was formulated by Viro, who intended to use these transformations in constructing new real algebraic curves. Another origin is the classical problem about the relationship between projective plane algebraic curves and 2-dimensional linear systems of quadratic hypersurfaces, which was solved by Tyurin [T] in the case of nonsingular curves. (Tyurin's result states that there is a one-to-one correspondence between the set of generic 2-dimensional linear systems of quadrics in $\mathbb{C}P^n$ and a Zariski open subset of the space of pairs (C, φ), where C is a nonsingular complex plane curve of degree $n+1$, and φ is a Spin-structure on C with Arf $\varphi = 0$.)

In this paper, we also exploit the latter relationship: transformations are studied via their *discriminant curves*, which, by definition, are the loci of the singular curves of the corresponding linear systems. For nonsingular transformations, which form a one-parameter family, we construct a complete invariant $\varepsilon \in \mathbb{R}$ (see Theorem 3.3), which is a ramification of the classical j-invariant of the discriminant curve (the latter being a nonsingular cubic in this case). Given a nonsingular real cubic C, transformations with discriminant curve isomorphic to C are in one-to-one correspondence with the real (i.e., conj-invariant) Spin-structures on the complexification of C. Besides, there are 24 discrete classes of degenerate transformations, which are classified by their discriminant curves (which are either singular plane cubics, or coincide with $\mathbb{R}P^2$), along with some natural stratification of these curves.

1991 *Mathematics Subject Classification.* Primary 14E09, 14J50.
Key words and phrases. Quadratic transformations, linear systems.

©1996, American Mathematical Society

In conclusion, I would like to express my gratitude to Professor O. Viro, who inspired this work.

§1. Some definitions

Throughout this section F denotes a fixed polynomial transformation $\mathbb{R}P^n \to \mathbb{R}P^n$ of degree d given in homogeneous coordinates $(x_1 : \ldots : x_n)$, $(y_1 : \ldots : y_m)$ by the formula $y_i = f_i(x_0, \ldots, x_n)$. The polynomials $f_0, \ldots, f_m$ are always assumed *linearly independent*.

1.1. In general, the transformation f may be undefined at some points of $\mathbb{R}P^n$, namely, at the common points of the zero sets of all the f_i's. These points will be called the *base points* of F, and the set of all base points will be denoted by $\mathcal{B}$. Sometimes, following Zariski [Z], we shall also consider infinitely close base points, i.e., those which lie in the exceptional divisor of some blowing-up of $\mathbb{R}P^n$.

We define the *Jacobian* $\mathcal{J}$ of F to be the projectivization of the critical point set of a linear lift $\mathbb{R}^{n+1} \to \mathbb{R}^{m+1}$ of F. (In terms of F itself, $\mathcal{J}$ consists of the critical points of the restriction $F|_{\mathbb{R}P^n \setminus \mathcal{B}}$ and, possibly, of some of the base points of F. If $m > n$, then $\mathcal{J} = \mathbb{R}P^n$. If $m = n$, then $\mathcal{J} \supset \mathcal{B}$.) As usual, the Jacobian is given by $\operatorname{rk}(\partial f_i / \partial x_j) < m + 1$. If $m = n$, the latter is equivalent to $\det(\partial f_i / \partial x_j) = 0$, so in this case the Jacobian is, in general, a hypersurface of degree $(d-1)^n$ in $\mathbb{R}P^n$ (though it may also coincide with the whole space $\mathbb{R}P^n$).

The Jacobian has a natural stratification defined by the types of critical points (see, e.g., [AVG]) and by the multiplicities and hierarchy of base points.

1.2. The transformation F obviously defines an m-dimensional linear system of hypersurfaces of degree d in $\mathbb{R}P^n$. (This system is spanned by the f_i's.) We shall call it the *adjoint linear system* of F and denote it by $\mathcal{L}$. In more precise terms, $\mathcal{L}$ is the image of the dual space $(\mathbb{R}P^m)^\vee$ of $\mathbb{R}P^m$ under the natural projective map $(\mathbb{R}P^m)^\vee \to \mathcal{C}_d\mathbb{R}P^n$ defined by F, where $\mathcal{C}_d\mathbb{R}P^n$ is the $(\binom{n+d}{d} - 1)$-dimensional projective space of all hypersurfaces of degree d in $\mathbb{R}P^n$. The set of all the singular hypersurfaces of $\mathcal{L}$ is called the *discriminant* of F and is denoted by Δ. (Δ either coincides with $\mathcal{L}$, or is a degree $(n+1)(d-1)^n$ hypersurface in $\mathcal{L}$.) The discriminant has a natural stratification defined by the types of degeneration of the hypersurfaces. The isotopy (rigid isotopy, etc.) types of nonsingular hypersurfaces define an additional structure on $\pi_0(\mathcal{L} \setminus \Delta)$.

There is the following useful connection between $\mathcal{J}$ and Δ: $\mathcal{J}$ is the locus of the singular points of the singular hypersurfaces of $\mathcal{L}$.

Obviously, the discriminant can be thought of as the intersection $\mathcal{L} \cap \operatorname{Discr}_d$, where $\operatorname{Discr}_d \subset \mathcal{C}_d\mathbb{R}P^n$ is the set of all the singular hypersurfaces of degree d in $\mathbb{R}P^n$. We shall need the following lemma:

1.3. LEMMA (Severi [S]). *Smooth points of* Discr_d *correspond to hypersurfaces with a single point of type* A_1 *(i.e., a nondegenerate double point). The tangent hyperplane to* Discr_d *at such a point consists of all the hypersurfaces passing through the above singular point.*

1.4. Define the *dual transformation* $\check{F} \colon (\mathbb{R}P^n)^\vee \to \mathbb{R}P^{D-(m+1)}$, where $D = \binom{n+d}{d} - 1$, in the following way: The system $\mathcal{L}$, which is an m-plane in $\mathcal{C}_d\mathbb{R}P^n$, defines its dual $(D - m - 1)$-plane $\mathcal{L}^\perp \subset (\mathcal{C}_d\mathbb{R}P^n)^\vee = \mathcal{C}_d(\mathbb{R}P^n)^\vee$, which, in turn, defines a transformation $(\mathbb{R}P^n)^\vee \to (\mathcal{L}^\perp)^\vee \cong \mathbb{R}P^{D-(m+1)}$. By definition, the latter is $\check{F}$. Obviously, two transformations are projectively equivalent if and only if their duals are.

Note that in the case of quadratic transformations $\mathbb{R}P^2 \to \mathbb{R}P^2$ (i.e., $d = 2$, $m = n = 2$), which we are mainly interested in, the dual transformation is of the same type.

§2. Main results

Up to projective equivalence, there exists a one 1-parameter family and 24 discrete types of quadratic transformations $\mathbb{R}P^2 \to \mathbb{R}P^2$. These types are presented in Table 1, which gives:
1. values of the parameters in the canonical representation of a transformation F (see below);
2. the codimension of the corresponding stratum in the 9-dimensional space of all quadratic transformations $\mathbb{R}P^2 \to \mathbb{R}P^2$;
3. the dual transformation $\check{F}$.

TABLE 1. Quadratic transformations $\mathbb{R}P^2 \to \mathbb{R}P^2$

Type	ε	codim	Dual
I1a	$\varepsilon < -8$	0	I1a'
I1b	$-8 < \varepsilon < 0$	0	I1b'
I1a'	$0 < \varepsilon < 1$	0	I1a
I1b'	$\varepsilon > 1$	0	I1b
I2a	1	1	I2$'a$
I2b	-1	1	I2$'b$
I2$'a$	1	1	I2a
I2$'b$	-1	1	I2b
I3	—	2	I3

Type	$(\varepsilon_0, \varepsilon_1, \varepsilon_2)$	codim	Dual
II1a	$(1, 1, 1)$	2	II$'$1a
II1b	$(-1, 1, 1)$	2	II$'$1b
II1c	$(-1, -1, 1)$	2	II$'$1c
II2$_1 a$	$(0, 1, 1)$	3	II$'$2$_1 a$
II2$_1 b$	$(0, -1, 1)$	3	II$'$2$_1 b$
II2$_2 a$	$(1, 0, 1)$	3	II$'$2$_2 a$
II2$_2 b$	$(-1, 0, 1)$	3	II$'$2$_2 b$
II3	$(0, 0, 1)$	4	II$'$3
II$'$1a	$(-1, 1, 1)$	2	II1a
II$'$1b	$(1, 1, 1)$	2	II1b
II$'$1c	$(-1, -1, 1)$	2	II1c
II$'$2$_1 a$	$(0, 1, 1)$	3	II2$_1 a$
II$'$2$_1 b$	$(0, -1, 1)$	3	II2$_1 b$
II$'$2$_2 a$	$(-1, 0, 1)$	3	II2$_2 a$
II$'$2$_2 b$	$(1, 0, 1)$	3	II2$_2 b$
II$'$3	$(0, 0, 1)$	4	II3
III	—	5	III
IV	$(1, 0, 0)$	7	IV$'$
IV$'$	$(1, 0, 0)$	7	IV

Below we present the canonical representations of the transformations.

Type I1: $\begin{cases} y_0 = x_2(x_2 - \varepsilon x_1), \\ y_1 = x_0(x_0 - x_2), \\ y_2 = x_1(x_1 - x_0), \end{cases}$
Type I2: $\begin{cases} y_0 = x_1^2 + \varepsilon x_2^2, \\ y_1 = x_0 x_2, \\ y_2 = x_1^2 - x_0 x_1, \end{cases}$

Type I2′: $\begin{cases} y_0 = (x_1 + x_2)^2, \\ y_1 = x_0 x_2, \\ y_2 = x_0^2 - \varepsilon x_1^2, \end{cases}$ Type I3: $\begin{cases} y_0 = (x_1 + x_2)^2, \\ y_1 = x_0 x_2, \\ y_2 = x_0^2 + x_0 x_1, \end{cases}$

Types II and IV: $\begin{cases} y_0 = \varepsilon_0 x_0^2 + \varepsilon_1 x_1^2 + \varepsilon_2 x_2^2, \\ y_1 = x_0 x_1, \\ y_2 = x_0 x_2, \end{cases}$ Types II′ and IV′: $\begin{cases} y_0 = \varepsilon_2 x_0^2 + \varepsilon_0 x_2^2, \\ y_1 = x_1 x_2, \\ y_2 = x_1^2 - \varepsilon_1 x_2^2, \end{cases}$

Type III: $\begin{cases} y_0 = x_2^2 + x_0 x_1, \\ y_1 = x_1 x_2, \\ y_2 = x_1^2. \end{cases}$

§3. Type I1: nondegenerate transformations

The classification of quadratic transformations $F \colon \mathbb{R}P^2 \to \mathbb{R}P^2$ is based on the consideration of the discriminant curve Δ. If F is given by

(3.1) $$y_k = \alpha_k^{ij} x_i x_j, \qquad i,j,k = 0,1,2,$$

then the well-known criterion for a conic to be degenerate gives the following equation for Δ:

(3.2) $$\det\left(\sum_{k=0}^{2} \check{y}_k \alpha_k^{ij}\right) = 0$$

(where $(x_0 : x_1 : x_2)$ are some coordinates in the domain $\mathbb{R}P^2$, $(y_0 : y_1 : y_2)$ are coordinates in the range $\mathbb{R}P^2$, and $(\check{y}_0 : \check{y}_1 : \check{y}_2)$ are the coordinates in $\mathcal{L}$ dual to $(y_0 : y_1 : y_2)$). So, in general, Δ is a cubic curve in $\mathbb{R}P^2$.

In this section we suppose that Δ is nonsingular.

3.3. THEOREM. (1) *Every nondegenerate quadratic transformation* $\mathbb{R}P^2 \to \mathbb{R}P^2$ *has a representation of the type*

(3.4) $$\begin{cases} y_0 = x_2(x_2 - \varepsilon x_1), \\ y_1 = x_0(x_0 - x_2), \\ y_2 = x_1(x_1 - x_0), \end{cases}$$

where $\varepsilon \in \mathbb{R} \setminus \{-8, 0, 1\}$.

(2) *The parameter ε is a complete invariant of the transformation, i.e., two transformations are equivalent if and only if the corresponding values of ε coincide.*

(3) *Given a nonsingular real cubic curve C with one (two) connected component(s), there is one (resp., three) quadratic transformation(s) with $\Delta \cong C$.*

REMARKS. 1. Straightforward calculations show that the dual transformation has the representation (3.4) with $\check{\varepsilon} = -8/\varepsilon$.

2. One can prove that any *complex* nondegenerate quadratic transformation $\mathbb{C}P^2 \to \mathbb{C}P^2$ also has the representation (3.4) above. But in this case ε depends on the choice of a representation, and a complete invariant of a transformation is $\lambda = (\varepsilon - 1)/\varepsilon^3(\varepsilon + 1) \in \mathbb{C} \setminus \{0\}$. The dual transformation has $\check{\lambda} = -16/\lambda$.

3. Statement (3) of the theorem agrees with the obvious real version of Tyurin's result [T]: given a nonsingular real cubic C, transformations with $\Delta \cong C$ are in a natural one-to-one correspondence with the real Spin-structures φ on the complexification of C with Arf $\varphi = 0$.

PROOF OF THEOREM 3.3.

3.5. LEMMA. *On any nonsingular real plane cubic curve C there are exactly two (up to cyclic reordering) triples of points (P_0, P_1, P_2) such that P_i lies on the tangent to C at P_{i+1} (where $P_3 = P_0$).*

PROOF. Fix an inflection point of C. This defines a group structure on C such that three points $P, Q, R \in C$ are concurrent if and only if $P + Q + R = 0$. Hence, the above three points must satisfy the system

$$2P_0 + P_2 = 2P_1 + P_0 = 2P_2 + P_1 = 0.$$

Since algebraically C is either $\mathbb{R}/\mathbb{Z}$ or $\mathbb{R}/\mathbb{Z} \oplus \mathbb{Z}_2$, this system has 9 solutions. Three of them (with $P_0 = P_1 = P_2$) correspond to the inflections points of C, and the other six form (up to reordering) the desired two triples. □

Fix one of the two triples (P_0, P_1, P_2) of Lemma 3.5 on the discriminant curve Δ. According to Lemma 1.3, P_0, P_1, P_2 are reducible conics with some singular points S_0, S_1, S_2 respectively such that P_i passes through S_{i+1} (where $S_3 = S_0$). Hence, if P_0, P_1, P_2 are the base points of the coordinate system $(\check{y}_0 : \check{y}_1 : \check{y}_2)$ in $\mathcal{L}$ (i.e., the points with coordinates $(1:0:0)$, $(0:1:0)$, $(0:0:1)$ respectively), and S_0, S_1, S_2 are the base points of the coordinate system $(x_0 : x_1 : x_2)$ in $\mathbb{R}P^2$, then, after a homogeneous coordinate change (i.e., a change of the type $x_i \mapsto \lambda_i x_i$, $y_j \mapsto \mu_j y_j$), the transformation has the representation (3.4). Its discriminant has the equation

(3.6) $$\varepsilon^2 \check{y}_0^2 \check{y}_1 + \check{y}_1^2 \check{y}_2 + \check{y}_2^2 \check{y}_0 + (\varepsilon - 4) \check{y}_0 \check{y}_1 \check{y}_2 = 0,$$

and an immediate calculation shows that this curve is nonsingular for any $\varepsilon \in \mathbb{R} \setminus \{-8, 0, 1\}$. This proves 3.3(1).

By a homogeneous coordinate change equation (3.6) can be converted to

(3.7) $$\check{y}_0^2 \check{y}_1 + \check{y}_1^2 \check{y}_2 + \check{y}_2^2 \check{y}_0 + u \check{y}_0 \check{y}_1 \check{y}_2 = 0,$$

where the new parameter u, defined up to multiplication by cubic roots of unity, is given by

(3.8) $$u^3 = (\varepsilon - 4)^3/e^2.$$

3.9. LEMMA. *The above u is an invariant of the (real) curve C given by (3.7).*

PROOF. By definition, u is determined by the curve and the choice of one of the two triples (P_0, P_1, P_2) of Lemma 3.5. We shall prove that the two triples can be converted to each other by a projective transformation of C. Let P be the inflection point which defines the group structure on C. Then the embedding $C \hookrightarrow \mathbb{R}P^2$ is the composition of the canonical map $C \hookrightarrow |3P|^\vee$ and some isomorphism $|3P|^\vee \cong \mathbb{R}P^2$. On the other hand, the two triples (P_0, P_1, P_2) are $(1/9, 7/9, 4/9)$ and $(8/9, 2/9, 5/9)$ (recall that the component of unity of C is $\mathbb{R}/\mathbb{Z}$), and the automorphism $\times(-1) \colon C \to C$ of C induces the desired transformation of $|3P|^\vee$ and $\mathbb{R}P^2$. □

Now it is clear that, given a curve C, there are at most three quadratic transformations with $\Delta \cong C$ (which correspond to the three solutions of (3.8)). On the other hand, the transformations dual to these three have, in general, distinct values of u, so they, and hence the original transformations, are pairwise distinct.

To prove statement (3), it suffices to notice that, if a cubic curve has one (two) component(s), then $u^3 > 27$ (resp., $u^3 < 27$), and (3.8) has one (resp., three) solution(s). □

§4. Types I2, I3: transformations with irreducible discriminant

4.1. THEOREM. *Let the discriminant Δ of a quadratic transformation F have a single singular point. Then F has one of the following representations:*

(4.2)
$$\begin{cases} y_0 = x_1^2 + \varepsilon x_2^2, \\ y_1 = x_0 x_2, \\ y_2 = x_1^2 - x_0 x_1, \end{cases}$$

(4.3)
$$\begin{cases} y_0 = (x_1 + x_2)^2, \\ y_1 = x_0 x_2, \\ y_2 = x_0^2 - \varepsilon x_1^2, \end{cases}$$

(4.4)
$$\begin{cases} y_0 = (x_1 + x_2)^2, \\ y_1 = x_0 x_2, \\ y_2 = x_0^2 + x_0 x_1, \end{cases}$$

where in the case of (4.2) and (4.3) $\varepsilon = +1$ (Δ has an isolated singular point) or -1 (Δ has a node), and in the case of (4.4) Δ has a cusp.

PROOF. Consider the case when Δ has a nondegenerate double point (i.e., either a node or an isolated singular point). Then it can be given by

(4.5)
$$\check{y}_0(\check{y}_1^2 + \varepsilon \check{y}_2^2) + \check{y}_1^2 \check{y}_2 = 0,$$

and in the corresponding coordinate system $P_0 = (1:0:0)$ is the singular point of Δ and $P_1 = (0:1:0)$ lies in the tangent to Δ through $P_2 = (0:0:1)$. Consider the following two cases.

Case 1: $P_0 = (1:0:0)$ is a tangency point of $\mathcal{L}$ and Discr_2. Let the singular points S_0, S_1, S_2 of the curves P_0, P_1, P_2 respectively be the vertices of the coordinate system $(x_0:x_1:x_2)$ in $\mathbb{R}P^2$. According to Lemma 1.3, every curve of $\mathcal{L}$ should pass through S_0. Besides, P_1 passes through S_2. Hence, the transformation has the representation

(4.6)
$$\begin{cases} y_0 = ax_1^2 + bx_1 x_2 + cx_2^2, \\ y_1 = x_0 x_2, \\ y_2 = x_1^2 - x_0 x_1, \end{cases}$$

and Δ is given by

(4.7)
$$\check{y}_0(a\check{y}_1^2 + c\check{y}_2^2) + \check{y}_1^2 \check{y}_2 - 2b\check{y}_0\check{y}_1\check{y}_2 = 0.$$

Comparing this and (4.5), one finds $a = 1$, $b = 0$, $c = \varepsilon$, which gives (4.2).

Case 2: $P_0 = (1:0:0)$ *is a singular point of* Discr_2. Let S_0, S_1, S_2 and coordinate system in $\mathbb{R}P^2$ be the same as in Case 1. (Now S_0 is *one* of the singular points of P_0, which is a double line.) Then the transformation has the representation

$$\text{(4.8)} \quad \begin{cases} y_0 = (x_1 + x_2)^2, \\ y_1 = x_0 x_2, \\ y_2 = ax_1^2 + bx_0 x_1 + cx_0^2, \end{cases}$$

Δ is given by

$$\text{(4.9)} \qquad \check{y}_0[c\check{y}_1^2 + (b^2 - 4ac)\check{y}_2^2] + c\check{y}_1^2 \check{y}_2 - 2b\check{y}_0\check{y}_1\check{y}_2 = 0,$$

and comparing this to (4.5) gives $a = -\varepsilon/4$, $b = 0$, $c = 1$, i.e., (4.3) (after an appropriate homogeneous coordinate change).

Now consider the case when Δ has a cusp. Then it can be given by

$$\text{(4.10)} \qquad \check{y}_0(\check{y}_1 - \check{y}_2)^2 + \check{y}_1^2 \check{y}_2 = 0.$$

4.11. LEMMA. *The cusp of* Δ *is a singular point of* Discr_2 *(i.e., a double line)*.

PROOF. Assume the contrary. Then the transformation can be represented by (4.6), and comparing (4.7) and (4.10) we obtain $a = c = 1$, $b = -2$. Hence, P_0 is given by $(x_1 - x_2)^2 = 0$, i.e., P_0 is a singular point of Discr_2. □

Due to the lemma, the transformation can be represented by (4.8). Then comparing (4.9) and (4.10) we obtain $a = 0$, $b = c = 1$, i.e., (4.4). □

§5. Pencils of conics

In this section we give a projective classification of pencils (i.e., one-dimensional linear systems) of plane conics, or, in other words, of quadratic transformations $\mathbb{R}P^2 \to \mathbb{R}P^1$.

5.1. THEOREM. (1) *Up to projective equivalence, a pencil of conics with nonsingular generic curve is determined by its set of base points (including complex and infinitely close points). The nine types are presented in Table* 2 *(where the notation* $r\mathbb{R}$ *(resp.,* $r\mathbb{C}$*) means that a system has a real (resp., complex) base point along with the infinitely close points up to the* $(r-1)$*st order).*

(2) *Any pencil of singular conics has one of the following representations*:

$$\text{(5.2)} \qquad \begin{cases} y_0 = x_0 x_1, \\ y_1 = x_0 x_2, \end{cases}$$

$$\text{(5.3)} \qquad \begin{cases} y_0 = x_1 x_2, \\ y_1 = x_1^2 - \varepsilon x_2^2, \quad \varepsilon = \pm 1, 0. \end{cases}$$

A pencil is said to be of type $\mathrm{II1}_1$ *if it has representation* (5.2), *and of type* $\mathrm{II1}_2 a$, $\mathrm{II1}_2 b$, *or* $\mathrm{II2}$ *if it has representation* (5.3) *with* $\varepsilon = 1$, -1, *or* 0 *respectively*.

PROOF. First consider the case in which a generic curve of the pencil is nonsingular (statement (1) of the theorem). Then the pencil has four (including complex and infinitely close) base points in general position (i.e., no three of these four points lie on a line), and, for dimensional reasons, it coincides with the complete linear system through these points. On the other hand, a set of four points in general position in

TABLE 2. Pencils of conics with nonsingular generic curve

Type	Base points	codim
I1a	$(\mathbb{R}, \mathbb{R}, \mathbb{R}, \mathbb{R})$	0
I1b	$(\mathbb{R}, \mathbb{R}, \mathbb{C}, \mathbb{C})$	0
I1c	$(\mathbb{C}, \mathbb{C}, \mathbb{C}, \mathbb{C})$	0
I2a	$(2\mathbb{R}, \mathbb{R}, \mathbb{R})$	1
I2b	$(2\mathbb{R}, \mathbb{C}, \mathbb{C})$	1
I3$_1$	$(3\mathbb{R}, \mathbb{R})$	2
I3$_2 a$	$(2\mathbb{R}, 2\mathbb{R})$	2
I3$_2 b$	$(2\mathbb{C}, 2\mathbb{C})$	2
I4	$(4\mathbb{R})$	3

$\mathbb{R}P^2$ is defined, up to projective equivalence, by its combinatorial characteristics, i.e., the infinite closeness relation and realness. The nine possible types are given in Table 2. (One should keep in mind that, if a pencil has a complex base point, it should also have the conjugate point with the same multiplicity.)

Now suppose that all the conics of the pencil are singular. Consider the following two cases:

Case 1: *the pencil does not have a singular base point.* Then, according to the Bertini theorem, all the conics have a common component, which in this case must be a line, say, $\{x_0 = 0\}$. The remaining components of the conics form a line pencil, whose base point does not belong to $\{x_0 = 0\}$ (since there is no common singular point), and the pencil can obviously be presented by (5.2).

Case 2: *all the conics have a common singular point.* Let $S = (1:0:0)$ be this point. Then eventually we have a pencil of homogeneous quadratic polynomials in $(x_1 : x_2)$, which, in an appropriate coordinate system, can be given by (5.3). □

§6. Type II: Δ has a linear component of type II1$_1$

In the following two sections we consider transformations $\mathbb{R}P^2 \to \mathbb{R}P^2$ whose discriminant set Δ is reducible. In this case Δ contains at least one line, which is a pencil of singular conics (i.e., a pencil of one of the types II1$_1$, II1$_2$, or II2 of §5). In this section we suppose that at least one of the line components of Δ is of type II1$_1$.

6.1. THEOREM. *Suppose that the adjoint linear system $\mathcal{L}$ of a transformation F contains a pencil of type* II1$_1$ *(see Theorem* 5.1*), and its discriminant set Δ does not coincide with* $\mathbb{R}P^2$. *Then F has the representation*

$$(6.2) \quad \begin{cases} y_0 = \varepsilon_0 x_0^2 + \varepsilon_1 x_1^2 + \varepsilon_2 x_2^2, \\ y_1 = x_0 x_1, \\ y_2 = x_0 x_2, \end{cases}$$

where the value of the triple of parameters $(\varepsilon_0, \varepsilon_1, \varepsilon_2)$ is one of those given in Table 1.

REMARK. The other values of $\varepsilon_i = \pm 1, 0$ are also possible, but the transformations obtained are equivalent to those given in Table 1.

PROOF. Choose a coordinate system $(y_0 : y_1 : y_2)$ in $\mathcal{L}$ so that its base points P_1, P_2 lie in (one of) the components of type II1$_1$ of Δ and P_0 is an arbitrary point not on this line. Then, in an appropriate coordinate system $(x_0 : x_1 : x_2)$ in $\mathbb{R}P^2$, the transformation can be given by

(6.3) $$\begin{cases} y_0 = ax_0^2 + bx_1^2 + cx_2^2 + 2dx_0x_1 + 2ex_0x_2 + 2fx_1x_2, \\ y_1 = x_0x_1, \\ y_2 = x_0x_2. \end{cases}$$

It is easy to see that there is a projective transformation

$$x_0 \mapsto \alpha x_0, \qquad (x_1, x_2) \mapsto (x_1, x_2) \cdot A,$$

where $\alpha \neq 0$ is a real, and A is a nondegenerate (2×2)-matrix,

$$y_0 \mapsto y_0, \qquad (y_1, y_2) \mapsto (\alpha y_1, \alpha y_2) \cdot A$$

which converts the first equation of (6.3) to

$$y_0 = \varepsilon_0 x_0^2 + \varepsilon_1 x_1^2 + \varepsilon_2 x_2^2 + 2\tilde{d}x_0x_1 + 2\tilde{e}x_0x_2, \qquad \varepsilon_i = \pm 1, 0,$$

while leaving the two other equations unchanged. Finally, the coordinate change $y_0 \mapsto y_0 + 2\tilde{d}y_1 + 2\tilde{e}y_2$ gives representation (6.2). The discriminant Δ then has the equation

(6.4) $$\check{y}_0(4\varepsilon_0\varepsilon_1\varepsilon_2\check{y}_0^2 - \varepsilon_2\check{y}_1^2 - \varepsilon_1\check{y}_2^2) = 0.$$

Now, the fact that
1. (6.2) is symmetric in $\varepsilon_1, \varepsilon_2$;
2. ε_1 and ε_2 cannot be both zero (since otherwise (6.4) would imply $\Delta = \mathbb{R}P^2$);
3. simultaneous change of the signs of all the ε_i's gives an equivalent transformation

shows that there are only eight essentially different values of $(\varepsilon_0, \varepsilon_1, \varepsilon_2)$, which are given in Table 1. $\square$

§7. Type II′: Δ has a 1-fold linear component of type II1$_2$ or II2

7.1. THEOREM. *Suppose that the discriminant set Δ of a transformation F has a 1-fold linear component of type* II1$_2$ *or* II2, *and* $\Delta \neq \mathcal{L}$. *Then F has the representation*

(7.2) $$\begin{cases} y_0 = \varepsilon_2 x_0^2 + \varepsilon_0 x_2^2, \\ y_1 = x_1 x_2, \\ y_2 = x_1^2 - \varepsilon_1 x_2^2, \end{cases}$$

where the value of the parameter triple $(\varepsilon_0, \varepsilon_1, \varepsilon_2)$ *is one of those given in Table* 1. (*Also see the remark after Theorem* 6.1.)

PROOF. Choose a coordinate system $(\check{y}_0 : \check{y}_1 : \check{y}_2)$ in $\mathcal{L}$ so that its base points P_1, P_2 lie in (one of) the 1-fold components of type II1$_2$ or II2 of Δ, and let P_0 be an arbitrary point not on this line. Then, in an appropriate coordinate system $(x_0 : x_1 : x_2)$ in $\mathbb{R}P^2$, the transformation can be given by

(7.3) $$\begin{cases} y_0 = ax_0^2 + bx_1^2 + cx_2^2 + 2dx_0x_1 + 2ex_0x_2 + 2fx_1x_2, \\ y_1 = x_1x_2, \\ y_2 = x_1^2 - \varepsilon_1 x_2^2. \end{cases}$$

We claim that $a \neq 0$. Let $a = 0$. Then the coordinate change $y_0 \mapsto y_0 + 2fy_1 + by_2$ converts the first equation of (7.3) to

$$y_0 = \tilde{c}x_2^2 + 2dx_0x_1 + 2ex_0x_2,$$

and Δ has the equation

(7.4) $$\check{y}_0^2[de\check{y}_1 + (d^2\varepsilon_1 - e^2)\check{y}_2 - \tilde{c}d^2\check{y}_0] = 0,$$

which shows that $\{\check{y}_0 = 0\}$ is an at least 2-fold component of Δ. This contradicts our choice of the coordinate system.

Hence, $a \neq 0$. In this case the coordinate change

$$x_0 \mapsto x_0 - \frac{d}{a}x_1 - \frac{e}{a}x_2$$

transforms the first equation to

$$y_0 = ax_0^2 + \tilde{b}x_1^2 + \tilde{c}x_2^2 + 2\tilde{f}x_1x_2,$$

and the transformation $y_0 \mapsto y_0 + 2\tilde{f}y_1 + \tilde{b}y_2$, followed by a homogeneous coordinate change, gives (7.2). The discriminant Δ then has the equation

(7.5) $$\varepsilon_2\check{y}_0(\check{y}_1^2 + 4\varepsilon_1\check{y}_2^2 - 4\varepsilon_0\check{y}_0\check{y}_2) = 0.$$

Now, taking into account the fact that $\varepsilon_2 \neq 0$ (since otherwise $\Delta = \mathcal{L}$) and that the simultaneous change of the signs of ε_0 and ε_2 gives an equivalent system, one sees that there are nine essentially different values of $(\varepsilon_0, \varepsilon_1, \varepsilon_2)$. Eight of them are given in Table 1, and the ninth one, namely $(1, -1, 1)$, is equivalent to II$'$1c (via the coordinate change $x_1 \mapsto x_2, x_2 \mapsto x_1; y_0 \mapsto y_0 + y_2$). □

§8. The other types (III, IV, IV$'$)

8.1. THEOREM. (1) *Any transformation F whose discriminant Δ is a 3-fold line has the representation*

(8.2) $$\begin{cases} y_0 = x_2^2 + x_0x_1, \\ y_1 = x_1x_2, \\ y_2 = x_1^2. \end{cases}$$

(2) *If Δ coincides with the adjoint linear system $\mathcal{L}$, then F has the representation*

(8.3) $$\begin{cases} y_0 = x_0^2, \\ y_1 = x_0x_1, \\ y_2 = x_0x_2, \end{cases} \quad or \quad \begin{cases} y_0 = x_2^2, \\ y_1 = x_1x_2, \\ y_2 = x_1^2 \end{cases}$$

(*which, in fact, are* (6.2) *or* (7.2) *with* $(\varepsilon_0, \varepsilon_1, \varepsilon_2) = (1, 0, 0)$).

PROOF OF STATEMENT (1). We shall prove that Δ is a pencil of type III$_2$ or II2 (actually, II2), and $a = 0$ in (7.3). Indeed, suppose that Δ is of type III$_1$. Then F can be given by (6.2), and Δ can be given by (6.4), which is never a triple line. Similarly, if Δ is of type III$_2$ or II2 (and, hence, F is represented by (7.3)), and $a \neq 0$, then Δ is given by (7.5) (after a coordinate change) and is not a triple line either.

Hence, F is given by (7.3) with $a = 0$, which can be converted to

$$(8.4) \quad \begin{cases} y_0 = \tilde{c}x_2^2 + 2dx_0x_1 + 2ex_0x_2, \\ y_1 = x_1x_2, \\ y_2 = x_1^2 - \varepsilon x_2^2 \end{cases}$$

(see the proof of Theorem 7.1), and Δ has equation (7.4). The condition that Δ be a triple line then yields $de = 0$, $d^2\varepsilon_1 - e^2 = 0$, $\tilde{c}d^2 \neq 0$, which implies $\tilde{c} \neq 0$, $d \neq 0$, $e = \varepsilon_1 = 0$, and (8.2) is obtained by a homogeneous coordinate change. □

PROOF OF STATEMENT (2). If $\mathcal{L}$ contains at least one pencil of type III$_1$, then F has representation (6.2), Δ is given by (6.4), and the condition $\Delta = \mathcal{L}$ implies $\varepsilon_1 = \varepsilon_2 = 0$, i.e., the first system of (8.3).

Now suppose that Δ contains a pencil of type III$_2$ or II2 and, hence, F is represented by (7.3). If $a \neq 0$, this can be converted to (7.2), equation (7.5) yields $\varepsilon_2 = 0$, and the coordinate change $y_2 \mapsto y_2 - \varepsilon_1 y_0$ gives the second system of (8.3). If $a = 0$, then F is represented by (8.4) with $de = d^2\varepsilon_1 - e^2 = \tilde{c}d^2 = 0$ (see (7.4)). Hence, either $d = e = 0$, or $d \neq 0$, $\tilde{c} = e = \varepsilon_1 = 0$. In the first case $\tilde{c} \neq 0$, and the coordinate change $y_0 \mapsto \tilde{c}y_0$, $y_2 \mapsto y_2 - \varepsilon_1 y_0$ shows that F is equivalent to the previous transformation. In the second case, the pencil $\{y_2 = 0\} \in \mathcal{L}$ is of type III$_1$, and, hence, the transformation is equivalent to that given by the first system of (8.3). □

References

[AVG] V. I. Arnold, S. M. Guseĭn-Zade, and A. N. Varchenko, *Singularities of differentiable maps*. I, "Nauka", Moscow, 1984; English transl., Birkhäuser, Basel, 1985.

[S] F. Severi, *Vorlesungen über die Algebraische Geometrie*, Teubner, Leipzig, 1921.

[T] A. Tyurin, *Geometry of the singularities of a generic quadratic form*, Izv. Akad. Nauk SSSR Ser. Mat. **44** (1980), 1200–1212; English transl. in Math. USSR-Izv. **15** (1981).

[Z] O. Zariski, *Algebraic surfaces*, Springer-Verlag, Heidelberg, 1971.

Translated by THE AUTHOR

Real Algebraic Curves and Link Cobordism. II

Patrick Gilmer

Dedicated to the memory of Professor D. A. Gudkov

Introduction

The purpose of this paper is to show how a number of important restrictions on the topology of real algebraic curves can be rederived from analogs of well-known restrictions on cobordisms between classical links. Let Q denote the projective tangent bundle of $\mathbb{R}P(2)$. Thus a point in Q consists of a point $p \in \mathbb{R}P(2)$ together with the direction of an unoriented line in $\mathbb{R}P(2)$ passing through p. Q can be identified with $S^3/\{\pm 1, \pm i, \pm j, \pm k\}$ (with the orientation reversed). By a Gauss map construction, one can define a link $L(D)$ lying above any collection D of simple closed smooth curves in general position in $\mathbb{R}P(2)$ [G1]. One has that $H_1(Q) \approx \mathbb{Z}_2 \times \mathbb{Z}_2$ with generators $\flat$ (represented by L (a line)) and $\mathfrak{f}$ (represented by a fiber of the map $Q \to \mathbb{R}P(2)$). In [G1], we showed how the complex locus $\mathbb{C}A$ of a real algebraic curve $\mathbb{R}F$ of degree m could be used to find a link cobordism of specified topological type between $L(\mathbb{R}F)$ and $L_m = L(\mathcal{L}_m)$, where $\mathcal{L}_m$ denotes m lines in general position.

In [G2], we developed the theory of signatures for links in rational homology 3-spheres. We proved analogs of the Murasugi–Tristram inequalities for nonorientable cobordisms. We also defined a notion of properness for links with extra data in a rational homology sphere, and defined the Arf invariant of proper links with this extra structure. We also studied the behavior of the Arf invariant under link cobordism. In particular, we generalized the fact that two proper links in the 3-sphere which are related by a planar cobordism must have the same Arf invariants.

In this paper, we will rederive the Gudkov conjecture [Gu, R1] and some related congruences by calculating the Arf invariants of the above links, and applying the above-mentioned result. We will also rederive the strengthened Arnold and strengthened Petrovskiĭ inequalities [A] by calculating the 2-signatures and 2-nullities of the above links and applying our generalized Murasugi–Tristram inequalities. See [W] for a good presentation of the strengthened Arnold and strengthened Petrovskiĭ inequalities as well as a discussion of the history of their refinement. In a later paper [G3], we will derive Fielder's congruence [F1], Zvonilov's inequality [V, 3.23], and the

---

1991 *Mathematics Subject Classification.* Primary 14P25, 14H99; Secondary 57M25.

This research was supported by a grant from the Louisiana Education Quality Support Fund.

Viro–Zvonilov inequality [**V**, 3.10] by studying links in the tangent circle bundle of $\mathbb{R}P(2)$.

Throughout this paper $\mathbb{R}F$ will stand for a nonsingular real algebraic curve in $\mathbb{R}P(2)$ of degree $2k$. C will denote a collection of oriented disjoint simple closed 2-sided curves in $\mathbb{R}P(2)$ called *ovals*. In §3 and §6, p, $p_\pm$, p_0, n, $n_\pm$, n_0, and l will denote the numerical characteristics of $\mathbb{R}F$ which are usually denoted in this way. See for instance [**V**, §2]. In §1, §2, §4, and §5, p, $p_\pm$, p_0, n, $n_\pm$, n_0, and l will denote these same numerical characteristics of C. Similarly an oval of C is called *even* (*odd*) if it is surrounded by an even (odd) number of other ovals. There is a very close relation between our proofs below of the Gudkov conjecture and related congruences and Marin's proofs [**M**, **A'C**] of these same results. However we believe our proof gives a new perspective on these results. It also allows a derivation without any appeal to Rokhlin's theorem on the signatures of spin 4-manifolds or a discussion of spin structures. There is also a close connection (but not as close) between Arnold's original derivation of the Arnold and strengthened Petrovskiĭ inequalities, and also Marin's derivation of the Petrovskiĭ inequalities [**M**, **A'C**] and our proofs of these same results. Our approach does allow a down to earth approach to a slightly weakened version of these results, which makes no use of covering spaces or the G-signature theorem. We also remark that knowing the 2-signatures of $L(C)$ may make it easier to calculate the not-yet-calculated p^r-signatures of $L(C)$ discussed in [**G1**,§9].

§1. Some spanning surfaces in Q

We wish to define a (usually nonorientable) spanning surface for $L(C)$. Consider the region B^+ consisting of those points lying on C or within an odd number of ovals of C. We draw a vector field $\vec{v}$ on B^+ with isolated zeros of index plus or minus one which is tangent to C along the boundary. To be specific, we choose to have a positive zero just inside each even oval and a negative zero just outside each odd curve. Thus we have p positive zeros and n negative zeros. This vector field induces a line segment field **l** with zeros of index ± 2. This line field produces a section of the bundle $Q \to \mathbb{R}P(2)$ over B^+ minus the zeros of $\vec{v}$. The image of this section may be completed to a surface F_C^+ (usually denoted just F^+) in Q homeomorphic to B^+ blown up at each zero of $\vec{v}$. Here we use "blow-up" in the sense of (real) algebraic geometry. Thus a surface blown up at a point is homeomorphic to the surface obtained by replacing a neighborhood of that point by a Möbius band. We can think of a surface blown up at a point as this same surface with the point replaced by the set of all lines through that point. In fact the projection of Q to $\mathbb{R}P(2)$ restricted to F^+ blows down these resolutions. The boundary of F^+ is $L(C)$. One can see what is going on above a neighborhood of a zero quite well. In Figure 1, we show in gray the solid torus which lies above a neighborhood of a zero of index one for $\vec{v}$. We show how F^+ intersects the boundary of the solid torus in black (dotted when hidden by the solid torus). The core of the solid torus is also the core of the Möbius band obtained by intersecting F^+ with the torus. Above a zero of index minus one, we will have a picture which looks like the reflection of Figure 1 through the plane of the page.

If C is oriented as the boundary of some orientation on B^+, then $\gamma(F^+) \in H_1(F^+)$ is represented by the circles lying over the blown up points. Thus $\gamma(F^+) = (p+q)\mathfrak{f} = l\mathfrak{f} \in H_1(Q)$. Since each component of $L(C)$ is null-homologous in Q, we have that $\gamma(F^+) \in H_1(Q) = l\mathfrak{f}$, no matter what orientation we assign to $\partial(F^+)$. Note $e(F^+) = -2l(C)$ using [**G1**, p. 39] and the definition [**G2**, §3].

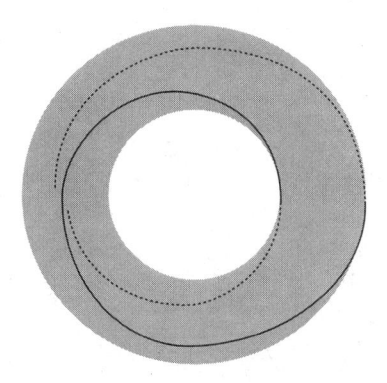

FIGURE 1

Give $\mathcal{L}_{2k}$ some specified orientation. Recall that this link with any orientation is isotopic to this link with any other. Also the isotopy type of this link does not depend on our choice of lines in general position. Partition our lines into k pairs and for each pair pick a pencil of lines through their point of intersection joining the pair of lines. For a given pencil $\mathcal{P}$, consider $\widetilde{\mathcal{P}} = \bigcup_{\mathcal{L} \in \mathcal{P}} L(\mathcal{L})$. $\widetilde{\mathcal{P}}$ is an annulus in Q with boundary L (the pair of lines joined by $\mathcal{P}$). If our lines are oriented, then we can choose our pencil so that $\gamma(\widetilde{\mathcal{P}}) \in H_1(Q)$ is either $\mathfrak{b}$ or zero. The union of the $\widetilde{\mathcal{P}}$'s over all our pairs of lines is then a spanning surface P_{2k} for L_{2k} which consists of a disjoint union of annuli. Depending on our choice of pencils, we can arrange that the γ-invariant of this surface is either $\mathfrak{b}$ or zero. Let P_{2k} denote such a surface with $\gamma(P_{2k}) = 0$. Let P'_{2k} denote such a surface with $\gamma(P'_{2k}) = \mathfrak{b}$. Note that the Euler numbers of these surfaces are zero [**G1**, p. 39].

§2. Some Arf invariants for $L(C)$ and L_{2k}

For $\varkappa, \theta \in \{0, 1\}$, let $q_{(\varkappa,\theta)}$ denote the quadratic refinement of the linking form on $H_1(Q)$ given by $q_{(\varkappa,\theta)}(x\mathfrak{b} + y\mathfrak{f}) = (1/2)(x^2\varkappa + xy + y^2\theta)$. By [**G2**, 6.2] and [**G1**, 3.1], we have:

PROPOSITION 2.1. $(L(C), \gamma, q)$ is proper for all (γ, q). $(L_{2k}, \gamma, q_{\varkappa,\theta})$ is proper only for the following $(\gamma, (\varkappa, \theta))$: $(0, (0, 0))$, $(0, (0, 1))$, $(\mathfrak{b}, (0, 0))$, $(\mathfrak{b}, (0, 1))$, $(\mathfrak{f}, (1, 0))$, $(\mathfrak{f}, (1, 1))$, $(\mathfrak{b} + \mathfrak{f}, (1, 0))$, $(\mathfrak{b} + \mathfrak{f}, (1, 1))$.

We will now calculate the Arf invariant $A(L, \gamma, q)$ defined in [**G**, §6].

PROPOSITION 2.2. If $(L_{2k}, a\mathfrak{b}, q)$ is proper for some a and q, then $A(L_{2k}, a\mathfrak{b}, q) = 0$. For any $\varkappa$,

$$A(L(C), l\mathfrak{f}, q_{(\varkappa,0)}) = p - q - l \quad \text{and} \quad A(L(C), l\mathfrak{f}, q_{(\varkappa,1)}) = q - p - l.$$

PROOF. The P_{2k} and P'_{2k} capped off are both planar, so $\beta(\varphi_q^{P_{2k}})$ and $\beta(\varphi_q^{P'_{2k}})$ when defined are both zero. Thus we obtain the above Arf invariants of L_{2k}. Let $x \in H_1(F^+)$ be a homology class representing the inverse image of a positive zero of $\vec{v}$ in F^+ with either orientation. Then $\mathcal{G}_{F^+}(x, x) = 1$. Let $y \in H_1(F^+)$ be a homology class representing the inverse image of a negative zero of $\vec{v}$ in F^+ with either orientation. Then $\mathcal{G}_{F^+}(y, y) = -1$. Thus $\varphi_{q_{(\varkappa,0)}}^{F^+}(x) = \varphi_{q_{(\varkappa,1)}}^{F^+}(y) = 1$, and $\varphi_{q_{(\varkappa,0)}}^{F^+}(y) = \varphi_{q_{(\varkappa,1)}}^{F^+}(x) = -1$. Since the blown up curves are all disjoint, and B^+ capped off is planar,

$\beta(\varphi_{q_{(\varkappa,0)}}^{F^+}) = p - q$. Similarly $\beta(\varphi_{q_{(\varkappa,1)}}^{F^+}) = q - p$. Thus $A(L(C), l\mathfrak{f}, q_{(\varkappa,0)}) = p - q - l$, and $A(L(C), l\mathfrak{f}, q_{(\varkappa,1)}) = q - p - l$. □

Let $\{\mathfrak{b}_2, \mathfrak{f}_2\}$ denote the basis for $H_1(Q, \mathbb{Z}_2)$ obtained by reducing $\mathfrak{b}, \mathfrak{f} \in H_1(Q)$ modulo two. Let $\{\mathfrak{b}^\#, \mathfrak{f}^\#\}$ denote the basis for $H^1(Q, \mathbb{Z}_2)$ dual to the basis $\{\mathfrak{b}_2, \mathfrak{f}_2\}$. $\mathfrak{f}^\#$ was denoted α in [**G1**]. We note that by [**G1**, p. 39], $\mathfrak{b}^\# \cdot \gamma = \gamma + \mathfrak{f}$ and $\mathfrak{f}^\# \cdot \gamma = \gamma + \mathfrak{b}$. Recall that for $\psi \in H^1(Q, \mathbb{Z}_2)$ we have $\psi \cdot q(x) = q(x) + 1/2(\psi(x))$. So $\mathfrak{b}^\# \cdot q_{\varkappa, \theta} = q_{\varkappa+1, \theta}$, and $\mathfrak{f}^\# \cdot q_{\varkappa, \theta} = q_{\varkappa, \theta+1}$. Here and below arithmetic modulo two is used in the subscripts of q.

PROPOSITION 2.3. $A(L_{2k}, \mathfrak{f}, q_{1,0}) = 0$.

PROOF. First we calculate $A(\varnothing, \mathfrak{f}, q_{1,0})$, where $\varnothing$ denotes the empty link. The Klein bottle $K_\mathcal{L}$, where $\mathcal{L}$ is a line in $\mathbb{R}P(2)$ discussed in [**G1**, p. 39], has $\gamma(K_\mathcal{L}) = \mathfrak{f}$ and is a spanning surface for the unlink. $H_1(K_\mathcal{L}, \mathbb{Z}_2)$ has a basis $\{b, f\}$, where b is represented by $L(\mathcal{L})$, and f is represented by the set of all unoriented directions through a point $p \in \mathcal{L}$. Thus $\varphi_{q_{(1,0)}}^{K_\mathcal{L}}(f) = 0$. So the Brown invariant of $\varphi_{q_{(1,0)}}^{K_\mathcal{L}}$ is zero. Thus $A(\varnothing, \mathfrak{f}, q_{1,0}) = 0$. Now

$$A(L_{2k}, \mathfrak{f}, q_{1,0}) = A(L_{2k}, \mathfrak{b}^\# \cdot 0, \mathfrak{b}^\# \cdot q_{0,0}) = A(L_{2k}, 0, q_{0,0}) + A(\varnothing, \mathfrak{f}, q_{1,0}),$$

where the last equal sign follows from [**G2**, (6.5)]. By Proposition 2.2 above we are done. □

These are all the Arf invariants we shall need. However, the rest may be easily calculated using the technique of Proposition 2.3 and results of Proposition 2.2. The following proposition is useful for this purpose. It is proved by calculating φ_q^F for all q, where F is one of the surfaces $K_\mathcal{L}$, $K_p = \mathcal{T}(K_\mathcal{L})$, $\mathcal{A} \circ \mathcal{T}(K_\mathcal{L})$. See [**G1**, p. 39] for the involutions $\mathcal{A}$ and $\mathcal{T}$. Let $A(\gamma, \varkappa, \theta)$ denote $A(\varnothing, \gamma, q_{(\varkappa, \theta)})$.

PROPOSITION 2.4.

$$A(\mathfrak{f}, 0, 0) = A(\mathfrak{f}, 1, 0) = A(\mathfrak{b}, 0, 0)$$
$$= A(\mathfrak{b}, 0, 1) = A(\mathfrak{b} + \mathfrak{f}, 1, 0) = A(\mathfrak{b} + \mathfrak{f}, 0, 1) = 0,$$
$$A(\mathfrak{f}, 0, 1) = A(\mathfrak{b}, 1, 0) = A(\mathfrak{b} + \mathfrak{f}, 0, 0) = 2,$$
$$A(\mathfrak{f}, 1, 1) = A(\mathfrak{b}, 1, 1) = A(\mathfrak{b} + \mathfrak{f}, 1, 1) = -2.$$

Turaev [**T**, (2.3)], [**G2**, (8.4)] showed that these numbers are actually the differences between the μ-invariants or Rokhlin invariants of different spin structures on Q. Let S_q denote the spin structure associated to a quadratic refinement q, and $\mu(\varkappa, \theta)$ denote $\mu(Q, S_{q_{(\varkappa, \theta)}})$. Using [**K**], for instance, one may calculate $\mu(1, 1) = -4$, and $\mu(\varkappa, \theta)$ is zero in the three other cases. Proposition 2.4 can be obtained in this way as well. The first derivation is more elementary.

§3. The Gudkov conjecture and related results

By [**G1**, (6.1)], there is a cobordism G in $I \times Q$ from $L(\mathbb{R}F)$ to L_{2k} with Euler number $2k^2 - 2l$. If $\mathbb{R}F$ is an M-curve, then the cobordism is a planar surface. We apply [**G2**, (6.7)] with $\gamma = l\mathfrak{f}$ and $q = q_{(k,0)}$ to this cobordism obtaining the congruence first conjectured by Gudkov [**Gu**]:

THEOREM 3.1 (Rokhlin [R1]). *For an M-curve,*
$$p - n \equiv k^2 \pmod{8}.$$

Since a dividing curve leads to an orientable cobordism, we have in the same manner:

THEOREM 3.2 (Arnold [A]). *For a dividing curve,*
$$p - n \equiv k^2 \pmod{4}.$$

If $\mathbb{R}F$ is an $(M-1)$-curve, then the cobordism G is a punctured projective plane, and $\gamma(G) = \mathfrak{f}$. Applying [G2, (6.7)], with the same γ and q, and noting that in this case $k \equiv l + 1 \pmod{2}$, we obtain:

THEOREM 3.3 (Gudkov–Krakhnov [GK], Kharlamov [Kh]). *For $(M-1)$-curves,*
$$p - n \equiv k^2 + \pm 1 \pmod{8}.$$

If we have an $(M-2)$-curve which is not dividing, then the resulting cobordism G in $I \times Q$ is a punctured Klein bottle with $\gamma = 0$. Applying [G2, (6.7)] as above, we obtain:

THEOREM 3.4 (Kharlamov [Kh]). *For nondividing $(M-2)$-curves,*
$$p - n \equiv k^2 + \begin{cases} 0 \\ \pm 2 \end{cases} \pmod{8}.$$

Equivalently, an $(M-2)$-curve satisfying $p - n \equiv k^2 + 4 \pmod{8}$ must be dividing.

§4. The Goeritz forms on F_C^+ and P_{2k}

We can orient the ovals of C so that whenever two ovals are nested, they are oriented in the same direction. Let $\{c_1, \ldots, c_n\}$ denote the odd ovals of C with orientations those chosen above. Let $\{x_1, \ldots, x_n\}$ denote the homology classes of $H_1(F^+)$ represented by components of $L(C)$ covering $\{c_1, \ldots, c_n\}$. Finally, let $\{y_1, \ldots, y_p\}$ denote the homology classes of the blown up positive zeros $\{a_1, \ldots, a_p\}$. Let $\{z_1, \ldots, z_q\}$ denote the homology classes of the blown up negative zeros $\{b_1, \ldots, b_q\}$. Here we orient the blown up zeros so that they link $L(C)$ positively if C is a surrounding oval. The union of these sets is a basis for $H_1(F^+)$. One calculates $\mathcal{G}_{F^+}(x_i, x_i) = 4$. Thus $\mathcal{G}_{F^+}(x_i, x_j) = 4$ if c_i and c_j are nested, and is zero otherwise. Also $\mathcal{G}_{F^+}(y_i, y_i) = 1$. Similarly $\mathcal{G}_{F^+}(x_i, y_j) = 2$ if c_i encloses a_j, and is zero otherwise. So $\mathcal{G}_{F^+}(z_i, z_i) = -1$, and $\mathcal{G}_{F^+}(x_i, z_j) = 2$ if c_i encloses b_j, and is zero otherwise. Moreover $\mathcal{G}_{F^+}(y_i, y_j) = \mathcal{G}_{F^+}(z_i, z_j) = \mathcal{G}_{F^+}(y_i, z_j) = 0$.

We wish to calculate the signature and nullity of this form. First note that if we change our basis by replacing x_i by $x_i' = x_i/2$, the matrix for this form with respect to the new basis will have all entries either zero, one or minus one. Next notice that the block with basis $\{y_1, \ldots, y_p\} \cup \{z_1, \ldots, z_q\}$ is diagonal with p ones and q minus ones down the diagonal. For $1 \leqslant i \leqslant n$, define

$$x_i'' = x_i' - \sum_{c_i \ni a_k} y_k + \sum_{c_i \ni b_k} z_k,$$

where $\ni$ means encloses. $\{x_1'', \ldots, x_n''\}$ is a basis for the orthogonal complement of the space spanned by $\{y_1, \ldots, y_p\} \cup \{z_1, \ldots, z_q\}$. Let A be the matrix of the Goeritz form restricted to this orthogonal complement with respect to $\{x_1'', \ldots, x_n''\}$. The entries of

A are given as follows. Let $E_{i,j}$ denote the number of even curves nested within both c_i and c_j. Let $O_{i,j}$ denote the number of odd curves nested within both c_i and c_j. We do not include either of c_i or c_j in these counts. Then $a_{i,j} = 1 + O_{i,j} - E_{i,j}$, if $i = j$ or if c_i and c_j are nested. Otherwise $a_{i,j}$ is zero.

PROPOSITION 4.1. $\text{Sign}(A) = n_+ - n_-$, and $\eta(A) = n_0$.

PROOF. Consider an innermost odd curve, and index the odd curves so that it is last. Then $a_{n,n}$ is positive, negative or zero according to whether this curve contributes to n_+, n_- or n_0. Moreover for all $i \neq n$, $a_{i,n} = a_{n,n}$ or $a_{i,n} = 0$ depending on whether the ith and nth n-curves are nested or not. Next notice that in the case $a_{n,n} \neq 0$, if we clear off diagonal entries of the nth row and column by row and column operations using $a_{n,n}$, then the matrix we obtain after deleting these cleared rows and columns is the A-matrix associated to the curve obtained by deleting this innermost odd curve from C. Finally in the case $a_{n,n} = 0$, the nth row and column are already zero, and if we delete them, we again obtain the A-matrix associated to C with this innermost odd curve deleted. Thus this proposition can be proved by induction on n. This proposition is trivially true if $n = 0$. □

We define $S_\gamma(C) = S(L(C), \gamma) + l$, and $\eta_\gamma(C) = \eta(L(C), \gamma)$. See [G2, §3] for the definitions of $S(L(C), \gamma)$ and $\eta(L(C), \gamma)$. We have added the term l so that we are really calculating the signature of $L(C)$ equipped with the framing described in [G1, §3]. Since F^+ has p components, by using Proposition 4.1 we see that:

PROPOSITION 4.2. $S_{l\mathfrak{f}}(C) = p - n_0 - 2n_-$ and $\eta_{l\mathfrak{f}}(C) = n_0 + p - 1$.

The Goeritz form on P_{2k} is identically zero. Thus we have:

PROPOSITION 4.3. $S_0(\mathcal{L}_{2k}) = S_\mathfrak{b}(\mathcal{L}_{2k}) = 0$ and $\eta_0(\mathcal{L}_{2k}) = \eta_\mathfrak{b}(\mathcal{L}_{2k}) = 2k - 1$.

§5. More spanning surfaces in Q and their Goeritz forms

We need to describe some more spanning surfaces for $L(C)$ and L_{2k} with other γ-invariants. Recall F^+ described above had $\gamma(F^+) = l\mathfrak{f}$. Consider an orientation reversing curve ω in B^-, and let K be the Klein bottle which is the inverse image of this curve in Q. Let $F'^+ = F^+ \cup K$. Then $\gamma(F'^+) = (l+1)\mathfrak{f}$. $L(\omega)$ represents a generator x'_0 of $H_1(K, \mathbb{Q}) \approx \mathbb{Q}$. If $\nu(\omega)$ represents the boundary of a neighborhood of ω in $\mathbb{R}P(2)$, then $\mathcal{G}_{F'^+}(x'_0, x'_0) = \mathcal{G}_K(x'_0, x'_0) = Lk(L(\omega), L(\nu(\omega))) = 1$. We also have that $\mathcal{G}_{F'^+}(x'_0, x'_i) = 1$ for $1 \leq i \leq n$. Let C'' denote C surrounded by a pair of nested ovals. Then the Goeritz form for F'^+ is isomorphic to the Goeritz form for $F^+_{C''}$, and $e(F'^+)/2 + l(C) = e(F^+_{C''})/2 + l(C'')$. Also F'^+ and $F^+_{C''}$ have the same number of components. Thus we have:

PROPOSITION 5.1. *We have*

$$S_{(l+1)\mathfrak{f}}(C) = \begin{cases} 1 & \text{if } C \text{ has no ovals,} \\ p - n_0 - 2n_- & \text{if } C \text{ has exactly one outer oval,} \\ p - n_0 - 2n_- - 1 & \text{if } C \text{ has more than one outer oval,} \end{cases}$$

$$\eta_{(l+1)\mathfrak{f}}(C) = \begin{cases} 0 & \text{if } C \text{ has no ovals,} \\ n_0 + p + 1 & \text{if } C \text{ has exactly one outer oval,} \\ n_0 + p & \text{if } C \text{ has more than one outer oval.} \end{cases}$$

Now consider a vector field defined on B^- with a negative zero just outside each even curve and a positive zero just inside each odd curve and one further positive zero outside every oval. Then we can construct a spanning surface F^- lying above B^- as in §1. Again we can find a basis for $H_1(F^-, \mathbb{Q})$ consisting of the curves over each of the $p + n + 1$ zeros, and curves lying over each of the even curves. Let C' denote C surrounded by a single oval. The Goeritz form on F^- is then isomorphic to the Goeritz form on $F_{C'}^+$, and

$$e(F^-)/2 + l(C) = e(F_{C'}^+)/2 + l(C').$$

Moreover F^- and $F_{C'}^+$ have the same number of components. $\gamma(F^-) = (l+1)\mathfrak{f} + \mathfrak{b}$. Consider now the image of F^- under the antipodal map $\mathcal{A}$ of the bundle Q over $\mathbb{R}P(2)$. Since $\mathcal{A}(\mathfrak{b}) = \mathfrak{f} + \mathfrak{b}$, we have $\gamma(\mathcal{A}(F^-)) = (l)\mathfrak{f} + \mathfrak{b}$. Moreover it is easy to see that $\mathcal{A}(L(C))$ is isotopic to $L(C)$. Putting all this together we obtain:

PROPOSITION 5.2.
$$S_\mathfrak{b}(C) = S_{\mathfrak{f}+\mathfrak{b}}(C) = n - p_0 - 2p_- + 1, \qquad \eta_\mathfrak{b}(C) = \eta_{\mathfrak{f}+\mathfrak{b}}(C) = p_0 + n.$$

We would like to find spanning surfaces for $\mathcal{L}_{2k}$ with γ-invariant $\mathfrak{f}$ and $\mathfrak{f} + \mathfrak{b}$. For this consider a Klein bottle K lying over a line ℓ_K in general position to the lines of $\mathcal{L}_{2k}$. Then K intersects $\mathcal{P}_{2k}$ in k connecting ribbon intersections. We resolve these ribbon intersections as is done in the proof of [G2, (4.3)] to obtain a new connected spanning surface $\mathcal{P}'_{2k}$ with $\gamma(\mathcal{P}'_{2k}) = \gamma(\mathcal{P}_{2k}) + \mathfrak{f}$. As $\gamma(\mathcal{P}_{2k})$ could be constructed to be either zero or $\mathfrak{b}$, we have realized the missing γ-invariants. $\beta_1(\mathcal{P}'_{2k}) = 2k + 1$. The Goeritz form on the first homology with rational coefficients of this surface is isomorphic to the direct sum of k hyperbolic planes and a single rank one positive definite space generated by $L(\ell_K)$. Thus we have:

PROPOSITION 5.3. $S_\mathfrak{f}(\mathcal{L}_{2k}) = S_{\mathfrak{f}+\mathfrak{b}}(\mathcal{L}_{2k}) = 1$ and $\eta_\mathfrak{f}(\mathcal{L}_{2k}) = \eta_{\mathfrak{f}+\mathfrak{b}}(\mathcal{L}_{2k}) = 0$.

§6. Strengthened Arnold and Petrovskiĭ inequalities

By [G1, (6.1)], there is a connected cobordism G from $L(\mathbb{R}F)$ to $\mathcal{L}_{2k}$ with $\chi(G) = k - 2k^2$, $e(G) = 2k^2 - 2l$, and $\gamma(G) = (l+k)\mathfrak{f}$. Let γ' denote $\gamma + (l+k)\mathfrak{f}$. Let X_γ be the two-fold branched covering space of $I \times Q$ along G which extends the covering space of Q along $\mathcal{L}_{2k}$ which is classified by γ ([G2]). Applying [G2, (4.8), (7.3), (7.4)], we obtain:

THEOREM 6.1. If RA is nonempty and $\gamma \in H_1(Q)$, then

$$|S_{\gamma'}(\mathbb{R}F) - S_\gamma(\mathcal{L}_{2k}) - k^2| + \eta_{\gamma'}(\mathbb{R}F) + \eta_\gamma(\mathcal{L}_{2k}) \leq 2k^2 - k + 2\Delta_\gamma,$$

where $\Delta_\gamma = \min\{\eta_{\gamma'}(\mathbb{R}F), \eta_\gamma(\mathcal{L}_{2k}), \dim H_1(I \times Q, G, \mathbb{Z}_2), \dim H_1(X_\gamma, \mathbb{Q})\}$.

The following lemma may be deduced from the proof of [G4, (1.5)].

LEMMA 6.2. Let $g: Y_1 \to Y_2$ be a map between finite CW complexes which induces a surjection $g_*: H_1(Y_1, \mathbb{Z}_2) \to H_1(Y_2, \mathbb{Z}_2)$. Let $\widetilde{Y}_2$ be a connected two-fold covering space of Y_2. Let $\widetilde{Y}_2$ be the pull back of $\widetilde{Y}_1$ under g. We have a map $\tilde{g}: \widetilde{Y}_1 \to \widetilde{Y}_2$ covering g. It follows that $\tilde{g}_*: H_1(\widetilde{Y}_1, \mathbb{Q}) \to H_1(\widetilde{Y}_2, \mathbb{Q})$ is surjective.

If Z is a space with an involution, we let $\bar{H}_k(Z)$ denote the minus one eigenspace for the action of the involution on $H_k(Z, \mathbb{Q})$.

LEMMA 6.3. *The plus one eigenspace for the action of the covering transformation on $H_1(X_\gamma, \mathbb{Q})$ is trivial, so $\bar{H}_1(X_\gamma)$ is isomorphic to $H_1(X_\gamma, \mathbb{Q})$. Let Y be a subspace of $(I \times Q) - G$ such that the inclusion induces an epimorphism $H_1(Y, \mathbb{Z}_2) \to H_1((I \times Q) - G, \mathbb{Z}_2)$. Let $\widetilde{Y}$ denote the inverse image of Y in X_g; then the inclusion induces an epimorphism $\bar{H}_1(\widetilde{Y}) \to H_1(X_\gamma, \mathbb{Q})$.*

PROOF. The transfer maps $H_1(I \times Q, \mathbb{Q})$ isomorphically to the plus one eigenspace of $H_k(X_\gamma, \mathbb{Q})$. But $H_1(I \times Q, \mathbb{Q}) = 0$. By Lemma 6.3, $\bar{H}_1(\widetilde{Y}) \to \bar{H}_1(X_\gamma - G, \mathbb{Q})$ is surjective. A Mayer–Vietoris sequence shows that the inclusion induces an isomorphism $\bar{H}_1(X_\gamma - G) \to \bar{H}_1(X_\gamma, \mathbb{Q}) = H_1(X_\gamma, \mathbb{Q})$. □

LEMMA 6.4. *Let $\mathcal{R}$ be a connected orientable component of $B^\pm$ with nonzero Euler characteristic. Let S denote the inverse image of $\mathcal{R}$ in $F^\pm$. If the 2-fold covering space of $Q - L(\mathbb{R}F)$ associated to γ' [G2, §7] is trivial when restricted to S, then $H_1(X_\gamma, \mathbb{Q}) = 0$.*

PROOF. If $s \in \mathcal{R}$ is a zero of the vector field used to construct $F^\pm$, let $f(s)$ be the curve on S lying over this zero. We may pick an orientation for $S' = S - \bigcup_{s \in \mathcal{R}}$(a neighborhood of $f(s)$). We may then orient each $f(s)$ parallel to the induced orientation on $\partial S'$. Let S_1 and S_2 be the two lifts of S in X_γ. Let $f(s)_i$ be the lift of $f(s)$ in S_i, oriented by lifting the orientation on $f(s)$. Then in $H_1(X_\gamma)$, we have $[f(s)_i] = [f(r)_i]$ if s and r are zeros of the same sign and $[f(s)_i] = -[f(r)_i]$ otherwise. To see this pick a path on $\mathcal{R}$ joining the zeros. Lying over this path we have a cylinder joining $f(s)$ and $f(r)$. The orientation on $f(s)$ and $f(r)$ will be the boundary of an orientation on this cylinder if and only if s and r are zeros of opposite sign. This cylinder is covered in X_γ by two cylinders which provide homologies between $f(s)_i$ and $\pm f(r)_i$. In fact the cover will necessarily be trivial on the union of S and this cylinder. By Lemma 6.3, $[f(s)_1] = -[f(s)_2]$ in $H_1(X_\gamma, \mathbb{Q})$, as $[f(s)_1] + [f(s)_2]$ is fixed by the involution. Now the number of zeros counted with sign is $\chi(\mathcal{R})$, which is nonzero. On the other hand, S_1 minus neighborhoods of the $f(s)_1$ provides a homology of $2\chi(R)[f(s)_1]$ to the homology class of the lift in X_γ of the boundary of S and this is fixed by the involution. Thus $f(s)_i$ represents zero in $H_1(X_\gamma, \mathbb{Q})$.

Let Y denote the union of the boundary of a tubular neighborhood of one component of L_{2k} in $Q \times \{0\}$, an arc on the tubular neighborhood of G in $I \times Q$ running from $Q \times \{0\}$ to $S \subset Q \times \{1\}$, an arc on S joining the end of the last arc to $f(s)$, and $f(s)$. Then the inclusion of Y into $(I \times Q) - G$ induces an isomorphism on homology with $\mathbb{Z}_2$ coefficients. Let $\widetilde{Y}$ denote the inverse image of Y in X_γ. By Lemma 6.3, the induced map $\bar{H}_1(\widetilde{Y}, \mathbb{Q}) \to H_1(X_\gamma, \mathbb{Q})$ is surjective. Above we showed that $f(s)_1$ and $f(s)_2$ represent zero in $H_1(X_\gamma, \mathbb{Q})$. As $\bar{H}_1(\widetilde{Y}, \mathbb{Q})$ is generated by $[f(s)_1] - [f(s)_2]$, we have $H_1(X_\gamma, \mathbb{Q}) = \bar{H}_1(X_\gamma, \mathbb{Q})$. □

LEMMA 6.5. *Let $\mathcal{R}$ be the connected nonorientable component of B^-. Let S denote the inverse image of $\mathcal{R}$ in F^-. Suppose the 2-fold covering space of $Q - L(\mathbb{R}F)$ associated to γ' [G2, §7] is trivial on the inverse image in S of an orientation reversing curve in $\mathcal{R}$, and is trivial on the inverse image in Q of any point in $\mathcal{R}$. Then $H_1(X_\gamma, \mathbb{Q}) = 0$.*

PROOF. We first observe that the inverse image of an orientation reversing curve in Q is a Klein bottle, which under our hypothesis will be covered by a pair of Klein bottles in X_γ. For $p \in R$, let $f(p)$ denote its inverse image in Q and $f'(p)$ denote a component of the inverse image of $f(p)$ in X_γ. Then $f'(p)$ is a fiber of a Klein bottle covering the Klein bottle associated to an orientation reversing curve passing through p. Thus $f'(p)$ is rationally null-homologous in this Klein bottle and so in X_γ. As above we let S_1 and S_2 denote the two components of the inverse image of S in X_γ. As the boundary of $\bar{S}_i$ is fixed by the involution and all the $f_i(s)$ are rationally null homologous in X_γ, we conclude that the image of $H_1(S_i, \mathbb{Q})$ in $H_1(X_\gamma, \mathbb{Q})$ is zero. Now let Y denote S union a meridian to $L(\mathbb{R}F)$, and $\widetilde{Y}$ its inverse image in $X_\gamma - G$. The inclusion of Y into $(I \times Q) - G$ will induce a map which is a surjection on $\mathbb{Z}_2$ homology. By Lemma 6.3, we only need to see that the image of the induced map $\bar{H}_1(\widetilde{Y}, \mathbb{Q}) \to H_1(X_\gamma, \mathbb{Q})$ is zero. Note that $\widetilde{Y}$ will consist of the union of a circle (the inverse image of the meridian) and S_1 and S_2. The homology class of the circle is fixed by the involution, and so must map to zero in $H_1(X_\gamma, \mathbb{Q})$. □

LEMMA 6.6. *Let $\mathcal{R}$ be the connected nonorientable component of B^-. Let S denote the inverse image of $\mathcal{R}$ in F^-. Suppose the 2-fold covering space of $Q - L(\mathbb{R}F)$ associated to γ' [G2, §7] is nontrivial on the inverse image in S of an orientation reversing curve in $\mathcal{R}$, and is trivial on the inverse image in S of any point in $\mathcal{R}$. Assume the Euler characteristic of $\mathcal{R}$ is nonzero. Then we have $H_1(X_\gamma, \mathbb{Q}) = 0$.*

PROOF. Let ω be an orientation reversing curve in $\mathcal{R}$ and ω' the inverse image of ω in S. Let $\widetilde{S}$ denote the connected inverse image of S in X_γ. Let $\widetilde{\omega}'$ denote the connected inverse image in $\widetilde{S}$ of ω'. $\widetilde{S} - \widetilde{\omega}'$ will have two components which we denote S_1 and S_2. Each projects homeomorphically to $S - \omega'$. The action of the involution on $\widetilde{\omega}'$ fixes the homology. Thus the involution fixes $H_1(\partial \bar{S}_1)$. For each zero $s \in \mathcal{R}$ of the vector field used to construct F^-, let $f(s)$ be the inverse image in Q and $f(s)_i$ the inverse image of $f(s)$ in S_i. As in the proof of Lemma 6.4, we have that S_i minus neighborhoods of the $f(s)_i$ provides a homology of $2\chi(R)[f(s)_i]$ to classes living on $\partial \bar{S}_1$. Thus $f(s)_i$ represents zero in $H_1(X_\gamma, \mathbb{Q})$. We conclude that the image of $H_1(S_i, \mathbb{Q})$ in $H_1(X_\gamma, \mathbb{Q})$ is zero.

Now let Y be the union of S, and a meridian of L (an exterior oval). Then the inclusion of Y into $(I \times Q) - G$ induces a surjection on homology with $\mathbb{Z}_2$ coefficients. Let $\widetilde{Y}$ denote the inverse image of Y in X_γ. $\widetilde{Y}$ consists of the union of $\bar{S}_1$, $\bar{S}_2$, and the lift of the meridian. By the above we see that the induced map $\bar{H}_1(\widetilde{Y}, \mathbb{Q}) \to H_1(X_\gamma, \mathbb{Q})$ is zero. On the other hand, by Lemma 6.3, the induced map $\bar{H}_1(\widetilde{Y}, \mathbb{Q}) \to H_1(X_\gamma, \mathbb{Q})$ is surjective. □

PROPOSITION 6.7. $\Delta_\gamma \leqslant 1$. *If $\Delta_\gamma = 1$, then $\mathbb{R}F$ is dividing and one of the following holds*:
  (a) $\gamma = \gamma' = 0$, k *is odd*, $n = n_0$ *and there is exactly one outer oval*.
  (b) $\gamma = \gamma' = \mathfrak{b}$, k *is even, and* $p = p_0$.

PROOF. As the components of L_{2k} represent $\mathfrak{b}$, we have $\dim H_1(I \times Q, G, \mathbb{Z}_2) \leqslant 1$. Thus $\Delta_\gamma \leqslant 1$. If $\mathbb{R}F$ is not dividing, $j^*(\mathfrak{f}^\#) \neq 0$ [G1, (6.1(2))], and so G has a curve representing $\mathfrak{f}$. Thus if $\mathbb{R}F$ is not dividing, $H_1(I \times Q, G, \mathbb{Z}_2) = 0$. Thus if $\Delta_\gamma = 1$, $\mathbb{R}F$ is dividing. We assume now that $\mathbb{R}F$ is dividing and that $\Delta_\gamma = 1$. As $\mathbb{R}F$ is dividing, we have $\gamma(G) = 0$ and $\gamma = \gamma'$. Moreover $\mathbb{R}F$ is not empty and so has at least one outer oval. We have that γ is not either $\mathfrak{f}$ or $\mathfrak{f} + \mathfrak{b}$, as otherwise we would have that $\eta_\gamma(L_{2k}) = 0$.

Assume $\gamma = \mathfrak{b}$. Note that the associated two-fold cover of $Q - L(\mathbb{R}F)$ is given by an element $\phi_\mathfrak{b} \in H^1(Q - L(\mathbb{R}F), \mathbb{Z}_2)$ which is defined by intersecting with a spanning surface F such that $\gamma(F) = \mathfrak{b}$. $\phi_\mathfrak{b}(x)$ is given by $Lk(x, L(\mathbb{R}F)) - 2\ell([x], \mathfrak{b})$ modulo two. It follows that this cover restricted to $F^+ - L(\mathbb{R}F)$ is trivial. By Lemma 6.4, every region of B^+ must have zero Euler characteristic, i.e., $p = p_0$. But then $n = p$, so l is even. As $\mathbb{R}F$ is dividing, we have $k \equiv l \pmod{2}$ by [R2, (3.2)] (or Theorem 3.2 above).

Assume $\gamma = \mathfrak{f}$. Note that the associated two-fold cover of $Q - L(\mathbb{R}F)$ is given by an element $\phi_\mathfrak{f} \in H^1(Q - L(\mathbb{R}F), \mathbb{Z}_2)$ which is defined by intersecting with a spanning surface F such that $\gamma(F) = \mathfrak{f}$. $\phi_\mathfrak{f}(x)$ is given by $Lk(x, L(\mathbb{R}F)) - 2\ell([x], \mathfrak{f})$ modulo two. Thus the cover is trivial on the inverse image in Q of any point of B^-. By Lemma 6.4, every orientable region of B^- must have zero Euler characteristic, i.e., $n = n_0$. Thus the number of ovals of $\mathbb{R}F$ is congruent modulo two to the number of outer ovals. Suppose the number of outer ovals is even, then the hypotheses of Lemma 6.5 hold. But the conclusion of Lemma 6.5 contradicts our hypothesis that $\Delta_\mathfrak{b} = 1$. Thus the number of outer ovals is odd. Thus the hypotheses of Lemma 6.6 hold, so the Euler characteristic of the nonorientable component of B^- is zero. In other words, there is one outer oval. Since we also have that $n = n_0$, we conclude that l and thus k must be odd. $\square$

We remark that [G2, (4.8)] alone yields $\Delta_\gamma \leqslant 2$. We mention this as [G2, (4.8)] is proved by more elementary means than [G2, (7.3), (7.4)]. In particular no mention of covering spaces or the G-signature theorem is made. In this way a slightly weakened version of the following theorem can be proved without using results from 6.2–6.7, and without [G2, §7]. This offers a down to earth approach to these inequalities.

We now prove the strengthened Arnold and strengthened Petrovskiĭ inequalities in the form obtained by Wilson [W, (7.4) and the Remark at the end of the first paragraph on p. 70]. Note [W, the first paragraph on p. 70] contains some misprints: (3) should be (4) and vice versa throughout this paragraph. Inequalities (1) and (3) are the strengthened Arnold inequalities. Inequalities (2) and (4) are the strengthened Petrovskiĭ inequalities. We remark that the proviso in the case $k = 1$ in inequality (3) and the proviso in the case $k \neq 1$ in inequality (4) are necessary. Wilson omits this proviso for inequality (3). We note that the results obtained as below, but letting $\gamma = \mathfrak{f}$ or $\mathfrak{f} + \mathfrak{b}$ instead, are weaker than those given below. This is due to the small nullities given in 5.3.

THEOREM 6.8. *We have*

(1)
$$p_- + p_0 \leqslant \frac{(k-1)(k-2)}{2} + \begin{cases} 1 & \text{if } p = p_0, k \text{ is even and } \mathbb{R}F \text{ is dividing,} \\ 0 & \text{otherwise,} \end{cases}$$

(2) $\quad n - p_- \leqslant \dfrac{3k(k-1)}{2},$

(3)
$$n_- + n_0 \leqslant \frac{(k-1)(k-2)}{2} + \begin{cases} 0 & \text{if } n = n_0, k \text{ is odd, } \mathbb{R}F \text{ is dividing} \\ & \text{and } \mathbb{R}F \text{ has exactly one outer oval,} \\ 0 & \text{if } k \text{ is even,} \\ 0 & \text{if } k = 1 \text{ and } \mathbb{R}F \text{ has no ovals,} \\ -1 & \text{otherwise,} \end{cases}$$

$$(4) \quad p - n_- \leq \frac{3k(k-1)}{2} + 1 + \begin{cases} -1 & \text{if } k \text{ is odd, } k \neq 1 \text{ and } \mathbb{R}F \\ & \text{has exactly one outer oval,} \\ 0 & \text{otherwise.} \end{cases}$$

PROOF. The result holds trivially if $\mathbb{R}F$ is empty. Thus we assume from now on that $\mathbb{R}F$ has at least one oval. We apply Theorem 6.1 with $\gamma = \mathfrak{b}$ and $\gamma' = (l+k)\mathfrak{f} + \mathfrak{b}$. Substituting in Propositions 4.3 and 5.2, we obtain:

$$|n - p_0 - 2p_- + 1 - k^2| + p_0 + n + 2k - 1 \leq 2k^2 - k + 2\Delta_\mathfrak{b}.$$

Making use of Proposition 6.7 and $|x| \geq \begin{cases} x, \\ -x, \end{cases}$ we obtain inequality (1) and inequality (2) with the proviso that inequality (2) may be wrong by one if $p = p_0$, k is even and $\mathbb{R}F$ is dividing. However if $p = p_0$, then $n = p$ and by inequality (1), we have that $n \leq (k-1)(k-2)/2 + 1$ and so inequality (2) holds trivially in this case, making use of the fact that $n = 0$ if $k = 1$.

We now apply Theorem 6.1 with $\gamma = 0$ and $\gamma' = (l+k)\mathfrak{f}$. Consider first the case in which k is even. Substituting in Propositions 4.2 and 4.3, we obtain:

$$|p - n_0 - 2n_- - k^2| + n_0 + p - 1 + 2k - 1 \leq 2k^2 - k + 2\Delta_0.$$

As above, we obtain inequalities (3) and (4) for k even.

Now considering the case in which k is odd, and $\mathbb{R}F$ has more than one outer oval, we obtain:

$$|p - n_0 - 2n_- - 1 - k^2| + n_0 + p + 2k - 1 \leq 2k^2 - k + 2\Delta_0.$$

This yields inequalities (3) and (4), for k odd when $\mathbb{R}F$ has more than one outer ovals.

Now considering the case k odd, and $\mathbb{R}F$ has exactly one outer oval, we obtain:

$$|p - n_0 - 2n_- - k^2| + n_0 + p + 1 + 2k - 1 \leq 2k^2 - k + 2\Delta_0.$$

This yields inequalities (3), for k odd when $\mathbb{R}F$ has exactly one outer oval. It also yields inequality (4) with the proviso that (4) may be wrong by one if in addition $n = n_0$, and $\mathbb{R}F$ is dividing. However if $n = n_0$ and $\mathbb{R}F$ has exactly one outer oval, then $p = n + 1$. So, by inequality (3), $p \leq 1 + (k-1)(k-2)/2$. This implies that inequality (4) follows trivially. $\square$

References

[A]   V. I. Arnold, *On the distribution of the ovals of real plane curves, involutions of 4-dimensional smooth manifolds, and the arithmetic of integer valued quadratic forms*, Funktsional. Anal. i Prilozhen. **5** (1971), no. 3, 1–9; English transl., Functional Anal. Appl. **5** (1971), no. 3, 169–176.

[A'C]   N. A'Campo, *Sur la première parte du seizième problème de Hilbert*, Séminaire Bourbaki **31** (1978–1979), no. 537, 1–20; Lecture Notes in Math., vol. 770, Springer-Verlag, Berlin and Heidelberg, 1980, pp. 208–227.

[F1]   T. Fiedler, *New congruences in the topology of real plane curves*, Dokl. Akad. Nauk SSSR **270** (1983), no. 1, 56–58; English transl., Soviet Math. Dokl. **27** (1983), 566–568.

[G1]   P. Gilmer, *Real algebraic curves and link cobordism*, Pacific J. Math. **153** (1992), no. 1, 31–69.

[G2]   _____, *Link cobordism in rational homology 3-spheres*, J. Knot Theory and Ramifications **2** (1993), 285–320.

[G3]   _____, *Real algebraic curves and link cobordism* III (to appear).

[G4]   _____, *Configuration of surfaces in 4-manifolds*, Trans. Amer. Math. Soc. **264** (1981), 353–380.

[Gu]   D. A. Gudkov, *Construction of a new series of M-curves*, Dokl. Akad. Nauk SSSR **200** (1971), no. 6, 1269–1272; English transl., Soviet Math. Dokl. **12** (1971), no. 5, 1559–1563.

[GK] D. A. Gudkov and A. D. Krakhnov, *On the periodicity of the Euler characteristic of real algebraic* $(M-1)$-*manifolds*, Funktsional. Anal. i Prilozhen. **7** (1973), no. 2, 15–19; English transl., Functional Anal. Appl. **7** (1973), no. 2, 82–98.

[K] S. J. Kaplan, *Constructing framed 4-manifolds with given almost framing*, Trans. Amer. Math. Soc. **254** (1979), 237–263.

[Kh] V. M. Kharlamov, *New congruences for the Euler characteristic of real algebraic manifolds*, Funktsional. Anal. i Prilozhen. **7** (1973), no. 2, 74–78; English transl., Functional Anal. Appl. **7** (1973), no. 2, 147–150.

[M] A. Marin, *Quelques remarques sur les courbes algebriques planes réélas*, Publ. Math. Univ. Paris VII **9** (1980), 51–68.

[R1] V. A. Rokhlin, *Congruences modulo 16 in sixteenth Hilbert problem*, Funktsional. Anal. i Prilozhen. **6** (1972), no. 4, 58–64; English transl. in Functional Anal. Appl. **6** (1972), no. 4, 301–306.

[R2] _____, *Complex topological characteristics of real algebraic curves*, Uspekhi Mat. Nauk **33** (1978), no. 5, 77–89; English transl., Russian Math. Surveys **33** (1978), no. 5, 85–98.

[T] V. G. Turaev, *Cohomology rings, linking coefficients and invariants of spin structures in three-dimensional manifolds*, Mat. Sb. **120** (1983), no. 1, 68–83; English transl., Math. USSR-Sb. **48** (1984), 65–79.

[V] O. Ya. Viro, *Progress in the topology of real algebraic varieties over the last six years*, Uspekhi Mat. Nauk **41** (1986), no. 3, 45–67; English transl., Russian Math. Surveys **41** (1986), no. 3, 55–82.

[W] G. Wilson, *Hilbert's sixteenth problem*, Topology **17** (1978), no. 1, 53–74.

DEPARTMENT OF MATHEMATICS, LOUISIANA STATE UNIVERSITY, BATON ROUGE, LOUISIANA 70803
E-mail address: gilmer@marais.math.lsu.edu

Morsifications of Rational Functions

V. V. Goryunov

ABSTRACT. The paper is devoted to enumerative problems of the theory of real singularities. We calculate the number of connected components of the set of rational functions on a real line having as many poles and critical points as possible. In the Appendix, in terms of snakes on Dynkin diagrams, we obtain the numbers of topologically different real morsifications of simple function-germs E_μ with μ distinct critical values.

A series of recent papers [1–3] by Arnold was devoted to the study of a new invariant of a real isolated function singularity, the number of its topologically different *very nice morsifications*. These are morsifications with as many real critical values as possible. Calculations done by Vakulenko [6] show that for an A_μ singularity the value of the invariant is equal either to an Euler number or to a tangent number. The values for the other infinite series, D, of the simple function-germs were obtained in [16] (see also [12, 13]). They turned out to be closely related to similar invariants for Laurent polynomials with a simple pole. In the present note, extending the area of study, we consider the space of rational functions on the real line. We calculate the number of connected components of a set of M-functions, the ones with as many real critical points and poles as possible. This numerical invariant is rougher than the invariant introduced by Arnold. Say a set of M-morsifications of an A_μ singularity, i.e., the space of M-polynomials of fixed degree, is connected. But for rational functions the situation is far from such simplicity.

In the Appendix we calculate the numbers of connected components of the set of very nice morsifications of E_6, E_7, and E_8 function singularities. These numbers are 82, 768, and 4056 respectively. The calculation is based on the bijection between the set of connected components of space of M-morsifications and different R-diagrams of a simple singularity.

Let $\Lambda_{\mu,\nu} = \mathbb{R}^{\mu+\nu}$ be the space of rational functions in one variable

$$r(x) = p(x)/q(x) = (x^\mu + \lambda_1 x^{\mu-1} + \cdots + \lambda_\mu)/(x^\nu + \lambda_{\mu+1} x^{\nu-1} + \cdots + \lambda_{\mu+\nu}).$$

For generic values of the parameters, the number of distinct complex critical points of the function is either $\mu + \nu - 1$ if $\mu \neq \nu$, or $2\mu - 2$ if $\mu = \nu$. This is the maximal

---

1991 *Mathematics Subject Classification*. Primary 58C27; Secondary 05A19, 11B68.
Supported by NSF grant DMS-9022140.

©1996, American Mathematical Society

possible number of critical points. In this case all of them are Morse, p and q are relatively prime and all the ν poles are simple.

DEFINITION. A rational function $r(x)$ is called a *rational M-function* if
(1) it has the above maximal number of critical points;
(2) all the critical points are real;
(3) all the poles of r are real too.

We denote the set of all rational M-functions in $\Lambda_{\mu,\nu}$ by $M_{\mu,\nu}$. Our goal is to calculate the number $m_{\mu,\nu}$ of connected components of this set.

EXAMPLES. 1. The set of all M-polynomials of fixed degree is connected, i.e., we have $m_{\mu,0} = 1$ [1].
2. The set of all Laurent M-polynomials with a single pole of order 1 has $m_{\mu,1} = \mu - 1$ components [16]. In this case, shifting the pole to the origin, we get the αth component formed by the functions with α critical points positive and $\mu - \alpha$ critical points negative, $0 < \alpha < \mu$.

For the general case we prove the following

THEOREM. *The number of connected components of the set $M_{\mu,\nu}$ of rational M-functions is equal to*

$$m_{\mu,\nu} = \begin{cases} \dfrac{(\mu - \nu)(\mu + 2\nu - 1)!}{(\mu + \nu)!\,\nu!}, & \text{if } \mu > \nu, \\ \dfrac{2(\nu - \mu)(\mu + 2\nu - 1)!}{(2\nu)!\,\mu!}, & \text{if } \mu < \nu, \\ \dfrac{4(3\nu + 1)!}{(2\nu + 2)!\,\nu!}, & \text{if } \mu = \nu. \end{cases}$$

Our enumeration of the connected components is based on the consideration of the curves $\gamma(r) = \{\text{Im}\, r(x) = 0\}$ in $\mathbb{C}$ up to orientation-preserving diffeomorphisms of $\mathbb{C} \simeq \mathbb{R}^2$, $\mathbb{R}$. In the simplest cases, for $\nu = 0$ or $\mu = 0$, all the rational M-functions r have equivalent curves $\gamma(r)$. We show them in Figure 1. The nodal points there are the critical points and the crosses are the poles.

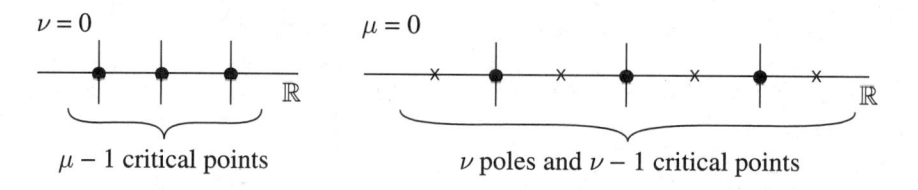

FIGURE 1. Inverse image $\gamma(r)$ of the real axis under a rational M-function r

In the other cases, up to the above diffeomorphisms, the following inductive surgery procedure is valid.

LEMMA 1. *The curve $\gamma(r) = \{\text{Im}\, r(x) = 0\}$ of a rational M-function $r \in M_{\mu,\nu}$, $\mu \neq \nu$, is obtained from the curve $\gamma(r') = \{\text{Im}\, r'(x) = 0\}$ of some rational M-function $r' \in M_{\mu-1,\nu-1}$ by replacing one of the $\mu + 2\nu - 3$ intervals into which the poles and the* *critical points of r' cut the real axis, by an elementary block with the following circular component*:

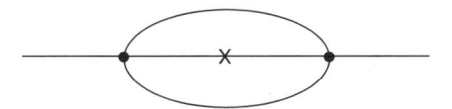

If $(\mu, \nu) \neq (2, 1)$, then among the intervals mentioned in Lemma 1 there are two infinite ones. If $(\mu, \nu) = (2, 1)$, the only interval we have is the whole real axis.

REMARK. The case $\mu = \nu$ is easily seen to give the same curves as $\nu = \mu + 1$.

EXAMPLES. The families represented in Figure 2 (see p. 88) have less than $\mu + \nu$ parameters due to linear transformations of the source and the target. The reduced parameter space is subdivided into several open regions (some of them are connected components of the set $M_{\mu,\nu}$ of rational M-functions) by a *bifurcation diagram*. The diagram is formed by the values of the parameters corresponding to functions having:

degenerate critical points (Σ_c),
nonsimple poles (Σ_p),
numerator and denominator with common roots (Σ_0),
a critical point at $x = \infty$ (Σ_∞, for $\mu = \nu$ only).

Each connected component of $M_{\mu,\nu}$ is marked with the corresponding curve γ.

The proof of Lemma 1 is very close to the proofs of similar statements in [2, 16]. We only point out the facts the proof is based on.

(1) The closure $\bar{\gamma}$ of the curve $\gamma(r)$ is smooth except for the nodes on $\mathbb{R}$ at the critical points of r.
(2) Exactly $2|\mu - \nu|$ branches of the curve $\gamma(r)$ go to infinity.
(3) Circulating by the gradient flow of r along $\bar{\gamma} \cup \{x = \infty\}$, we can make a cycle only having passed through a pole. Here $x = \infty$ is regarded as a pole if $\mu > \nu$.
(4) All the equivalence classes of the curves are realizable.

For (4) one constructs the corresponding mapping between the Riemann spheres as in [2, §2].

REMARK. The surgery of the Lemma relates a particular pole to each circular component of γ.

The mapping in (4), i.e., a rational M-function, depends continuously on the possible choice of the critical values and on the positions of the critical points and the poles while we stay in the same equivalence class of the curves γ. As in the polynomial case [2, §2] and the Laurent case [16], the set of M-functions r having equivalent curves $\gamma(r)$ is easily seen to be connected and, moreover, contractible (see the end of the main part of this note). Thus, we get

COROLLARY 2. *The number $m_{\mu,\nu}$, $\mu \neq \nu$, of connected components of the set of rational functions is equal to the number of equivalence classes of the curves $\gamma(r)$ of Lemma 1.*

Now we consider the three cases of the Theorem. In what follows by "a curve γ" we mean its equivalence class. We say that a node (resp. pole) is *free* if it has two infinite whiskers (resp. is not contained in a disk bounded by a circular component of γ). The curve γ has

for $\mu > \nu$: ν circles, $\mu - \nu - 1$ free nodes and no free poles;
for $\mu < \nu$: μ circles, $\nu - \mu - 1$ free nodes and $\nu - \mu$ free poles.

$\mu = 1, \nu = 2, r(x) = (x + \alpha_1)/(x^2 + \alpha_2)$

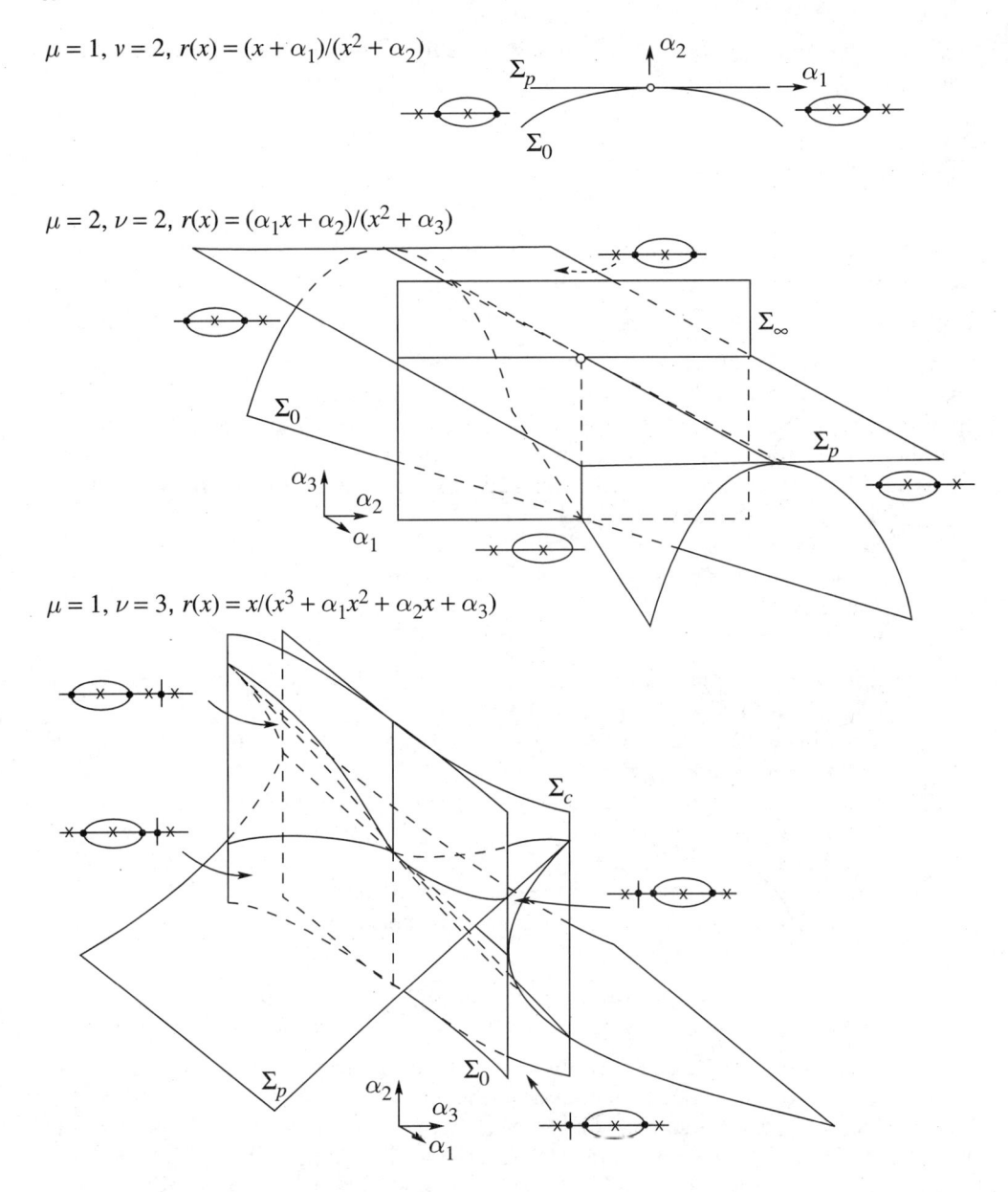

$\mu = 2, \nu = 2, r(x) = (\alpha_1 x + \alpha_2)/(x^2 + \alpha_3)$

$\mu = 1, \nu = 3, r(x) = x/(x^3 + \alpha_1 x^2 + \alpha_2 x + \alpha_3)$

FIGURE 2. Bifurcation diagrams and sets of rational M-functions for low values of μ

An *exterior circle* of γ is its circular component not contained in the disk bounded by any other circular component.

$\mu > \nu$. Let us slightly change the notation to indicate the number $\beta = \mu - \nu - 1$ of free nodes:

$$\phi_{\beta,\nu} = m_{\nu+\beta+1,\nu}, \qquad \beta, \nu \geqslant 0.$$

Thus, we need to show that
$$\phi_{\beta,\nu} = \frac{(\beta+1)(3\nu+\beta)!}{(2\nu+\beta+1)!\,\nu!}.$$

Let us start with $\beta = 0$: no free poles, no free nodes, all the 2ν critical points participate in the ν circles. Denote by $n_{\nu,s}$ the number of curves γ with exactly s exterior circles.

LEMMA 3. *We have*
$$n_{\nu,s} = \sum_{\alpha \geq 1} \sum_{k \geq 0} (k+1) n_{\alpha-1,k} n_{\nu-\alpha,s-1}, \qquad \nu \geq 1.$$

PROOF. Take an exterior circle of γ, the first from the right on the real axis. Let $\alpha - 1$ be the number of circles inside it, and k the number of exterior circles in this subconfiguration of $\alpha - 1$ circles. There are $k + 1$ possibilities for the position of the pole corresponding to the above circle of γ (see our latter remark) with respect to the subconfiguration.

Let us introduce the generating function
$$N(x, y) = \sum_{\nu,s \geq 0} n_{\nu,s} x^\nu y^{s+1}.$$

COROLLARY 4. *We have* $N(x,y) = xy \dfrac{\partial N}{\partial y}(x,1) N(x,y) + y$.

Let us take the partial y-derivative of this relation and evaluate both the derivative and the initial relation at $y = 1$. Introducing
$$\phi(x) = N(x,1) = \sum_{\nu \geq 0} \phi_{0,\nu} x^\nu, \qquad \psi(x) = \frac{\partial N}{\partial y}(x,1),$$
we obtain
$$\phi = x\psi\phi + 1, \qquad \psi = x\psi(\phi + \psi) + 1.$$
Excluding ψ, we obtain

COROLLARY 5. $x\phi^3 - \phi + 1 = 0$.

REMARK. Replacing here the degree 3 by 2, we get the generating function for the Catalan numbers.

The assertion of the Theorem for $\mu = \nu + 1$ is equivalent to the recurrence relation
$$\phi_{0,\nu+1} = \frac{3(3\nu+1)(3\nu+2)}{2(\nu+1)(2\nu+3)} \phi_{0,\nu} \quad \text{with} \quad \phi_{0,0} = 1.$$

This means that the function $\phi(x)$ must satisfy the differential equation
$$2(2x\partial_x + 1) x\partial_x \phi = 3x(3x\partial_x + 1)(3x\partial_x + 2) \phi$$

being its unique solution at $x = 0$ with $\phi(0) = 1$. Elementary calculations show that this follows from the last Corollary.

Now let $\mu - \nu - 1 = \beta > 0$. Let
$$\Phi(x, z) = \sum_{\beta \geqslant 0} \sum_{\nu \geqslant 0} \phi_{\beta,\nu} x^\nu z^\beta$$
be a generating function. Then

LEMMA 6. $\Phi = \phi/(1 - z\phi)$.

PROOF. For $\beta > 0$ let us consider the free node of γ, the first from the right on the real axis. Let α be the number of circles to the right of this node. Then
$$\phi_{\beta,\nu} = \sum_{\alpha \geqslant 0} \phi_{0,\alpha} \phi_{\beta-1,\nu-\alpha}.$$
For the generating functions, this means $\Phi(x, z) = z\phi(x)\Phi(x, z) + \phi(x)$.

Thus
$$\Phi(x, z) = \sum_{\beta \geqslant 0} z^\beta \phi^{\beta+1}(x).$$
To prove the theorem for $\mu > \nu$, it remains to show that
$$\phi^{\beta+1}(x) = \sum_{\nu \geqslant 0} \frac{(\beta + 1)(3\nu + \beta)!}{(2\nu + \beta + 1)!\,\nu!} x^\nu.$$
This is equivalent to the requirement that the function $\phi^{\beta+1}(x)$ is analytic at the origin and satisfies the differential equation
$$(2x\partial_x + \beta + 1)(2x\partial_x + \beta)x\partial_x \psi$$
$$= x(3x\partial_x + \beta + 3)(3x\partial_x + \beta + 2)(3x\partial_x + \beta + 1)\psi, \qquad \psi(0) = 1.$$
Again, the fact that this is exactly the case follows from the cubic equation on ϕ.

$\mu < \nu$. Let $\varepsilon = \nu - \mu > 0$ be the number of free poles of γ. To each of the free poles we attach a complex-conjugate-symmetric pair of whiskers going to infinity and intersecting neither γ nor the other pairs added. We get a curve γ' of some rational M-function $(x^{\nu+\varepsilon} + \ldots)/(x^{\nu-\varepsilon} + \ldots)$.

The mapping $\gamma \mapsto \gamma'$ is obviously one-to-one. Indeed, the inverse mapping is constructed as follows. Each curve γ' has exactly $2\varepsilon - 1$ free nodes. Order these nodes: $x_1 > x_2 > \cdots > x_{2\varepsilon-1}$. Now omit the whiskers starting at all x_{odd} and declare all x_{odd} to be poles. We obtain

LEMMA 7. $m_{\nu-\varepsilon,\nu} = m_{\nu+\varepsilon,\nu-\varepsilon}$.

This is exactly the statement of the Theorem for $\mu < \nu$.

REMARK. It is not too difficult to see that the function $\psi(x) = (\partial N/\partial y)(x, 1)$ above is the generating function $\sum_{\mu \geqslant 0} m_{\mu,\mu+1} x^\mu$.

$\mu = \nu$. The assertion of the Theorem for this case follows from

LEMMA 8. $m_{\nu,\nu} = 2m_{\nu-1,\nu}$.

PROOF. The function
$$r(x) = (x^\nu + \lambda_1 x^{\nu-1} + \ldots)/(x^\nu + \lambda_{\nu+1} x^{\nu-1} + \ldots)$$
is a rational M-function if and only if $\lambda_1 \neq \lambda_{\nu+1}$ and the function
$$(r(x) - 1)/(\lambda_1 - \lambda_{\nu+1}) = (x^{\nu-1} + \ldots)/(x^\nu + \ldots)$$
is a rational M-function. According to the two possible signs of $(\lambda_1 - \lambda_{\nu+1})$, we obtain $m_{\nu,\nu} = 2 m_{\nu-1,\nu}$.

Thus the Theorem is proved.

The decreasing order of critical points of a function on the real line induces an ordering of the critical values. Suppose $\mu \neq \nu$. Each equivalence class of γ imposes a certain system of $\mu + \nu - 2$ inequalities on the ordered set of the $\mu + \nu - 1$ critical values of a rational M-function r from the corresponding connected component R_γ. The same inequalities define a simplicial cone $\mathcal{R}_\gamma$ in Euclidean space $\mathbb{R}^{\mu+\nu-2} = \{z_1 + \cdots + z_{\mu+\nu-1} = 0\} \subset \mathbb{R}^{\mu+\nu-1}$ equipped with the set of diagonals $z_i = z_j$ (mirrors of the reflection group $A_{\mu+\nu-2}$). As in [2, 3, 16], we have
 (1) R_γ is homeomorphic to $\mathcal{R}_\gamma \times \mathbb{R}^2$, in particular R_γ is contractible;
 (2) the number of chambers in the cone $\mathcal{R}_\gamma$ coincides with the number of connected components of the set of very nice (= with all critical values different) rational functions contained in R_γ, and each of these components is contractible too.

For $\mu = \nu$, we must replace $\mu + \nu$ by $2\nu - 1$ everywhere and $\mathbb{R}^2$ for $\mathbb{R}^3$ in (1).

OPEN COMBINATORIAL QUESTION. Calculate the number of chambers in $\mathcal{R}_\gamma$.

Related problems. For recent progress in problems closely related to the topic of the present paper, see [4, 5, 7–15, 18]. For example, [10] establishes a direct correspondence between very nice morsifications of A_μ function singularities and lemniscate configurations of complex polynomials. The existence of such a relation has been suggested by the observation that in both situations the sequence of numbers of distinct objects was one and the same, namely, the Euler-tangent sequence [1, 11]

$$1, 1, 2, 5, 16, 61, 272, 1385, 7936, 50521, \ldots.$$

Probably a similar correspondence exists for rational functions as well.

In the quantum settings, two sequences starting like the Euler–tangent one were recently discovered. According to Kirillov [17], the number of conjugacy classes in the group of nondegenerate upper-triangular $n \times n$-matrices over $\mathbb{Z}_2$ is

$$1, 2, 5, 16, 61, 275, 1430, 8506, 57205, \ldots.$$

Zograf's calculations [19] of volumes of moduli spaces of punctured spheres produced

$$1, 5, 61, 1379, 49946, \ldots.$$

It would be very exciting to find, as suggested by Arnold, a direct relation between these quantum objects and, say, classes of very nice morsifications of function-germs.

Appendix. Very nice morsifications of E_μ singularities

In the introduction to the main part of this note, we mentioned two invariants of a real function singularity: the number of connected components of the set of M-morsifications and of very nice morsifications. For simple singularities of the series A and D, these invariants have been calculated in [1–5]. Here we present a different approach, universal for at least all the simple functions, and carry out the calculations for the remaining simple germs E_6, E_7, and E_8.

1. M-components. Consider any real simple function-germ X_μ, $X = A, D, E$, on a plane, with Milnor number μ. The base of its truncated miniversal deformation [4] contains an open set of M-morsifications of X_μ (functions with μ real critical points). According to [7], each connected component of this set (shortly, M-component) contains *sabirifications* of X_μ, functions with all saddle points on the same level.

A sabirification f defines an R-diagram of X_μ. This is an analog of a Dynkin diagram with extra information about the indices of the critical points of f. Each critical point of f corresponds to a vertex of the R-diagram marked by a plus (maximum of f) or by a minus (minimum) or considered neutral (saddle). The minima and maxima are taken on in the regions of the plane bounded by a saddle level curve. We join a plus-vertex and a minus-vertex by an edge if the corresponding regions are separated by an interval of the saddle level. We join a plus-vertex or a minus-vertex with a neutral vertex if the closure of the region contains the saddle point.

All the possible R-diagrams of the simple functions are listed, for example, in [7]. Their number $d(X_\mu)$ is given by the table:

X_μ	A_μ	D_{2k}^+	D_{2k}^-	D_{2k+1}	E_6	E_7	E_8
$d(X_\mu)$	1	$k-1$	k	k	2	4	5

For a fixed simple type X_μ of a real singularity, the R-diagrams have the same number of neutral vertices and the difference of the diagrams is equivalent to the difference of the numbers of the plus- (equivalently, minus-) vertices. Since the numbers of minima, saddles and maxima are the same for M-morsifications from the same M-component, for the numbers $m(X_\mu)$ of M-components of X_μ we have

PROPOSITION A1. $m(X_\mu) \geqslant d(X_\mu)$.

Actually, a stronger statement holds:

THEOREM A1. *The mapping relating an R-diagram of X_μ to an M-component of X_μ is one-to-one.*

COROLLARY A2. $m(X_\mu) = d(X_\mu)$.

PROOF OF THEOREM A1. For $X = A, D$, this is presented in [1, 5]. Thus, we need to study the E case only. In Figure A1 we show all R-diagrams of D_5 and E_μ singularities [7]. The numbers appearing there were explained above. In what follows it is convenient to set $E_5 = D_5$. We also do not distinguish between the '+' and '−' cases.

LEMMA A3. *Each M-component of E_μ contains a connected component of the stratum $E_{\mu-1}$ in its closure.*

PROOF. We consider only $\mu = 6$. The two other cases are similar.

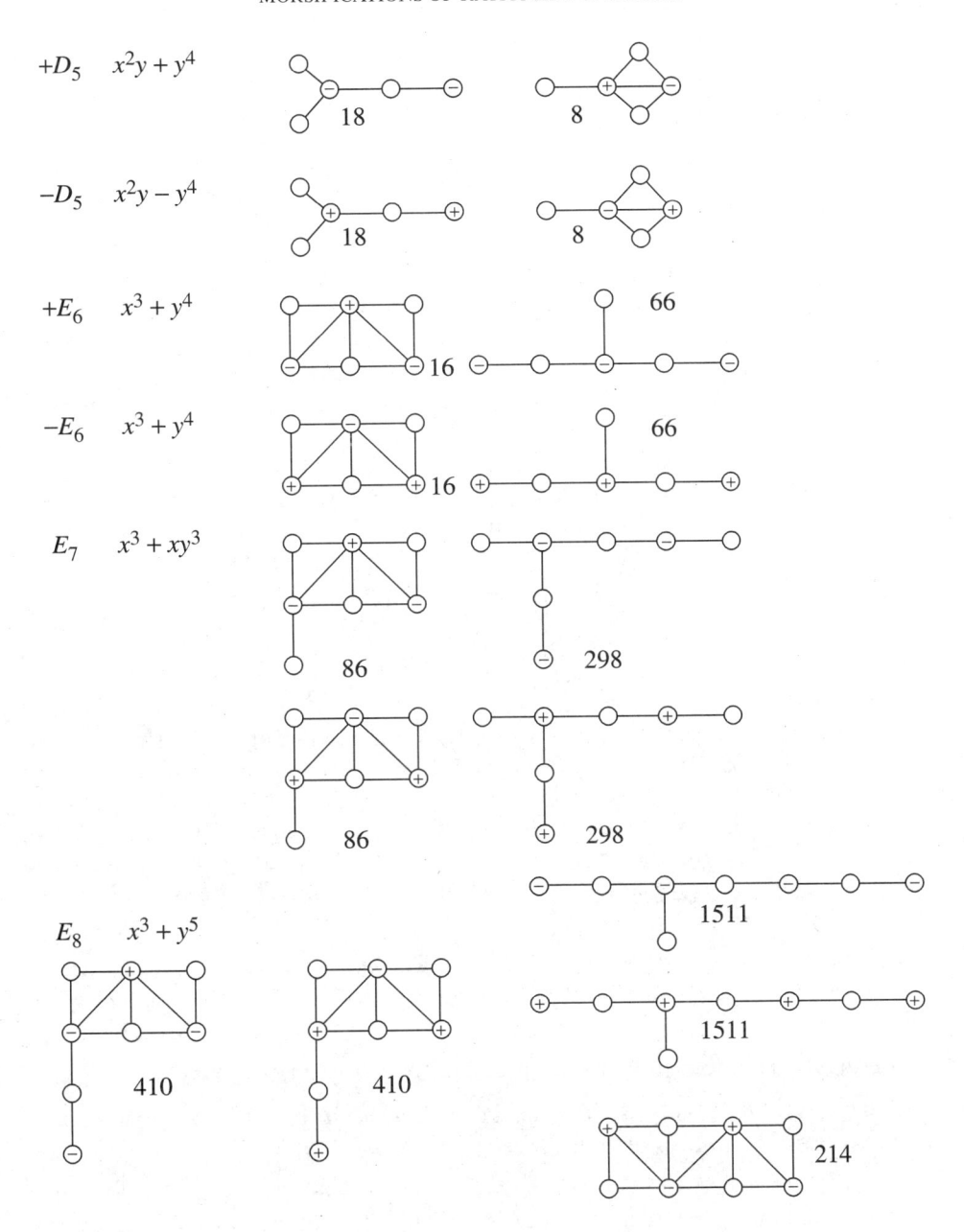

FIGURE A1. Normal forms, R-diagrams and the numbers of very nice components of D_5 and E_μ singularities

Let us take a sabirification of E_6 and consider its R-diagram (one of those given in Figure A1). Note that by omitting the lower right vertex of the diagram, together with all the edges starting at this vertex, we get a D_5 R-diagram. Let us fix the critical value of our sabirification corresponding to the lower right vertex and continuously deform all the other critical values, making them equal. According to [7], this deformation of the critical values induces a continuous deformation of the initial sabirification ending with a function with two critical points, A_1 and D_5, on different levels.

Easy direct computations imply

LEMMA A4. *The stratum $E_{\mu-1}$ in the base of a truncated miniversal deformation of E_μ is homeomorphic to a real line punctured (for E_μ itself omitted) at the origin.*

COROLLARY A5. $m(E_\mu) \leqslant 2m(E_{\mu-1})$.

Now we consider the three cases.

E_6. According to [5], $m(D_5) = 2$. Thus, by the last Corollary, $m(E_6) \leqslant 4$. But there are two different ways to deform a sabirification of E_6 into a function with a D_5 point (take the left vertex in the proof of Lemma A3 instead of the right one). This means that the closure of each of the M-components of E_6 must contain both half-branches of the stratum D_5. Thus $m(E_6) \leqslant 2 = d(E_6)$.

E_7. By Proposition A1 and Corollary A5: $4 = d(E_7) \leqslant m(E_7) \leqslant 2m(E_6) = 4$.

E_8. The E_7 stratum in the truncated versal deformation of E_8 is a one-parameter family $x^3 + y^5 + txy^3$, $t \neq 0$. In addition to the E_7 point, such a function has a Morse point on the other critical level. This point is a local minimum for $t > 0$ and local maximum for $t < 0$. By a slightly more precise argument than in the proof of Lemma A3, it is easy to see that the closures of M-components of E_8 corresponding to the three R-diagrams with both numbers of plus- and minus-vertices positive, contain both half-branches of the stratum E_7. Thus, $m(E_8) \leqslant 2m(E_7) - 3 = 5 = d(E_8)$.

2. Very nice components.

The base of a truncated miniversal deformation contains a *Maxwell stratum* Σ_m, i.e., a hypersurface corresponding to functions with coinciding values at at least two real critical points. The Maxwell stratum subdivides each M-component into a certain number of open regions each of which consists of morsifications with all the μ critical values different. Such morsifications are called *very nice*. Connected components of the set of very nice morsifications will be called *very nice components*.

THEOREM A2. *The number of very nice components in an M-component of an E_μ function singularity is the number given in Figure A1 alongside the R-diagram of the M-component.*

Taking the sum over all R-diagrams of a particular germ, we get

COROLLARY A6. *The numbers of connected components of the set of very nice morsifications of E_6, E_7, and E_8 are respectively 82, 768, and 4056.*

PROOF OF THEOREM A2. Consider an M-component $M_\mathcal{D}$ of E_μ corresponding to a certain R-diagram $\mathcal{D}$. Order the vertices of $\mathcal{D}$ in an arbitrary way (in what follows we consider the diagram with this extra ordering). Take a particular sabirification $f \in M_\mathcal{D}$ and deform it inside $M_\mathcal{D}$. This induces a deformation of the ordered set of critical values. By [7], staying in $M_\mathcal{D}$, we can move two critical values independently relatively to each other unless the corresponding vertices are connected by an edge. If there is an edge, the inequality between the critical values remains the same as for the critical values of f. Thus the edges of $\mathcal{D}$ give a system of necessary and sufficient inequalities on the critical values $v_1, \ldots, v_\mu$ of a very nice morsification from the M-component $M_\mathcal{D}$.

The same inequalities define a cone $C_\mathcal{D}$ in $\mathbb{R}^{\mu-1} = \{z_1 + \cdots + z_\mu = 0\} \subset \mathbb{R}^\mu$ equipped with the set W of all diagonals $z_i = z_j$. As in [2], shifting the critical values v_i by their arithmetical mean, we get a diffeomorphism between the pairs $(M_\mathcal{D}, M_\mathcal{D} \cap \Sigma_m)$ and $(C_\mathcal{D}, C_\mathcal{D} \cap W)$. As in [2], this diffeomorphism can be extended to a homeomorphism of the closures.

Thus we obtain the contractibility of each $M_\mathcal{D}$ and of each very nice component. We see that the number of very nice components in $M_\mathcal{D}$ coincides with the number of Weyl chambers of the reflection group $A_{\mu-1}$ in $C_\mathcal{D}$. The Weyl chambers in $C_\mathcal{D}$ are in one-to-one correspondence with the $\mathcal{D}$-snakes.

DEFINITION. A $\mathcal{D}$-snake is a permutation $\{i_1, \ldots, i_\mu\}$ of the numbers $\{1, \ldots, \mu\}$ subject to the system of inequalities corresponding to $\mathcal{D}$.

EXAMPLE. The diagram

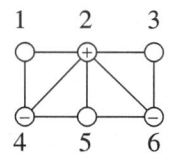

gives rise to the snakes $\{i_1, \ldots, i_6\}$ such that

$$
\begin{array}{ccc}
i_1 < i_2 > i_3 \\
\vee \quad \vee \quad \vee \\
i_4 < i_5 > i_6
\end{array}
$$

To calculate the number of $\mathcal{D}$-snakes, we use the following inductive procedure. Consider a vertex, say, the jth one, of $\mathcal{D}$ for which we may have $i_j = \mu$ (of course, there may be several options for j). Delete this vertex from $\mathcal{D}$, together with all the edges going from it, obtaining a diagram $\mathcal{D}_j$ with connected components $\mathcal{D}_{j,1}, \ldots, \mathcal{D}_{j,r_j}$. Then $\mathcal{D}_{j,k}$ is an R-diagram of a function with a Milnor number $\mu_{j,k} < \mu$, $\mu_{j,1} + \cdots + \mu_{j,r_j} = \mu - 1$. Suppose we already know the numbers $s_{j,k}$ of $\mathcal{D}_{j,k}$-snakes. Consider the product

$$ s_{j,1} \cdots s_{j,r_j} \frac{(\mu - 1)!}{\mu_{j,1}! \cdots \mu_{j,r_j}!}. $$

PROPOSITION A7. *The sum of such products over all the options for j is the number of $\mathcal{D}$-snakes.*

EXAMPLE. For the diagram of the previous example, there is only one option $j = 2$. Deleting the vertex 2, we get the R-diagram of A_5. The number of A_5-snakes is 16 [1], the third tangent number.

In a similar way, using the numerical data from [1, 5], we get the numbers of very nice components for all the other R-diagrams. Of course, the same inductive procedure works for the A and D series of simple function-germs (cf. [1, §5] and [5, 2.4]).

Acknowledgment

The author would like to thank MSRI, Berkeley for its support and hospitality during the Year on Algebraic Geometry.

References

1. V. I. Arnold, *Bernoulli–Euler updown numbers associated with function singularities, their combinatorics and arithmetics*, Duke Math. J. **63** (1991), 537–555.
2. _____, *Springer numbers and morsification spaces*, J. Alg. Geom. **1** (1992), 197–214.
3. _____, *The calculus of snakes and the combinatorics of Bernoulli, Euler and Springer numbers of Coxeter groups*, Uspekhi Mat. Nauk **47** (1992), no. 1, 1–51; English transl. in Russian Math. Surveys **47** (1992), no. 1.
4. _____, *Congruences for Euler, Bernoulli and Springer numbers of Coxeter groups*, Izv. Akad. Nauk SSSR Ser. Mat. **56** (1992), no. 5, 1129–1133; English transl., Math. USSR-Izv. **41** (1993), no. 2, 389–393.
5. _____, *Nombres d'Euler, de Bernoulli et de Springer pour les groupes de Coxeter et les espaces de Morsifications: Le calcul des serpents*, Leçons de Mathématiques d'Aujourd'hui, Univ. Bordeaux-1, Talence (1994).
6. V. I. Arnold, V. V. Goryunov, O. V. Lyashko, and V. A. Vassiliev, *Singularities*, Encyclopaedia of Math. Sci., vols. 6 and 39 (Dynamical systems, nos. 6 and 8), Springer-Verlag, Berlin, 1993.
7. S. A. Barannikov, *On the space of real polynomials without multiple critical values*, Funktsional. Anal. i Prilozhen. **26** (1992), no. 2, 10–17; English transl., Functional Anal. Appl. **26** (1992), no. 2, 84–90.
8. _____, *The complements of the resultant and discriminant sets in $\mathbb{C}^n$ are M-manifolds*, Funktsional. Anal. i Prilozhen. **27** (1993), no. 3, 1–4; English transl., Functional Anal. Appl. **27** (1993), no. 3, 155–157.
9. I. Bauer and F. Catanese, *Generic lemniscates of algebraic functions*, Preprint (1992), Università di Pisa.
10. F. Catanese and P. Frediani, *Configurations of real and complex polynomials*, Astérisque **218** (1993), 61–93.
11. F. Catanese and M. Paluszny, *Polynomial-lemniscates, trees and braids*, Topology **30** (1991), 623–640.
12. M. R. Èntov, *On real Morsifications of D_μ singularities*, Dokl. Akad. Nauk SSSR **325** (1992), no. 1, 32–36; English transl. in Soviet Math. Dokl. **46** (1993), no. 1, 25–29.
13. _____, *Number of components of the set of real morsifications of D_μ singularities*, Funktsional. Anal. i Prilozhen. **27** (1993), no. 2, 85–86; English transl., Functional Anal. Appl. **27** (1993), no. 2, 151–154.
14. _____, *Real preimages of a real point under Lyashko–Looijenga covering for simple singularities*, Izv. RAN Ser. Mat. **57** (1993), no. 5, 186–196; English transl., Russian Acad. Sci. Izv. Math. **43** (1994), no. 2, 186–196.
15. A. A. Glutsyuk, *An analogue of Caley's theorem for cyclically symmetric connected graphs with one cycle associated to generalized Lyashko–Looijenga covers*, Uspekhi Mat. Nauk **48** (1993), no. 2, 182–183; English transl. in Russian Math. Surveys **48** (1993), no. 2.
16. V. V. Goryunov, *Subprincipal Springer cones and morsifications of Laurent polynomials and D_μ singularities*, Singularities and Bifurcations (V. I. Arnold, ed.), Adv. Sov. Math., vol. 21, 1994, pp. 163–188.
17. A. A. Kirillov, *On the combinatorics of coadjoint orbits*, Funktsional. Anal. i Prilozhen. **27** (1993), no. 1, 73–76; English transl., Functional Anal. Appl. **27** (1993), no. 1, 62–64.
18. A. G. Kuznetsov, I. M. Pak, and A. E. Postnikov, *Increasing trees and alternating permutations*, Uspekhi Mat. Nauk (to appear).
19. P. Zograf, *The Weil–Petersson volume of the moduli space of punctured spheres*, Contemp. Math. **150** (1993), 367–372.
20. Yu. S. Chislenko, *Decompositions of simple singularities of real functions*, Funktsional. Anal. i Prilozhen. **22** (1988), no. 4, 52–67; English transl., Functional Anal. Appl. **22** (1988), no. 4, 297–310.

Translated by THE AUTHOR

Real Algebraic Curves with Real Cusps

Ilia Itenberg and Eugeniĭ Shustin

ABSTRACT. The classical question on the possible number of prescribed singularities of a plane algebraic curve of given degree is answered completely only for curves with nodes. We study curves with ordinary cusps and establish the following results on the maximal number $k(m)$ of real cusps of a real curve of degree m:

$$k(6) = 7, \quad k(7) = 10, \quad 14 \leqslant k(8) \leqslant 15$$

(for $m \leqslant 5$ the answer is known). We note that $k(6)$ differs from the maximal number 9 of complex cusps of a complex curve of degree 6.

§1. Introduction

The question on the possible number of prescribed singularities of a plane algebraic curve of a given degree is classical. A complete solution is known only for curves with nondegenerate double points: namely, for arbitrary positive integers d, m satisfying the inequality

$$d \leqslant \tfrac{1}{2}(m-1)(m-2),$$

there exists an irreducible plane complex (real) algebraic curve of degree m with d (real) nondegenerate double points (see [1, 2]). For the next singularity (in the sense of complexity), namely the cusp (type $x^2 + y^3 = 0$), only some constructions ([3, 4]) and some upper estimates ([5–8]) for complex curves are known.

In the present paper we study the number $k(m)$ defined as

$$\max\{n \mid \text{there exists an irreducible plane real algebraic curve of}$$
$$\text{degree } m \text{ with } n \text{ real cusps and without other singularities}\}.$$

It is well known (see [10]) that

(1.1) $\quad k(1) = k(2) = 0, \quad k(3) = 1, \quad k(4) = 3 \ [9], \quad k(5) = 5.$

We prove the following statement.

1991 *Mathematics Subject Classification*. Primary 14P25; Secondary 14H20.

©1996, American Mathematical Society

THEOREM 1.1. *We have*

(1.2) $$k(6) = 7,$$
(1.3) $$k(7) = 10,$$
(1.4) $$14 \leqslant k(8) \leqslant 15.$$

REMARKS. The inequality $k(7) \leqslant 10$ in (1.3) and the upper estimate in (1.4) are given (for complex curves) in [8]. A curve of degree 6 has no more than 9 cusps. Plücker's formulas show that a real curve of degree 6 with 9 cusps is dual to a nonsingular real curve of degree 3 having exactly 3 real inflection points (see, for example, [11]). Hence the above mentioned curve of degree 6 has only 3 real cusps. Thus, to prove the theorem, it suffices to construct curves of degree 6, 7, 8 with the required numbers of real cusps (§§2, 3) and to prove the nonexistence of degree 6 curves with 8 real cusps (§4).

§2. Gluing of singular algebraic curves

Let $z \in \mathbb{C}P^2$ be an isolated singular point of a real algebraic curve F, and let D be a sufficiently small ball in $\mathbb{C}P^2$ centered at z. The topological type of the pair $(D, D \cap F)$ if $z \notin \mathbb{R}P^2$ or the topological type of the triple $(D; D \cap F, D \cap \mathbb{R}P^2)$ if $z \in \mathbb{R}P^2$ is called the *type of* the singular point z of the curve F. In [12] (see also [4, 13]) the invariant $b(z)$ of a singular point is introduced. This invariant is equal to 1 for the cusp.

Let $F(x, y)$ be a polynomial with a nondegenerate Newton polygon Δ, and let ρ be an edge of Δ. The *restriction of $F(x, y)$ to the edge ρ* is the sum of monomials of $F(x, y)$ corresponding to the edge ρ. If the restriction of F to each edge of its Newton polygon Δ does not have multiple factors except x, y, then F is said to be a *peripherally nondegenerate polynomial* (PN-*polynomial*). For a PN-polynomial F, let $S(F)$ be the set of singularities of the curve $\{F = 0\} \setminus \{xy = 0\}$. Let

$$F_k(x, y) = \sum_{(i,j) \in \Delta_k} a_{ij} x^i y^j, \quad k = 1, \ldots, N,$$

be real PN-polynomials with Newton polygons $\Delta_1, \ldots, \Delta_N$. Suppose that
   a) $\Delta_1 \cup \cdots \cup \Delta_N = \Delta$, where Δ is a convex polygon;
   b) $\Delta_k \cap \Delta_l$ ($k \neq l$) is empty or is a common vertex or is a common edge;
   c) there exists a convex piecewise linear function $\Delta \to \mathbb{R}$ which is linear on each polygon Δ_i and is not linear on each union $\Delta_k \cup \Delta_l$ ($k \neq l$).

Suppose also that if the polygons Δ_k and Δ_l have a common edge, then the restrictions of the polynomials F_k and F_l to the common edge coincide.

Let Γ be the graph of adjoinings of the polygons $\Delta_1, \ldots, \Delta_N$ oriented in such a way that it does not contain oriented cycles.

In [4] a modification for singular curves of Viro's method of gluing of real algebraic varieties (see [14, 15]) is suggested. The following theorem holds.

THEOREM 2.1. *For each $k = 1, \ldots, N$ let the curve $\{F_k = 0\} \setminus \{xy = 0\}$ be irreducible and*

$$\sum_{\delta \in S(F_k)} b(\delta) < \sum_{\gamma_k} (\text{card}\,(\gamma_k \cap \mathbb{Z}^2) - 1),$$

where γ_k runs over all edges of the polygon Δ_k that do not correspond to arcs of the graph Γ going inside Δ_k.

Then there exists a real PN-polynomial F with Newton polygon Δ for which the set $S(F)$ coincides with an arbitrary given subset of $S(F_1) \sqcup \cdots \sqcup S(F_N)$.

The polynomial F is said to be obtained *by gluing* the polynomials $F_1, \ldots, F_N$.

§3. Constructions

Curves of degree 6. Let us introduce the following notation. Let $F \in \langle k \rangle$ (or $F \in \Delta \langle k \rangle$) mean that the set $S(F)$ of the real PN-polynomial F consists of k real cusps (and F has the Newton polygon Δ).

LEMMA 3.1. *There exists a polynomial F_1 which has the Newton triangle*
$$\{(0;2),(0;6),(6;0)\}$$
and satisfies $F_1 \in \langle 5 \rangle$.

PROOF. A real algebraic curve C_5 of degree 5 with 5 real cusps and with 5 points of inflection is constructed in [10]. Choose projective coordinates $(x_0:x_1:x_2)$ so that the point $(0:0:1)$ is a point of inflection and the line $x_0 = 0$ is the tangent to the curve C_5 at this point. Then the curve C_5 is defined by a PN-polynomial $\Phi_1(x,y)$ with the Newton quadrangle
$$\{(0;0),(0;4),(3;2),(5;0)\}$$
in the coordinates $x = x_1/x_0$, $y = x_2/x_0$ of the affine plane $x_0 \neq 0$.

Let $\Phi_2(x,y)$ be a PN-polynomial such that
i) its Newton polygon coincides with the triangle $\{(3;2),(5;0),(6;0)\}$;
ii) $S(\Phi_2) = \varnothing$;
iii) its restriction to the edge $[(3;2),(5;0)]$ of the Newton polygon coincides with the restriction of the polynomial Φ_1 to this edge.

We can apply Theorem 2.1 to the polynomials Φ_1 and Φ_2, and obtain a polynomial Φ_3 having the Newton triangle $\{(0;0),(0;4),(6;0)\}$ and satisfying $\Phi_3 \in \langle 5 \rangle$. Then put
$$F_1(x,y) = y^6 \Phi_3(x/y, 1/y). \quad \square$$

LEMMA 3.2. *There exists a polynomial F_2 with the Newton triangle*
$$\{(0;0),(0;2),(6;0)\},$$
with the property $F_2 \in \langle 2 \rangle$, and with a given restriction to the edge $[(0;2),(6;0)]$.

PROOF. Let us consider the polynomials
$$\Psi_1(x,y) = y^2 + (x^2-1)^3, \qquad \Psi_2(x,y) = y^2 - (x^2-1)^3.$$
It is easy to see that each set $S(\Psi_1)$, $S(\Psi_2)$ consists of two real cusps. To obtain a given restriction to the edge $[(0;2),(6;0)]$, one can apply the transformation $x \mapsto \alpha x$, $y \mapsto \beta y + \gamma x^3$ to the polynomial Ψ_1 (or Ψ_2) and multiply the polynomial obtained by a suitable real number. $\quad \square$

The following proposition is a corollary of Lemmas 3.1 and 3.2.

PROPOSITION 3.3. *There exists a real algebraic curve of degree 6 with 7 real cusps.*

PROOF. The desired curve of degree 6 with 7 real cusps is obtained by gluing the polynomials F_1 and F_2 constructed in Lemmas 3.1 and 3.2. $\quad \square$

REMARK. Applying Theorem 2.1 to the polynomials F_1 and F_2, we can also obtain curves of degree 6 with $1, \ldots, 6$ real cusps.

Curves of degree 7. Let us divide the triangle with vertices $(0;0)$, $(7;0)$, $(0;7)$ into the polygons

$$\tau_1 = \{(0;0), (3;0), (0;3), (3;3)\},$$
$$\delta_1 = \{(0;3), (0;7), (3;4), (3;3)\},$$
$$\delta_2 = \{(3;0), (7;0), (4;3), (3;3)\},$$
$$\delta_3 = \{(4;3), (3;4), (3;3)\}.$$

This subdivision satisfies conditions a)–c) of §2.

LEMMA 3.4. *There exists a polynomial G_1 with the property $G_1 \in \tau_1\langle 4 \rangle$ whose restrictions to the edges $[(3;0), (3;3)]$, $[(0;3), (3;3)]$ are products of real linear factors.*

PROOF. Consider a curve C_4 of degree 4 with 3 real cusps (see [9]) and choose a coordinate system $(x_0 : x_1 : x_2)$ in such a way that
  i) the point $(1:0:0)$ does not belong to the curve C_4;
  ii) the points $(0:1:0)$, $(0:0:1)$ belong to the curve C_4 and are nonsingular;
  iii) the line $x_0 = 0$ is a tangent to the curve C_4 at a real point not coinciding with $(0:1:0)$ or with $(0:0:1)$,
  iv) each line $x_1 = 0$, $x_2 = 0$ intersects the curve C_4 at 4 real points.
Then the image C_4^* of the curve C_4 under the transformation

$$x_0 = x_1' x_2', \qquad x_1 = x_0' x_2', \qquad x_2 = x_0' x_1'$$

is a curve of degree 6 with 4 real cusps and 2 ordinary triple points (see [11]). Let us choose a coordinate system $(y_0 : y_1 : y_2)$ so that the points $(0:1:0)$ and $(0:0:1)$ are ordinary triple points of the curve C_4^* and $(1:0:0) \notin C_4^*$. Then the curve C_4^* is defined by the desired polynomial G_1 in the coordinates

$$x = y_1/y_0, \qquad y = y_2/y_0$$

of the affine plane $y_0 \neq 0$. □

LEMMA 3.5. *There exist polynomials G_2, G_3 with the properties*

$$G_2 \in \delta_1\langle 3 \rangle, \qquad G_3 \in \delta_2\langle 3 \rangle,$$

such that the restriction of G_2 to the edge $[(0;3), (3;3)]$ and the restriction of G_3 to the edge $[(3;0), (3;3)]$ are given products of real linear factors.

PROOF. Choose a coordinate system $(x_0 : x_1 : x_2)$ so that the points $(1:0:0)$, $(0:0:1)$ do not belong to the curve C_4 (see Lemma 3.4) and the point $(0:1:0)$ is a nonsingular point of C_4. Let the polynomial $G(x, y)$ define the curve C_4 in the coordinates $x = x_1/x_0$, $y = x_2/x_0$ of the affine plane $x_0 \neq 0$.
Put

$$G_2(x, y) = y^3 G(x, y), \qquad G_3(x, y) = x^3 G(y, x).$$

The condition on the restrictions of the polynomials G_2 and G_3 means that the line $x_2 = 0$ intersects the curve C_4 at 4 real points with a given anharmonic ratio. Let us show that we can choose the line $x_2 = 0$ to obtain an arbitrary anharmonic ratio. The cusps divide the real branch of the curve C_4 into 3 nonsingular arcs. Let a be the arc with endpoints p, q. Consider the tangent L to the arc a at some point z ($z \neq p$,

q) and the intersection point z^* of the lines L and (pq). Let Π be a pencil of lines such that $\Pi = [L, (pq)]$ and each line of this pencil intersects the curve C_4 only at real points. Then for an arbitrary positive number ρ there exists a line l from the pencil Π such that the anharmonic ratio of the points $l \cap C_4$ equals ρ. □

PROPOSITION 3.6. *There exists a real algebraic curve of degree 7 with 10 real cusps.*

PROOF. We can obtain curves of degree 7 with $1, \ldots, 10$ real cusps by gluing the polynomials G_1, G_2, G_3 (constructed in Lemmas 3.4 and 3.5), and a polynomial G_4 with the property $G_4 \in \delta_3 \langle 0 \rangle$. □

Curves of degree 8. Let us divide the triangle with vertices $(0;0), (0;8), (8;0)$ into the quadrangles

$$\tau_1 = \{(0;0), (3;0), (0;3), (3;3)\},$$
$$\pi_1 = \{(0;3), (0;8), (4;4), (3;3)\},$$
$$\pi_2 = \{(3;0), (8;0), (4;4), (3;3)\}.$$

This subdivision, obviously, satisfies conditions a)–c) of §2.

LEMMA 3.7. *There exist polynomials H_1 and H_2 with the properties*

$$H_1 \in \pi_1 \langle 5 \rangle, \qquad H_2 \in \pi_2 \langle 5 \rangle,$$

such that the restriction of H_1 to the edge $[(0;3), (3;3)]$ and the restriction of H_2 to the edge $[(3;0), (3;3)]$ are given products of real linear factors.

PROOF. Let C_5 be the curve of degree 5 with 5 real cusps constructed in [10]. Choose a coordinate system $(x_0 : x_1 : x_2)$ so that
  i) the points $(1:0:0), (0:0:1)$ do not belong to the curve C_5;
  ii) the point $(0:1:0)$ belongs to C_5 and is nonsingular;
  iii) the line $x_2 = 0$ is tangent to the curve C_5 at the point $(0:1:0)$.
Let $H(x, y)$ be the polynomial defining the curve C_5 in the coordinates

$$x = x_1/x_0, \qquad y = x_2/x_0$$

of the affine plane $x_0 \neq 0$. Put

$$H_1(x, y) = y^3 H(x, y), \qquad H_2(x, y) = x^3 H(y, x).$$

The condition on the restrictions means that the line $\{x_2 = 0\}$ intersects the curve C_5 only at four real points with the given anharmonic ratio. The curve C_5 has only one real branch (see [10]). This branch is divided by the cusps into 5 nonsingular arcs. It is shown in [10] that at least one of these arcs (which we shall denote by a) has the following property: the straight line (p, q) passing through the endpoints p and q of a meets a at one nonsingular point and divides a into two arcs a_p (with endpoint p) and a_q (with endpoint q); see Figure 1. There exists exactly one point belonging to the arc a_q such that the tangent at this point to the arc a_q contains the cusp p. Denote this point by s. It is easy to see that there is a point of inflection t in the interval (p, s) of the arc a. If we slightly shift the point z along the arc a from the point s towards the point t, then the corresponding tangent Λ_z to the arc a intersects the curve C_5 at

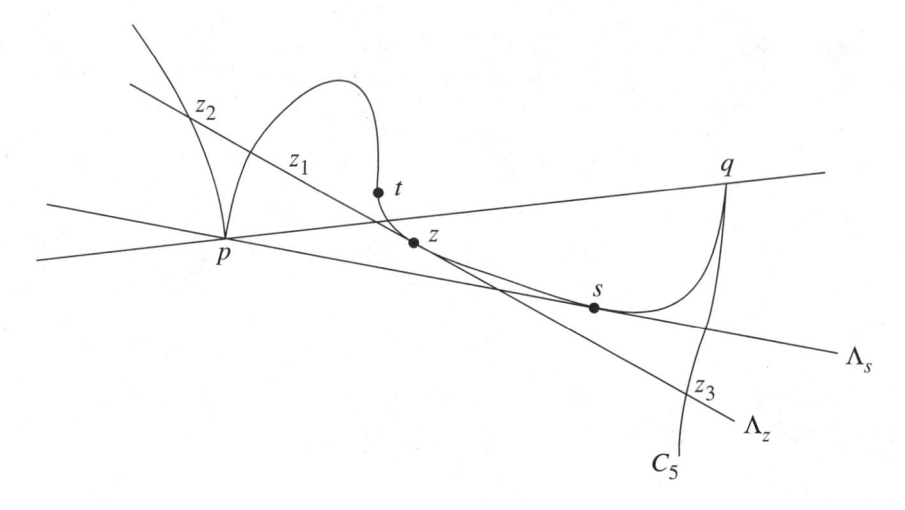

FIGURE 1

4 real points that we denote by z, z_1, z_2, z_3, as shown in Figure 1. Two situations are possible.

1) Each line Λ_z (where $z \in (t, s) \subset a$) intersects the curve C_5 at 4 real points. Then, if z tends to s, the points z_1, z_2 tend to p, and if z tends to t, then the point z_1 tends to t as well. This means that the anharmonic ratio of the points z, z_1, z_2, z_3 ranges over the interval $(0, \infty)$ when the point z runs over the arc (s, t).

2) A line Λ_τ (where $\tau \in (t, s)$) contains a cusp or this line is a double tangent; each line Λ_z (where $z \in (\tau, s)$) intersects the curve C_5 at 4 real points. Then similarly, if z tends to s, the points z_1, z_2 tend to p, and if z tends to τ, the points z_2, z_3 tend to one point. Again this means that the anharmonic ratio of z, z_1, z_2, z_3 ranges over $(0, \infty)$ when z runs over the arc (s, τ).

In both situations, for an arbitrary positive number ρ, we can find a line Λ_z (where $z \in (t, s)$) such that the anharmonic ratio of the points z, z_1, z_2, z_3 is equal to ρ. Choosing this line as the coordinate axis $x_2 = 0$, we complete the proof. □

PROPOSITION 3.8. *There exists a curve of degree* 8 *with* 14 *real cusps.*

PROOF. Consider the polynomial G_1 (see Lemma 3.4) and the corresponding polynomials H_1, H_2 (see Lemma 3.7). We can obtain coinciding restrictions of the polynomials H_1 and H_2 to the edge $[(3; 3), (4; 4)]$ by the change $(x, y) \mapsto (\lambda x, y)$ in the polynomial H_2 and by multiplying the polynomial obtained by a suitable real number. Curves of degree 8 with $1, \ldots, 14$ real cusps can be obtained by gluing the polynomials G_1, H_1, and H_2. □

§4. Upper bound

This section is entirely devoted to curves of degree 6. It was proved in §3 that $k(6) \geq 7$. To finish the proof of statement (1.2) of Theorem 1.1, we shall show that a real plane projective curve of degree 6 cannot have more than 7 real cusps.

Suppose that there exists a curve A' of degree 6 with 8 real cusps and without other singularities. We can smooth all the cusps of this curve to obtain a nonsingular curve A of degree 6 with 8 additional components of the real point set. Thus, the curve A has at least 9 ovals (the connected components of the set of real points of a nonsingular curve of even degree are called *ovals*).

A nonsingular curve is an *M-curve* if it has the maximal possible number of connected components of the set of real points for the given degree. This number is equal to 11 for degree 6. Thus, the curve A has 9, 10, or 11 ovals.

Two nonsingular curves of the same degree are called *rigidly isotopic* if two points corresponding to these curves in the space of all curves of the given degree can be connected by a path whose points correspond only to nonsingular curves.

The rigid isotopy type of a nonsingular curve of degree 6 is determined (see [16]) by the disposition scheme of the ovals (this scheme is called the *real scheme* of the curve) and the type (I or II) of the curve, which is defined as follows: if the set of real points of a curve divides the set of complex points, then the curve is *of type* I, if not, the curve is *of type* II.

The list of all rigid isotopy types of nonsingular curves of degree 6 with at least 9 ovals is shown in Figure 2 (we use the system of notations for real schemes of curves suggested by O. Viro [17]). Our goal is to show, for each described rigid isotopy type, that the curve A cannot be of this type. This would give, evidently, a proof of the inequality $k(6) \leq 7$.

We shall divide the rigid isotopy types presented in Figure 2 into three groups. For the first group, the desired statement can be obtained by considering complex orientations (Proposition 4.1); for the second one, by studying the possible gluings of ovals in the situation when the complex orientations do not give sufficient information (Proposition 4.2); and for the third group, we shall need some more delicate considerations (Proposition 4.3). The proof of Proposition 4.3 is more general than the proofs of Propositions 4.1 and 4.2. It works for any rigid isotopy type of curves of degree 6.

Complex orientations. Let $\overset{\circ}{A}$ be a curve of type I. Then its set of real points $\mathbb{R}\overset{\circ}{A}$ divides the complex locus $\mathbb{C}\overset{\circ}{A}$ in two parts exchanged by the involution of complex conjugation of $\mathbb{C}P^2$. The natural orientations of these parts define two opposite orientations of $\mathbb{R}\overset{\circ}{A}$ which are called the *complex orientations* of the curve $\overset{\circ}{A}$. This notion was introduced by V. Rokhlin [18, 19].

Two ovals of a curve form an *injective pair* if one of these ovals lies inside the other one. An injective pair of a curve of type I is called *positive* (resp. *negative*), if the orientations of the ovals of this pair given by the complex orientation of the curve can be induced (resp. cannot be induced) by some orientation of the annulus bounded by these ovals.

Let Π_+ (resp. Π_-) denote the number of positive (resp. negative) injective pairs of a curve.

The following statement is called the Rokhlin formula (see [19]):

For a curve of even degree m of type I with l ovals,

$$2(\Pi_+ - \Pi_-) = l - m^2/4.$$

For example, for an M-curve of degree 6, we have $\Pi_+ - \Pi_- = 1$.

PROPOSITION 4.1. *The rigid isotopy types* $\langle 1 \cup 1\langle 9 \rangle \rangle$, $\langle 6 \cup 1\langle 2 \rangle \rangle$, $\langle 4 \cup 1\langle 4 \rangle \rangle$ (*type* I), $\langle 2 \cup 1\langle 6 \rangle \rangle$, $\langle 1\langle 8 \rangle \rangle$ (*type* I) *of nonsingular curves of degree 6 do not contain the curve A*.

PROOF. Suppose that the curve A is of one of the rigid isotopy types mentioned in the statement of the proposition. Then the set of real points $\mathbb{R}A'$ of the curve A' has 3

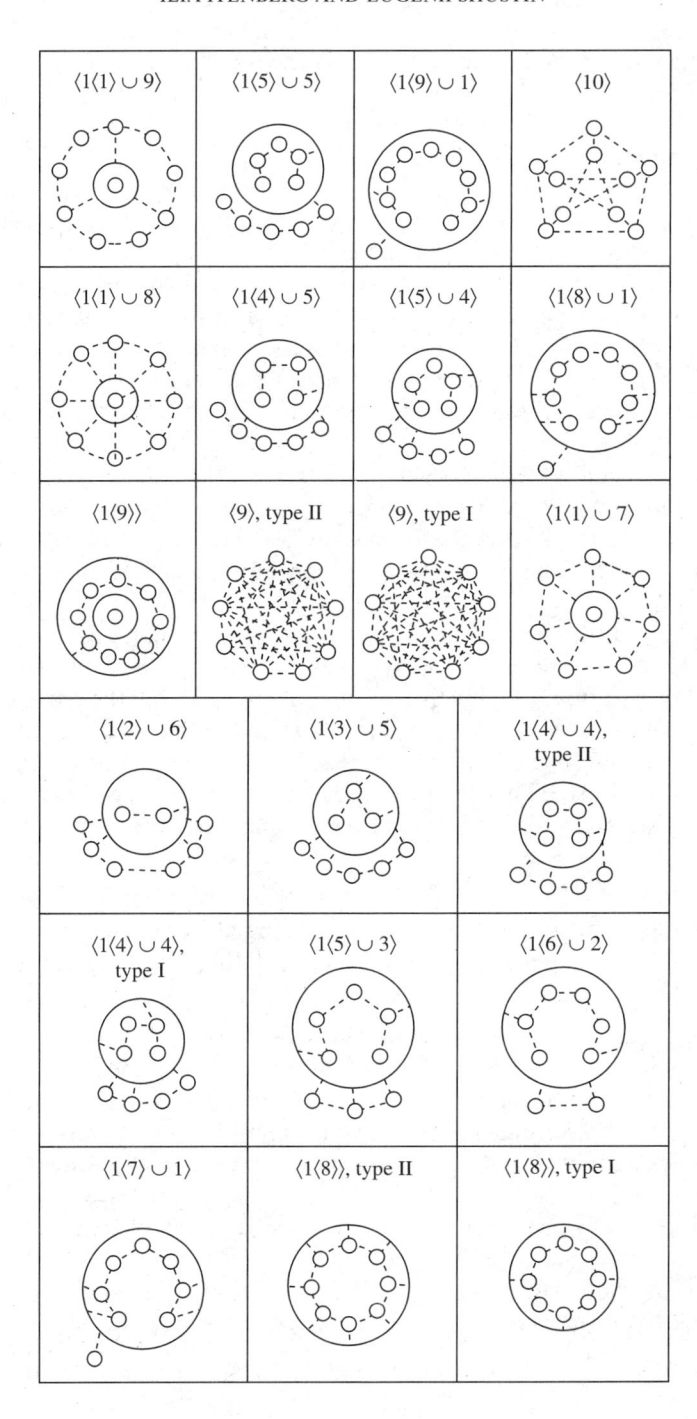

FIGURE 2

connected components corresponding to the nonempty oval and two empty ovals of A, if A is of type $\langle 1 \cup 1\langle 9\rangle\rangle$, and $\mathbb{R}A'$ has one connected component corresponding to the nonempty oval of A in the other cases.

An interior oval (i.e., an oval lying inside the nonempty one) of the curve A is called *positive* if it forms a positive injective pair with the nonempty oval, and *negative* in the opposite case.

Using the Rokhlin formula, we see that the curve A has
  5 positive and 4 negative ovals, if A is of type $\langle 1 \cup 1\langle 9\rangle\rangle$;
  1 positive and 1 negative oval, if A is of type $\langle 6 \cup 1\langle 2\rangle\rangle$;
  2 positive and 2 negative ovals, if A is of type $\langle 4 \cup 1\langle 4\rangle\rangle$ (type I);
  3 positive and 3 negative ovals, if A is of type $\langle 2 \cup 1\langle 6\rangle\rangle$;
  4 positive and 4 negative ovals, if A is of type $\langle 1\langle 8\rangle\rangle$ (type I).

We make two simple observations.

1. Only negative ovals of A can be glued with the nonempty one to produce a cusp of A' (a singular curve obtained as a result of gluing two ovals of A also must have a complex orientation). Two interior ovals of A can be glued to produce a cusp of A' only if one of them is positive and the other one is negative.

2. It is a corollary of the Bézout theorem that an interior oval of the curve A can be glued with no more than 2 other interior ovals (let us choose a point inside an empty oval of A lying outside of the nonempty one; the rotation of a line around the chosen point gives a linear order, defined up to inversion, on the interior ovals; this order is invariant up to rigid isotopies).

Using these remarks, it is easy to see that if the curve A belongs to one of the rigid isotopy types listed in Proposition 4.1, then the curve A' cannot have more than 7 real cusps. This contradiction completes the proof. □

Gluing ovals of curves of degree 6. The consideration of complex orientations is not sufficient (or does not work at all as, for example, for curves of type II) for the other 16 rigid isotopy types shown in Figure 2. Nevertheless, one can try to obtain information on the possible gluings of ovals of degree 6 curves. We use the technique developed by V. Kharlamov [20] and V. Nikulin [16].

Two polytopes will be associated with a nonsingular curve of degree 6. We follow the construction described in [16, 21, 22].

Let A_6 be a nonsingular curve of degree 6, and $\mathbb{C}A_6$ be the set of complex points of this curve. The set $\mathbb{R}A_6$ of real points of the curve A_6 divides the real projective plane $\mathbb{R}P^2$ in two subsets with common boundary $\mathbb{R}A_6$. Exactly one of these sets is nonorientable. Denote it by $\mathbb{R}P^2_-$ and the other by $\mathbb{R}P^2_+$.

Consider the two-sheeted branched covering space of the complex projective plane $\mathbb{C}P^2$ with the branch locus $\mathbb{C}A_6$. The variety X obtained is a K3 surface. The involution of complex conjugation acting on $\mathbb{C}P^2$ can be lifted to X in two different ways. Denote the antiholomorphic involution on X with the set of fixed points lying above $\mathbb{R}P^2_-$ by conj.

Consider the quadruple (L, B, conj_*, h), where
(1) L is the homology group $H_2(X; \mathbb{Z})$;
(2) B is the intersection form on L;
(3) conj_* is the involution on L induced by conj;
(4) h is the element of L corresponding to a hyperplane section.

Denote by L_+ and by L_- the eigenspaces of the involution conj_* corresponding to the eigenvalues 1 and -1 respectively. It is known (see, for example, [23]) that
(a) the rank of L is equal to 22;
(b) the diagonal forms over $\mathbb{Q}$ of the restrictions of form B to L_+ and L_- have one and two positive squares respectively;

(c) $h \in L_-$, $B(h, h) = 2$.

Denote the orthogonal complement of h in L_- by L_{-h}. Let

$$\mathcal{L}_- = \{x \mid x \in L_{-h} \otimes \mathbb{R}, \ B_R(x, x) > 0\}/\mathbb{R}^*,$$

where B_R is the bilinear extension of the form B to $L \otimes \mathbb{R}$. Then $\mathcal{L}_-$ is a Lobachevskiĭ space. The group generated by reflections with respect to the hyperplanes in $\mathcal{L}_-$ orthogonal to elements v of L_{-h} with the square -2 acts in $\mathcal{L}_-$. Let Ω_- be a polytope which is a fundamental domain of this group.

Similarly, define a polytope Ω_+ in the Lobachevskiĭ space $\mathcal{L}_+$, where

$$\mathcal{L}_+ = \{x \mid x \in L_+ \otimes \mathbb{R}, \ B_R(x, x) > 0\}/\mathbb{R}^*.$$

Let C_- (resp. C_+) be the Coxeter scheme (see, for example, [24]) of the polytope Ω_- (resp. Ω_+). Obviously, the schemes C_+ (resp. C_-) of rigidly isotopic nonsingular curves of degree 6 are isomorphic.

Sometimes it is difficult to calculate the schemes C_+, C_- (for example, they can be infinite). However, it is always possible to calculate two important subschemes K_+, K_- of C_+, C_-. The subschemes K_+, K_- are defined as follows. Remove all thick and dotted edges of the schemes C_+ and C_- (two vertices of a scheme are connected by thick (or dotted) edge if the value of the bilinear form B is greater than or equal to 2 on the corresponding vectors) to obtain the schemes C'_+ and C'_-. Now consider the connected components C''_+, C''_- of C'_+, C'_- such that each contains at least two vertices (if all components of the scheme C'_+ (resp. C'_-) contain only one vertex, then the scheme K_+ (resp. K_-) is by definition a single vertex). It is shown in [21, 22] that the vertices of C''_+, C''_- correspond to the simplest degenerations of the curve A_6. For each pair of ovals of A_6, let us do the following: if there exist vertices corresponding to the gluings of two given ovals, remove all these vertices but one (the edges incident to the removed vertices should also be removed); the resulting schemes are by definition the schemes K_+, K_-.

The schemes K_+, K_- are calculated in [22] for all rigid isotopy types of nonsingular curves of degree 6. As explained in [21, 22], using these calculations one can obtain a complete description of the possible gluings of ovals of nonsingular curves of degree 6. This information is presented by dotted lines in Figure 2 for the curves with at least 9 ovals (the exact statement is the following: for any rigid isotopy type presented in Figure 2 and for any curve of this type, there exists a bijection between the ovals of the curve and the ovals of the scheme, such that two ovals of the curve can be glued if and only if the corresponding ovals of the scheme are connected by a dotted line).

To explain the correspondence between the schemes of possible gluings of ovals and K_+, K_-, let us take the scheme of Figure 2, replace each empty oval and each dotted line by a vertex, and connect two vertices if one of them is obtained from an oval and the other one is obtained from a dotted line incident to this oval. The graph described is isomorphic to the disjoint union of the schemes K_+, K_- (see [21, 22]).

The information on the possible gluings of ovals presented in Figure 2 yields the following statement.

PROPOSITION 4.2. *The rigid isotopy types* $\langle 9 \cup 1\langle 1\rangle\rangle$, $\langle 5 \cup 1\langle 5\rangle\rangle$, $\langle 10\rangle$, $\langle 5 \cup 1\langle 4\rangle\rangle$, $\langle 4 \cup 1\langle 5\rangle\rangle$, $\langle 1 \cup 1\langle 8\rangle\rangle$, $\langle 1\langle 9\rangle\rangle$, $\langle 5 \cup 1\langle 3\rangle\rangle$, $\langle 4 \cup 1\langle 4\rangle\rangle$ (*type* II), $\langle 3 \cup 1\langle 5\rangle\rangle$, $\langle 1 \cup 1\langle 7\rangle\rangle$ *of nonsingular curves of degree* 6 *do not contain the curve* A.

The schemes of possible gluings of ovals (or the schemes K_+, K_-) do not forbid degenerations with 8 real cusps for curves of the 5 remaining rigid isotopy types

$\langle 8 \cup 1\langle 1\rangle\rangle$, $\langle 9\rangle$ (type I), 9 (type II), $\langle 7 \cup 1\langle 1\rangle\rangle$, $\langle 1\langle 8\rangle\rangle$ (type II). In these cases we need to have more detailed information on the Coxeter schemes C_+, C_- of the polytopes Ω_+, Ω_-.

The last 5 rigid isotopy types. An arbitrary subset of the set of vertices of a Coxeter scheme with all edges connecting the vertices of this subset is called a *subscheme of a Coxeter scheme*.

Recall that we assume A' to be a curve of degree 6 with 8 real cusps and without other singularities and A to be a smoothing of the curve A' having 8 additional ovals. The following statements can be obtained from the results of [21, 22]:

The disjoint union of the schemes C_+ and C_- of the curve A must contain a subscheme isomorphic to the scheme S shown in Figure 3.

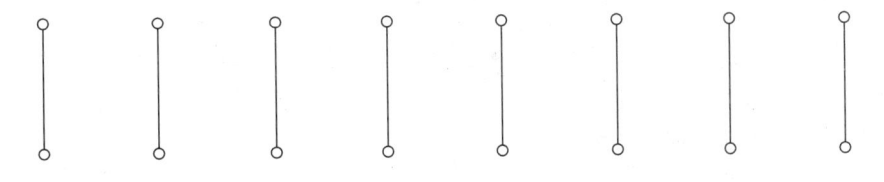

FIGURE 3

PROPOSITION 4.3. *The rigid isotopy types* $\langle 8 \cup 1\langle 1\rangle\rangle$, $\langle 9\rangle$ *(type* I*),* 9 *(type* II*),* $\langle 7 \cup 1\langle 1\rangle\rangle$, $\langle 1\langle 8\rangle\rangle$ *(type* II*) of nonsingular curves of degree 6 do not contain the curve A.*

PROOF. Let us denote by U the disjoint union of the Coxeter schemes C_+, C_-. As mentioned above, to prove the assertion of Proposition 4.3, it suffices to show that the scheme U never contains a subscheme isomorphic to S for curves of the listed rigid isotopy types.

Each subscheme of U isomorphic to S is a subscheme of the union U'' of the schemes C''_+, C''_- (defined in the previous subsection). The inverse statement is not true, because we remove certain thick and dotted edges of C_+, C_- in the process of constructing C''_+, C''_-.

We claim that a subscheme of U'' isomorphic to the scheme S cannot be a subscheme of U (in other words, any subscheme of the union of C''_+ and C''_- isomorphic to S contains a pair of vertices connected by a thick or a dotted edge in U).

We shall prove this here for the rigid isotopy type $\langle 8 \cup 1\langle 1\rangle\rangle$. For the other types the proof is similar.

The schemes C''_+, C''_- for the rigid isotopy type $\langle 8 \cup 1\langle 1\rangle\rangle$ are presented in Figure 4 (the calculation is done in [22]; one can easily draw the schemes C''_+, C''_- of any rigid isotopy type mentioned in Proposition 4.3 in the following way: take the corresponding scheme from Figure 2, replace each empty oval by a vertex, replace each dotted line by a vertex or by a collection of vertices, and connect two vertices if one of them is obtained from an oval and the other one is obtained from a dotted line incident to this oval).

It follows from the results of [22] that only vertices of degree 1 (a vertex is *of degree* 1 if there exists only one edge incident to it) of the scheme U'' of the rigid isotopy type $\langle 8 \cup 1\langle 1\rangle\rangle$ can be connected by thick and dotted edges in U.

Let S_u be a subscheme of U isomorphic to S. Suppose that no vertices of S_u are connected by thick or dotted edges in U. We can also suppose that the vertices with the numbers 1, 5, 16, 19, 20 (see Figure 4) belong to S_u. Consider the elements of L_h

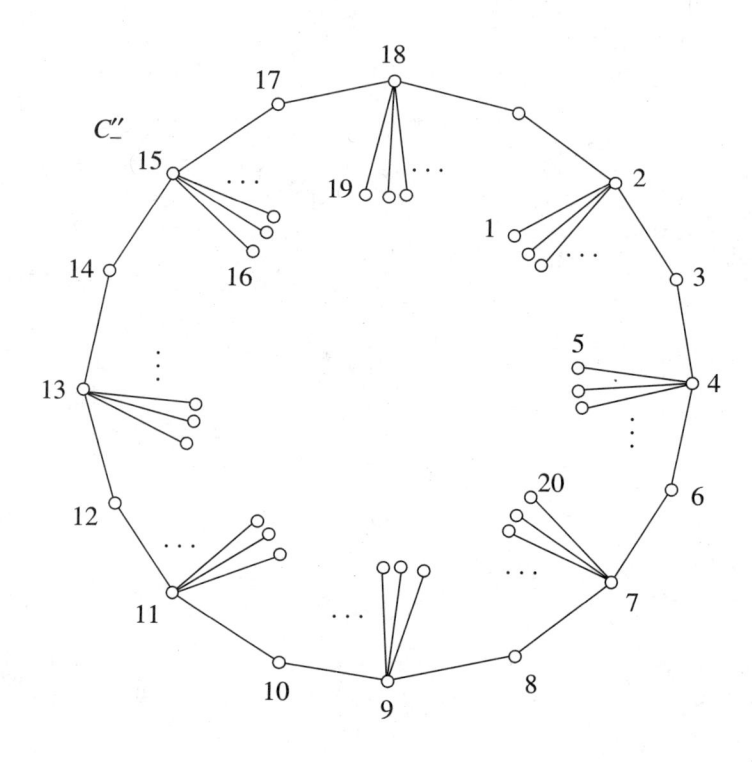

FIGURE 4

corresponding to the vertices with the numbers $1, \ldots, 19$. It is easy to verify that they are linearly independent, which contradicts the fact that the rank of L_h is equal to 18 in our case. □

References

1. F. Severi, *Vorlesungen über die Algebraische Geometrie*, Teubner, Leipzig, 1921.
2. L. Brusotti, *Sulla "piccola variazione" di una curva algebrica piana reali*, Rend. Rom. Ac. Lincei (5) **30** (1921), 375–379.
3. M. Gradolato and E. Mezzetti, *Curves with nodes, cusps and ordinary triple points*, Ann. Univ. Ferrara. Sez. VII **31** (1985), 23–47.
4. E. I. Shustin, *Gluing of singular algebraic curves*, Methods of Qualitative Theory, Gor'ky Univ. Press, Gor'ky, 1985, pp. 116–128. (Russian)
5. A. N. Varchenko, *Asymptotic behavior of integrals, and Hodge structures*, Itogi Nauki i Tekhniki: Sovremennye Problemy Mat., vol. 22, VINITI, Moscow, 1983, pp. 130–166; English transl. in J. Soviet Math. **27** (1984), no. 3.
6. F. Hirzebruch, *Singularities of algebraic surfaces and characteristic numbers*, Contemp. Math. **58** (1986), 141–155.
7. K. Ivinskis, *Normale Flächen und die Miyaoka–Kobayashi Ungleichung*, Diplomarbeit, Bonn, 1985.
8. A. Hirano, *Constructions of plane curves with cusps*, Saitama Math. J. **10** (1992), 21–24.
9. H. G. Zeuthen, *Sur les différentes formes des courbes planes du quadrième ordre*, Math. Ann. (1893), 408–432.
10. D. A. Gudkov, *On a fifth-order curve with five cusps*, Funktsional. Anal. i Prilozhen. **16** (1982), no. 3, 54–55; English transl., Functional Anal. Appl. **16** (1982), no. 3, 201–202.

11. R. J. Walker, *Algebraic curves*, Princeton Univ. Press, Princeton, NJ, 1950.
12. E. I. Shustin, *On manifolds of singular algebraic curves*, Selecta Math. Soviet. **10** (1991), no. 1, 27–37.
13. _____, *A new M-curve of degree eight*, Mat. Zametki **42** (1987), no. 6, 180–186; English transl., Math. Notes **42** (1987), no. 1–2, 606–610.
14. O. Ya. Viro, *Gluing of algebraic hypersurfaces, smoothing of singularities and construction of curves*, Proc. Leningrad Intern. Topological Conf. (Leningrad, Aug. 1983), "Nauka", Leningrad, 1983, pp. 149–197. (Russian)
15. _____, *Gluing of plane real algebraic curves and construction of curves of degree 6 and 7*, Lecture Notes in Math., vol. 1060, Springer-Verlag, Berlin, 1984, pp. 187–200.
16. V. V. Nikulin, *Integral symmetric bilinear forms and some of their geometric applications*, Izv. Akad. Nauk SSSR Ser. Mat. **43** (1979), 111–177; English transl., Math. USSR-Izv. **14** (1980), no. 1, 103–167.
17. O. Ya. Viro, *Curves of degree 7, curves of degree 8, and the Ragsdale conjecture*, Dokl. Akad. Nauk SSSR **254** (1980); English transl., Soviet Math. Dokl. **22** (1980), no. 2, 566–569.
18. V. Rokhlin, *Complex orientations of real algebraic curves*, Funktsional. Anal. i Prilozhen. **8** (1974), no. 4, 71–75; English transl. in Functional Anal. Appl. **8** (1974).
19. _____, *Complex topological characteristics of real algebraic curves*, Uspekhi Mat. Nauk **33** (1978), no. 5, 77–89; English transl., Russian Math. Surveys **33** (1978), no. 5, 85–98.
20. V. M. Kharlamov, *Classification of nonsingular surfaces of degree 4 in $\mathbb{R}P^2$ with respect to rigid isotopies*, Funktsional. Anal. i Prilozhen. **18** (1984); English transl., Functional Anal. Appl. **18** (1984), 39–45.
21. I. V. Itenberg, *Curves of degree 6 with one nondegenerate double point and groups of monodromy of nonsingular curves*, Real Algebraic Geometry, Proceedings, Rennes 1991, Lecture Notes in Math., vol. 1524, Springer-Verlag, Berlin, 1992, pp. 267–288.
22. _____, *Plane projective real curves of degree 6 with a nondegenerate double point*, Ph. D. Thesis, St. Petersburg Univ., 1991. (Russian)
23. I. R. Shafarevich, *Algebraic surfaces*, Trudy Mat. Inst. Steklov **75** (1965); English transl. in Proc. Steklov Inst. Math. (1967), Amer. Math. Soc., Providence, RI.
24. E. B. Vinberg, *The groups of units of certain quadratics forms*, Mat. Sb. **87** (1972), 17–35; English transl. in Math. USSR-Sb. **16** (1972).

Translated by THE AUTHORS

IRMAR, Université de Rennes 1, Campus de Beaulieu, F-35042 Rennes cedex, France

School of Mathematical Sciences, Tel Aviv University, Ramat Aviv, Tel Aviv 69978, Israel

Towards the Maximal Number of Components of a Nonsingular Surface of Degree 5 in $\mathbb{R}P^3$

V. Kharlamov and I. Itenberg

§1. Introduction

The problem of determining the maximal number of connected components of a surface of given degree m in $\mathbb{R}P^3$ was posed by Hilbert in 1900 (see the 16th problem of his famous list). Despite developments in the last decades in the topology of real algebraic varieties, the answer is still unknown, except in the trivial cases $m \leqslant 3$ and the case $m = 4$. In this last case the maximal number of components is equal to 10 (surfaces with 10 components were constructed by Rohn [**Ro**] in 1886; a proof of the maximality was given by Kharlamov [**Kh1**] in 1972).

It is well known that to determine the maximal number of components it suffices to consider nonsingular surfaces: by a small variation, any singular surface can be replaced by a nonsingular one with at least the same number of components.

A standard application of the Smith and Comessatti inequalities (see §3) gives the following estimate: the number of components of a nonsingular surface of degree m in $\mathbb{R}P^3$ is less than or equal to $(5m^3 - 18m^2 + 25m)/12$. In particular, it cannot be more than 25 for $m = 5$.

Kharlamov [**Kh2**] constructed a surface of degree 5 in $\mathbb{R}P^3$ with 21 components. (The surface constructed is an M-surface: the total $\mathbb{Z}_2$-homology group has the same rank as that of its complexification; see §3)

In the present paper we construct a nonsingular surface of degree 5 in $\mathbb{R}P^3$ with 22 components. We follow the scheme of [**Kh2**] and use, in addition, some elements of Itenberg's recent construction [**It**] of counter-examples to the Ragsdale conjecture (see [**Ra**]).

§2. Construction

2.1. An equivariant analog of Horikawa's theorem. By a real algebraic (or analytic) manifold we mean a complex manifold supplied with complex conjugation. For a real variety X we denote the set of its real points by $\mathbb{R}X$ and the set of complex points by $\mathbb{C}X$.

1991 *Mathematics Subject Classification*. Primary 14P25, 14J99.

Let Σ_2 be the standard nonsingular model of the real cone defined in P^3 by the equation $x_0^2 + x_1^2 = x_2^2 + x_3^2$. Following Horikawa [**Ho**], consider an irreducible curve B on Σ_2 satisfying the following conditions:
  (i) its intersection number with a linear generator of Σ_2 is equal to 6;
  (ii) its intersection number with the inverse image of the vertex of the cone is equal to 1;
  (iii) it has only two singular points, these points are ordinary triple points and they both lie on the same linear generator L.

We shall call it the *Horikawa curve*.

Denote by $\widetilde{W}$ the surface obtained from Σ_2 by blowing-up the two singular points of the Horikawa curve B, and by $\widetilde{L}$ and $\widetilde{B}$ the proper transforms of L and B under this blowing-up. Then take a double covering $\widetilde{S} \to \widetilde{W}$ with branch locus $B \cup L$. Such a covering exists because of (i)–(iii) and is unique.

The inverse image of L is a nonsingular rational curve with self-intersection number -1. Contracting it to a point, we get a nonsingular surface; denote it by S.

If the Horikawa curve is real, the surface $\widetilde{S}$ acquires, in the usual way, two canonical real structures. They are liftings of the complex conjugation of $\widetilde{W}$. They differ by the covering transformation and both can be projected to S; we also call *canonical* the two resulting real structures on S.

PROPOSITION 1. *Let the Horikawa curve B be real and let S be supplied with one of its canonical real structures. Then there exists an equivariant deformation of S to a nonsingular surface of degree 5 in $\mathbb{R}P^3$.*

PROOF. Take a versal deformation $p: L \to M$ of S. By Horikawa's theorem (see [**Ho**, Theorem 3]), M consists of two smooth irreducible components M_0 and M_1 intersecting normally; $\dim_\mathbb{C} M_0 = \dim_\mathbb{C} M_1 = 40$, $\dim_\mathbb{C} M_0 \cap M_1 = 39$. Points of $M_0 \setminus M_0 \cap M_1$ correspond to quintic surfaces and points of $M_1 \setminus M_0 \cap M_1$ to coverings of $\mathbb{P}^1 \times \mathbb{P}^1$. The standard versality arguments show that the deformation may be made equivariant. It remains to notice that the corresponding antiholomorphic involution does not interchange irreducible components of M (they are of different nature) and that $\mathbb{R}M_0$ and $\mathbb{R}M_1$, as fixed point sets of an antiholomorphic involution on smooth complex manifolds, are smooth connected manifolds; they intersect normally and

$$\dim_\mathbb{R} M_0 = \dim_\mathbb{R} M_1 = 40, \qquad \dim_\mathbb{R} M_0 \cap M_1 = 39. \qquad \square$$

2.2. A special case of Viro's theorem. Let P be a convex polygon in $\mathbb{R}^2$ with integer vertices that verifies the following condition: it is contained in the triangle

$$\Delta = \{x \geqslant 0,\ y \geqslant 0,\ x + y \leqslant m\}, \qquad m \in \mathbb{N},$$

and it contains the vertices $x = 0$, $y = m$ and $x = m$, $y = 0$. In the sequel, such a polygon will serve as a Newton polygon of curves of degree m with a singularity at the origin prescribed by the Newton polygonal line $\Gamma(P)$, which is, by definition, the union of sides of P facing the origin.

Suppose that P is triangulated, that the vertices of the triangles are integer and that some distribution of signs, $a_{i,j} = \pm$, at the vertices of the triangulation is given. Then there arises a naturally associated piecewise-linear curve L in $\mathbb{R}P^2$.

The construction of L is the following.

Take the copies

$$P_x = s_x(P), \quad P_y = s_y(P), \quad P_{xy} = s(P),$$
$$\Delta_x = s_x(\Delta), \quad \Delta_y = s_y(\Delta), \quad \Delta_{xy} = s(\Delta)$$

of P and Δ, where $s = s_x \circ s_y$ and s_x, s_y are reflections with respect to the coordinate axes. Extend the triangulation of P to a symmetric triangulation of $P_* = P \cup P_x \cup P_y \cup P_{xy}$ and extend the distribution of signs to a distribution at the vertices of the extended triangulation so that it verifies the modular property: $g^*(a_{i,j} x^i y^j) = a_{g(i,j)} x^i y^j$ for $g = s_x, s_y$, and s (in other words, the sign at a vertex is the sign of the corresponding monomial in the quadrant containing the vertex).

If a triangle of the triangulation has vertices of different signs, select a midline separating them. If a midline comes to $\Gamma(P)$ at a point b, select also the segment joining b to the origin. Denote by L_* the union of the selected midlines and segments. It is contained in $T_* = \Delta \cup \Delta_x \cup \Delta_y \cup \Delta_{xy}$. Glue the sides of T_* by s. The resulting space T is homeomorphic to $\mathbb{R}P^2$. Take the curve L to be the image of L_* in T.

A pair (T, L) is called a *chart* of a real plane algebraic curve A if there exists a homeomorphism $(T, L) \to (\mathbb{R}P^2, \mathbb{R}A)$ that maps each segment of L joining the origin and an edge of $\Gamma(P)$ to the branch of A corresponding to this edge.

A curve A is called *regular* if it does not have singular points outside of the origin.

Let us introduce two additional assumptions: the triangulation of P considered is *primitive* and *convex*. The first condition means that all triangles are of area $1/2$ (or, equivalently, that all integer points of P are vertices of the triangulation). The second one means that there exists a convex piecewise-linear function $P \to \mathbb{R}$ which is linear on each triangle of the triangulation and not linear on the union of two triangles.

The following statement is a special case of Viro's theorem [**Vi1**, Theorem 1.4]:

PROPOSITION 2. *Under the assumptions made above concerning the polygon P and its triangulation, there exists a real regular curve A in $\mathbb{R}P^2$ with chart (T, L).*

2.3. A lemma. We say that a real curve in P^2 is *of class H*, if
  (i) its Newton diagram is the pentagon Π presented in Figure 1,
  (ii) there are three branches corresponding to the edge BC of the diagram, they are smooth and each of them is tangent with simple inflection to another.

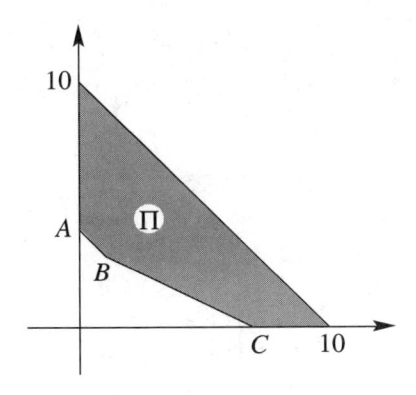

FIGURE 1

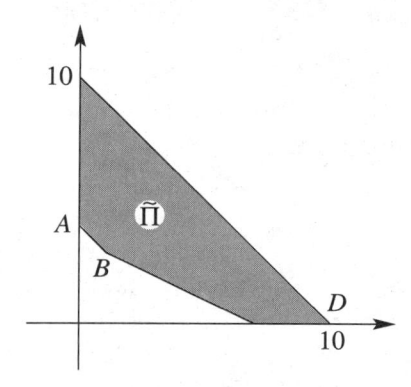

FIGURE 2

We say that a real curve in P^2 is *of class $\widetilde{H}$*, if
 (i) its Newton diagram is the quadrangle $\widetilde{\Pi}$ presented in Figure 2,
 (ii) the truncation f_{BD} of the polynomial f defining the curve is equal to
$$\gamma x(y - ax^3)(y - bx^3)(y - cx^3),$$
where $\gamma \neq 0$, a, b, c are real numbers and a, b, c are pairwise different.

Curves of class $\widetilde{H}$ have 4 branches at the origin, three of them are smooth and pairwise tangent with simple inflection; the same is true for that class H. (The principal difference between these two classes is that a curve of class H is degenerate at the origin with respect to the Newton diagram and a curve of class $\widetilde{H}$ is nondegenerate.)

LEMMA 3. *Up to a homeomorphism of the plane, each real regular curve $\widetilde{C}$ of class $\widetilde{H}$ is equivalent to a real regular curve C of class H. Moreover, a homeomorphism may be chosen to transform three tangent branches of $\widetilde{C}$ to three tangent branches of C.*

PROOF. Let
$$Q(x, y) = \sum_{(i,j)\in\widetilde{\Pi}} a_{i,j} x^i y^j$$
be a polynomial which defines a real regular curve of class $\widetilde{H}$, and let
$$\Gamma(x, y) = \gamma x(y - ax^3)(y - bx^3)(y - cx^3)$$
be the truncation of Q on BD.

Take a linear function v in two variables with integer coefficients vanishing at each point of BD and positive at the other points of $\widetilde{\Pi}$ and put
$$Q'_t(x, y) = \sum_{(i,j)\in\widetilde{\Pi}} a_{i,j} x^i y^j t^{v(i,j)}$$
$$+ \gamma x(y - x^2 - ax^3)(y - x^2 - bx^3)(y - x^2 - cx^3) - \Gamma(x, y).$$

For any real t, the curve $Q'_t = 0$ is of class H. Following the same lines as in [**Vi2**] in the proof of the theorem on the smoothing of quasi-homogeneous singularities, one verifies that for any sufficiently small positive value of t there exist two radii, $r_2 > r_1 > 0$, such that the curve $Q'_t = 0$ is approximated
 (a) inside of the disc D_1 of radius r_1 centered at the origin, by the curve
$$\gamma x(y - x^2 - ax^3)(y - x^2 - bx^3)(y - x^2 - cx^3) = 0;$$

(b) outside of the disc D_2 of radius r_2 centered at the origin, by the curve $Q = 0$;

(c) in the annulus $D_2 \setminus D_1$, by the curve $\Gamma = 0$.

Thus for a sufficiently small positive t the curve $Q'_t = 0$ is regular and topologically equivalent to the initial curve $Q = 0$ and a homeomorphism of the plane, mapping one into another, may be chosen to transform tangent branches into tangent branches. □

2.4. The curve.

PROPOSITION 4. *There exists a real regular curve of class H of the isotopy type represented in Figure 3 (the letters a, b, c mark the three branches with common tangent).*

PROOF. By Lemma 3, it suffices to realize the given isotopy type by a real regular curve of class $\widetilde{H}$.

Any convex primitive triangulation of a convex part of a convex polygon is extendible to a convex primitive triangulation of the polygon. Inside the part $BKMN$ of the quadrangle $\widetilde{\Pi}$, take the convex primitive triangulation shown in Figure 4 and extend it to $\widetilde{\Pi}$.

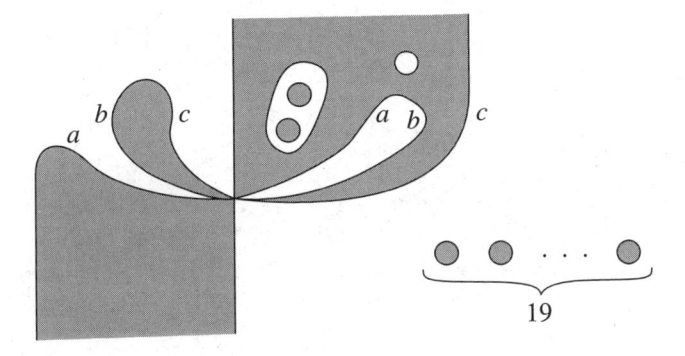

FIGURE 3

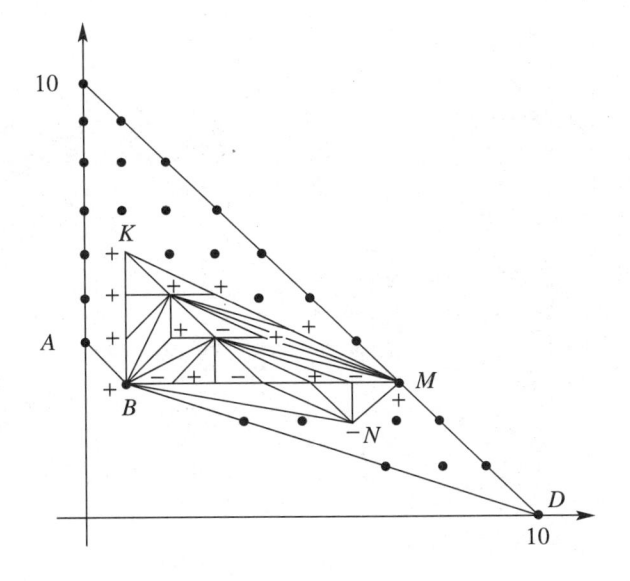

FIGURE 4

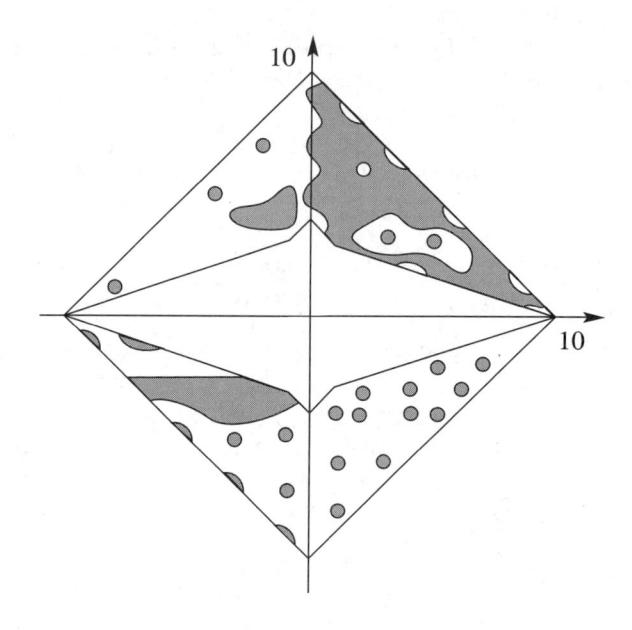

FIGURE 5

To apply Proposition 2, we need to choose signs on the vertices in $\widetilde{\Pi}$. Inside $BKMN$ put signs according to Figure 4, outside, use the following rule: the vertex (i, j) acquires sign "$-$" if i, j are even, and the sign "$+$" otherwise.

The corresponding piecewise-linear curve L is of the required isotopy type (see Figure 5) and Proposition 2 gives the desired result. □

2.5. The surface.

THEOREM 5. *There exists a nonsingular surface of degree 5 in $\mathbb{R}P^3$ having 21 connected components homeomorphic to the sphere and one component homeomorphic to the sphere with 7 Möbius bands.*

PROOF. To each real regular curve of class H corresponds a Horikawa curve: make two consecutive blowing-ups at the origin, the second one corresponding to the direction of the tangent line l to the parabolic branches, and then contract the proper transform of l to a point; thus we get Σ_2 and a Horikawa curve on it.

Let us start with the curve A constructed in 2.4. Then, applying Proposition 1, we obtain a nonsingular surface of degree 5 in $\mathbb{R}P^3$ homeomorphic to $\mathbb{R}S$ (see 2.1). The surface $\mathbb{R}S$ is the real part of a nonsingular real model of the two-sheeted covering Y of P^2, ramified along A. Choosing the appropriate real structure on Y from the two canonical ones (namely, take the one for which $\mathbb{R}S$ is situated over the dark regions in Figure 3), we get, according to Proposition 4, exactly 21 connected components homeomorphic to the sphere and one additional component. It now suffices to note that this component is not orientable and to calculate its Euler characteristic by retracing blowing-ups:
$$\chi = 2 - 1 + 2(1 - 2) - 3 + 0 + (1 - 2) = -5$$
(on the nonsingular model $\mathbb{R}S$ of $\mathbb{R}Y$, the singular point is replaced by a wedge of two circles). □

§3. Limits of the method

3.1. Known restrictions on the topological type of a real surface. We mention three well-known results (see, for example, the survey articles [**Wi, Kh3**]): for a nonsingular real projective surface X,

(a) $\chi(\mathbb{R}X) \leqslant h^{1,1}(\mathbb{C}X) - 2(\rho - 1)$ (Comessatti inequality),

where χ is the Euler characteristic and ρ is the number of linearly independent real algebraic classes in $H_2(\mathbb{C}X; \mathbb{R})$;

(b) $\beta_*(\mathbb{R}X) \leqslant \beta_*(\mathbb{C}X) - 2\phi$ (Smith inequality),

where β_* is the rank of the total $\mathbb{Z}_2$-homology group and $\rho + \phi$ is the number of linearly independent algebraic classes (not only real ones) in $H_2(\mathbb{C}X; \mathbb{R})$;

(c) if $\beta_*(\mathbb{R}X) = \beta_*(\mathbb{C}X)$, then

$$\chi(\mathbb{R}X) \equiv \sigma(\mathbb{C}X) \bmod 16 \qquad \text{(Rokhlin congruence).}$$

3.2. Application to surfaces of degree 5. If X is a nonsingular surface of degree 5, then
$$h^{1,1}(\mathbb{C}X) = 45, \quad \beta_*(\mathbb{C}X) = 55, \quad \text{and} \quad \sigma(\mathbb{C}X) = -35.$$
Thus, according to the Smith and Comessatti inequalities, the number of components of a surface of degree 5 in $\mathbb{R}P^3$ is not greater than 25.

PROPOSITION 6. *The real part $\mathbb{R}S$ of a Horikawa surface S cannot have more than 24 connected components. If the singular points of a Horikawa curve B are real, then $\mathbb{R}S$ has no more than 23 components.*

PROOF. First, consider the case when the singular points are real.

Then the surface $\widetilde{W}$ has at least 4 independent real algebraic cycles: the inverse image of the vertex of the cone, the inverse images of the singular points of B and the hyperplane section. So this is also the case for $\widetilde{S}$. Thus

$$\chi(\mathbb{R}S) = 1 + \chi(\mathbb{R}\widetilde{S}) \leqslant 1 + (h^{1,1}(\mathbb{C}\widetilde{S}) - 2 \cdot 3) = h^{1,1}(\mathbb{C}S) - 4 = 41,$$
$$\beta_0(\mathbb{R}S) = (\chi(\mathbb{R}S) + \beta_*(\mathbb{R}S))/4 \leqslant 24.$$

Moreover, if $\beta_0(\mathbb{R}S) = 24$, then $\beta_*(\mathbb{R}S) = 55$ and $\chi(\mathbb{R}\widetilde{S}) = 41$. The last combination contradicts the Rokhlin congruence.

If the singular points are imaginary, then $\rho \geqslant 3$ and $\phi \geqslant 1$. Thus

$$\chi(\mathbb{R}S) \leqslant 1 + (h^{1,1}(\mathbb{C}\widetilde{S}) - 2 \cdot 2) = 43, \qquad \beta_*(\mathbb{R}S) \leqslant \beta_*(\mathbb{C}S) - 2 = 53$$

and we obtain the bound $\beta_0(\mathbb{R}S) \leqslant 24$ again. □

3.3. Concluding remarks. A. It was conjectured by V. Arnold (see [**Vi3**]) that a nonsingular surface of degree m in $\mathbb{R}P^3$ has at most

$$(m^3 - m + 3((-1)^{m+1} + 1))/6$$

components. Viro [**Vi3**] showed that for any even $m \geqslant 6$ the conjecture is not true. The surface constructed in the present paper provides a counter-example for $m = 5$ (for $m \leqslant 4$ the conjecture is true).

B. Real double planes $\mathbb{R}Y$ ramified along real plane curves constructed by Itenberg in [**It**] have more than $(2 + h^{1,1}(\mathbb{C}Y))/2$ components. For a surface X of degree 5 in $\mathbb{R}P^3$

$$(2 + h^{1,1}(\mathbb{C}X))/2 = 23.5,$$

and one may expect that a clever direct application of Viro's construction can give examples of surfaces of degree 5 with at least 24 components.

C. The case of M-surfaces, $\beta_*(\mathbb{R}X) = \beta_*(\mathbb{C}X)$, is always of particular interest. By 3.1, an M-surface of degree 5 in $\mathbb{R}P^3$ may have 5, 9, 13, 17, 21, or 25 connected components. Examples with 5, 94, 13, 17, and 21 components were constructed by Kharlamov [**Kh2**]. If M-surfaces with 25 components really exist, then, again according to 3.1, they may be only of the following topological types:

$$24S \amalg P(2), \quad 23S \amalg S(1) \amalg P(1), \quad 23S \amalg S(2) \amalg P, \quad 22S \amalg S(1) \amalg S(1) \amalg P,$$

where S is the sphere, P is the projective plane, $S(q)$ and $P(q)$ are the sphere and the projective plane with q handles. The two last types are not realizable (the fourth was excluded by Viro, the third by Kharlamov; see [**Kh4**]). The problem of the existence of M-surfaces of degree 5 of the two other topological types $24S \amalg P(2)$, $23S \amalg S(1) \amalg P(1)$ is open.

D. Taking the other canonical real structure (see 2.1) on the Horikawa surface constructed in 2.5, one gets a surface with real part homeomorphic to $S \amalg S(2) \amalg P(20)$. In particular, here $\beta_1 = 45 = h^{1,1}$. The same value is given by M-surfaces having 5 components (see C above). It would be interesting to construct surfaces of degree 5 with larger β_1.

References

[Ho] E. Horikawa, *On deformations of quintic surfaces*, Invent. Math. **31** (1975), 43–85.

[It] I. V. Itenberg, *Contre-exemples à la conjecture de Ragsdale*, C. R. Acad. Sci. Paris Sér. I Math. (1993), no. 317, 277–282.

[Kh1] V. M. Kharlamov, *Maximal number of components of a 4th degree surface in $\mathbb{R}P^3$*, Funktsional. Anal. i Prilozhen. **6** (1972), no. 4, 101; English transl., Functional Anal. Appl. **6** (1972), no. 4, 345–346.

[Kh2] _____, *On a number of components of a M-surface of degree 5 in $\mathbb{R}P^3$*, Proc. of the XVI Soviet Algebraic Conference, Leningrad, 1981, pp. 353–354. (Russian)

[Kh3] _____, *Real algebraic surfaces*, Proc. Internat. Congress Math., Helsinki, vol. 1, 1978, pp. 421–428. (Russian)

[Kh4] _____, *Estimates of Betti numbers in topology of real algebraic surfaces*, Notes International Topological Conference, Oberwolfach, 1987, pp. 12–13.

[Ra] V. Ragsdale, *On the arrangement of the real branches of plane algebraic curves*, Amer. J. Math. **28** (1906), 377–404.

[Ro] K. Rohn, *Flächen vierter Ordnung*, Preisschriften der Fürstlich Jablonowski-schen Gesellschaft, Leipzig, 1886.

[Vi1] O. Ya. Viro, *Gluing of plane real algebraic curves and construction of curves of degree 6 and 7*, Lecture Notes in Math., vol. 1060, Springer-Verlag, Berlin, 1984, pp. 187–200.

[Vi2] _____, *Real algebraic plane curves: constructions with controlled topology*, Algebra i Analiz **1** (1989), no. 5, 1–73; English transl., Leningrad Math. J. **1** (1990), no. 5, 1059–1134.

[Vi3] _____, *Construction of multicomponent real algebraic surfaces*, Dokl. Akad. Nauk SSSR **248** (1979), no. 2 pages 279–282; English transl., Soviet Math. Dokl. **20** (1979), no. 5, 991–995.

[Wi] G. Wilson, *Hilbert's sixteenth problem*, Topology **17** (1978), no. 1, 53–74.

Translated by THE AUTHORS

Stable Equivalence of Real Projective Configurations

Sergeĭ I. Khashin and Vladimir F. Mazurovskiĭ

ABSTRACT. A nonsingular $(2k + 1; k)$-configuration is defined as a finite collection of k-dimensional linear pairwise disjoint subspaces of $\mathbb{R}P^{2k+1}$. An isotopy of such a collection is called rigid if it consists of nonsingular $(2k + 1; k)$-configurations. Viro defined a suspension construction of real projective configurations. We shall say that two real projective configurations are stably equivalent if their s-fold suspensions are rigidly isotopic for some s. The main result of the present paper is that two nonsingular $(2k+1; k)$-configurations are stably equivalent if and only if they have the same linking numbers of their subspaces. For the class of nonsingular isotopy join configurations of lines in 3-dimensional real projective space, we prove that such configurations are rigidly isotopic if and only if they have the same collections of linking numbers.

§0. Introduction

Recently knots and links were investigated actively not only in the traditional spaces $\mathbb{R}^3$ and S^3, but in other manifolds as well. Links in $\mathbb{R}P^3$ appear, in particular, as finite collections of projective pairwise disjoint lines. For such collections of lines together with ordinary link isotopies one can also consider rigid isotopies. By rigid isotopy one means an isotopy in the process of which the lines of the collection remain pairwise disjoint. Along with the rigid isotopies of configurations of lines of $\mathbb{R}P^3$ one can study rigid isotopies of finite collections of projective subspaces of any dimension in $\mathbb{R}P^n$. In the present paper the rigid isotopies of configurations of k-dimensional linear pairwise disjoint subspaces of $\mathbb{R}P^{2k+1}$ are investigated.

An *ordered (unordered) real projective $(n; k)$-configuration of degree m* is defined as an ordered (respectively unordered) collection of m linear k-dimensional subspaces of $\mathbb{R}P^n$. We associate with each configuration its *upper* and *lower ranks*, i.e., the dimensions of the projective hull and intersection, respectively, of all the subspaces of the configuration. The *combinatorial characteristic* of a configuration is, by definition, the list of the upper and lower ranks of all its subconfigurations. The *combinatorial type* is defined as the set of all the configurations with the same combinatorial characteristics. Two configurations are called *rigidly isotopic* if they can be joined by an isotopy

---

1991 *Mathematics Subject Classification.* Primary 51A20.

Key words and phrases. Real projective configurations, vector configurations, isotopies and rigid isotopies, homology equivalence, join, stable equivalence.

The research described in this publication was made possible in part by Grant No. 94-1.1-144 from the State Committee of Russian Federation for Higher Education.

©1996, American Mathematical Society

which consists of configurations of the same combinatorial type. It is obvious that the property of being rigidly isotopic is an equivalence relation. The corresponding equivalence class of a configuration is called its *rigid isotopy type*.

A configuration is said to be *nonsingular* if all its subspaces are in general position. It is clear that the nonsingular configurations constitute an open dense subset in the configuration space. Viro (see [8]) defined a suspension construction of real projective configurations. The suspension operator increases the dimension of the ambient space by 4 and the dimension of the subspaces by 2. Viro also put forward the conjecture that, to some extent, this construction should realize a one-to-one correspondence between the set of the rigid isotopy types of $(2k+1;k)$-configurations and the set of the rigid isotopy types of $(2k+5;k+2)$-configurations. In [3] it was shown that such a stabilization takes place for nonsingular configurations.

The main result of the present paper is that two nonsingular projective $(2k+1;k)$-configurations of degree m belong to the same stable type if and only if they have the same linking numbers of their subspaces (Theorem 1.7.6 and Theorem 2.25). The authors proved this statement independently, using different methods. The paper is divided into three parts. The first part was written by Mazurovskiĭ and contains a geometric proof of the main result. The second part is due to Khashin and contains an algebraic proof of the main theorem. At the beginning of the third part Khashin proves that the homology trivial nonsingular $(3;1)$-configuration is trivial (Theorem 3.1). Then we prove a joint result that a nonsingular isotopy join $(3;1)$-configuration is determined up to rigid isotopy by the linking numbers of its lines (Theorem 3.2).

§1. Geometric proof of the main result

1.1. Rigid isotopies of real projective configurations. A family $Y(t)$, $t \in [0,1]$, of subsets of a topological space X is called an *isotopy* of $Y = Y(0)$ if the *graph* of the family $\Gamma_{Y(t)} = \{(x,t) \in X \times [0,1] \mid x \in Y(t)\}$ is fiberwise homeomorphic to the cylinder $Y \times [0,1]$. Let $A = \{A_1, \ldots, A_m\}$ be an ordered (unordered) configuration of subspaces of $\mathbb{R}P^n$. An ordered (respectively unordered) collection $A(t) = \{A_1(t), \ldots, A_m(t)\}$ of isotopies of the subsets $A_1, \ldots, A_m$ in $\mathbb{R}P^n$ is called an *isotopy of the configuration* A if the family $\bigcup_{i=1}^{m} A_i(t)$ of subsets of $\mathbb{R}P^n$ ($t \in [0,1]$) is an isotopy of $\bigcup_{i=1}^{m} A_i$. An isotopy of real projective configuration is said to be *rigid* if for any $t \in [0,1]$ the sets $A_1(t), \ldots, A_m(t)$ are linear subspaces and $A(t)$ belongs to the same combinatorial type as A. The rigid isotopy $A(t) = \{A_1(t), \ldots, A_m(t)\}$ of a configuration A is called a *rigid ε-isotopy* if $A_i(t) \subset U_i$ for any $t \in [0,1]$ and $i = 1, \ldots, m$, where U_i is the ε-neighborhood of A_i in $\mathbb{R}P^n$ (in this case we consider $\mathbb{R}P^n$ as a metric space).

1.2. Degeneration and perturbation. Let X be the space of $(n;k)$-configurations of degree m and $s : [0,1] \to X$ be a path such that the restriction $s|_{[0,b]}$ is a rigid isotopy for any $0 < b < 1$. If the configurations $A = s(0)$ and $A' = s(1)$ belong to distinct combinatorial types, then s is called a *degeneration* of A, and s^{-1} is called a *perturbation* of A'. The perturbation s^{-1} is called an *ε-perturbation* if the restriction $s^{-1}|_{[a,1]}$ is a rigid ε-isotopy for any $0 < a < 1$.

A configuration A of m linear k-dimensional subspaces of $\mathbb{R}P^n$ is said to be *1-singular* if all the configurations rigidly isotopic to A form a codimension 1 subset in the space of $(n;k)$-configurations of degree m. It is easy to see that any perturbation of a 1-singular configuration gives a nonsingular configuration. It is also clear that a

$(2k + 1; k)$-configuration is 1-singular iff only two of its subspaces intersect and their intersection is a point. Ordered and unordered, 1-singular and nonsingular $(2k+1; k)$-configurations are the main objects of our investigation. In what follows we do not specify whether a configuration is ordered or unordered if this is not essential.

Let two ordered nonsingular $(2k + 1; k)$-configurations $B^1 = \{B_1^1, B_2^1\}$ and $B^2 = \{B_1^2, B_2^2\}$ be obtained from an ordered 1-singular $(2k + 1; k)$-configuration $B = \{B_1, B_2\}$ by ε-perturbations, where ε is so small that the union of the ε-neighborhoods of B_1 and B_2 in $\mathbb{R}P^{2k+1}$ is a regular neighborhood of $B_1 \cup B_2$. In this case we say that B^1 and B^2 are obtained from B by ε-perturbations of the same type if they can be joined by a rigid isotopy in this regular neighborhood.

1.2.1. LEMMA. *In the above situation there exist exactly two types of ε-perturbations of a 1-singular $(2k + 1; k)$-configuration B.*

As a consequence of Lemma 1.2.1 we obtain the following lemma.

1.2.2. LEMMA. *Let $S = \{S_1, \ldots, S_{k+1}, S_{k+2}\}$ be an ordered (unordered) $(2k + 1; k)$-configuration, let the subconfiguration $\{S_1, \ldots, S_{k+1}\}$ of S be 1-singular (nonsingular), and let the subspaces S_i and S_{k+2} intersect at a point for some $i \in \{1, \ldots, k + 1\}$, and $S_i \cap S_j \cap S_{k+2} = \varnothing$ for any $j = 1, \ldots, k + 1$, $j \neq i$. Let $\widetilde{S} = \{S_1, \ldots, S_{k+1}, \widetilde{S}_{k+2}\}$ be a 1-singular (nonsingular) configuration which is obtained from S by ε-perturbation. Consider a rigid isotopy*
$$S(t) = \{S_1(t), \ldots, S_{k+1}(t), S_{k+2}(t)\}$$
of S. Suppose that ε is so small that for any $t \in [0, 1]$ the union of the ε-neighborhoods of $S_i(t)$ and $S_{k+2}(t)$ in $\mathbb{R}P^{2k+1} \times \{t\}$ is a regular neighborhood of $S_i(t) \cup S_{k+2}(t)$. Then there exists a rigid isotopy
$$\widetilde{S}(t) = \{S_1(t), \ldots, S_{k+1}(t), \widetilde{S}_{k+2}(t)\}$$
of $\widetilde{S}$ such that:

1) $\widetilde{S}(t)$ *coincides with $S(t)$ on the subconfiguration $\{S_1, \ldots, S_{k+1}\}$;*
2) *for any $t \in [0, 1]$ the subconfiguration $\{S_i(t), \widetilde{S}_{k+2}(t)\}$ of $\widetilde{S}(t)$ is obtained from the subconfiguration $\{S_i(t), S_{k+2}(t)\}$ of $S(t)$ by an ε-perturbation.*

Let $C = \{C_1, \ldots, C_m\}$ be a nonsingular (1-singular) $(2k + 1; k)$-configuration, and let C_i, C_j be two subspaces of C ($C_i \cap C_j = \varnothing$). We say that C_i and C_j can be *moved up to intersection at a point*, if there exists a degeneration $C(t) = \{C_1(t), \ldots, C_m(t)\}$ of C such that:

a) the restrictions of $C(t)$ to the subconfigurations $\{C_1, \ldots, C_{i-1}, C_{i+1}, \ldots, C_m\}$ and $\{C_1, \ldots, C_{j-1}, C_{j+1}, \ldots, C_m\}$ of C are rigid isotopies;
b) the subspaces $C_i(1)$ and $C_j(1)$ intersect at a point.

Consider a perturbation $[C(1)](t) = \{[C_1(1)](t), \ldots, [C_m(1)](t)\}$ of $C(1)$ for which the restriction to the subconfigurations $\{C_1(1), \ldots, C_{i-1}(1), C_{i+1}(1), \ldots, C_m(1)\}$ and $\{C_1(1), \ldots, C_{j-1}(1), C_{j+1}(1), \ldots, C_m(1)\}$ of $C(1)$ are rigid isotopies. Let $0 < \delta \leqslant 1$, $0 < \eta \leqslant 1$. Consider the perturbations
$$[C^\delta(1)](t) = \{C_1(1 - \delta t), \ldots, C_m(1 - \delta t)\},$$
$$[C^\eta(1)](t) = \{[C_1(1)](\eta t), \ldots, [C_m(1)](\eta t)\}$$
of $C(1)$. Suppose that $[C^\delta(1)](t)$ and $[C^\eta(1)](t)$ are ε-perturbations of $C(1)$, and the union of the ε-neighborhoods of $C_i(1)$ and $C_j(1)$ in $\mathbb{R}P^{2k+1}$ is a regular neighborhood

of $C_i(1) \cup C_j(1)$. Suppose also that the restrictions of $[C^\delta(1)](t)$ and $[C^\eta(1)](t)$ to the subconfiguration $\{C_i(1), C_j(1)\}$ of $C(1)$ belong to distinct types of ε-perturbations of $\{C_i(1), C_j(1)\}$. In this case we shall say that the configuration

$$[C(1)](1) = \{[C_1(1)](1), \ldots, [C_m(1)](1)\}$$

is obtained from C by *passing across* the subspaces C_i and C_j.

1.3. Adjacency graph. The set of all nonsingular configurations of the same rigid isotopy type is called a *chamber*, and the set of all 1-singular configurations of the same rigid isotopy type is called a *wall*. The mutual position of the chambers in the configuration space can be described by means of the *adjacency graph* (see [1]), whose vertices and edges are in one-to-one correspondence with the chambers and walls respectively, and two vertices representing some chambers are connected by an edge if and only if these chambers are adjacent to the wall corresponding to this edge.

1.4. Linking numbers. In the following three subsections we describe some constructions due to Viro (see [7, 8]).

Let $B = \{B_1, B_2, B_3\}$ be an ordered nonsingular configuration of three k-dimensional subspaces of the oriented space $\mathbb{R}P^{2k+1}$. Consider the canonical projection pr: $\mathbb{R}^{2k+2} \setminus \{0\} \to \mathbb{R}P^{2k+1}$. The orientation of $\mathbb{R}P^{2k+1}$ induces an orientation of the vector space $\mathbb{R}^{2k+2}$. Let $\overline{B}_i = \text{pr}^{-1}(B_i) \cup \{0\}$, $i = 1, 2, 3$. It is clear that $\overline{B}_1$, $\overline{B}_2$, $\overline{B}_3$ are $(k+1)$-dimensional vector subspaces of $\mathbb{R}^{2k+2}$. Let $\overline{B}_1^*, \overline{B}_2^*, \overline{B}_3^*$ be the same subspaces equipped with some orientations. To every ordered pair $(\overline{B}_i^*, \overline{B}_j^*)$, $i, j = 1, 2, 3$, $i \neq j$, we assign an integer $\text{lk}(\overline{B}_i^*, \overline{B}_j^*)$, which is equal to $+1$ if the orientation of $\mathbb{R}^{2k+2}$ coincides with that of $\overline{B}_i^* \oplus \overline{B}_j^*$, and equal to -1 otherwise. The product $\text{lk}(\overline{B}_2^*, \overline{B}_3^*) \text{lk}(\overline{B}_3^*, \overline{B}_1^*) \text{lk}(\overline{B}_1^*, \overline{B}_2^*)$, denoted by $\text{lk}(B_1, B_2, B_3)$, is called the *linking number* of the triple B_1, B_2, B_3. It is easy to see that $\text{lk}(B_1, B_2, B_3)$ does not depend on the choice of the orientations of $\overline{B}_i$, $i = 1, 2, 3$, is preserved under rigid isotopies of B, and changes its sign when the orientation of $\mathbb{R}P^{2k+1}$ is reversed. It can also be seen that, if k is odd, then $\text{lk}(B_1, B_2, B_3)$ does not depend on the choice of the order of the subspaces of B and is preserved under isotopies of B (in this case $\text{lk}(\overline{B}_i^*, \overline{B}_j^*)$ coincides with the doubled linking number of the oriented cycles B_i and B_j in the oriented manifold $\mathbb{R}P^{2k+1}$).

Two ordered nonsingular $(2k+1; k)$-configurations $C = \{C_1, \ldots, C_m\}$ and $C' = \{C'_1, \ldots, C'_m\}$ are said to be *homology equivalent* if for a fixed orientation of $\mathbb{R}P^{2k+1}$

$$\text{lk}(C_i, C_j, C_l) = \text{lk}(C'_i, C'_j, C'_l) \quad \text{for any } i, j, l = 1, \ldots, m, i < j < l.$$

Two ordered 1-singular $(2k+1; k)$-configurations are called *homology equivalent* if after their perturbations two pairs of homology equivalent ordered nonsingular configurations are obtained. Two unordered 1-singular (nonsingular) $(2k+1; k)$-configurations are called *homology equivalent* if they can be ordered in such a way that the obtained ordered configurations are homology equivalent.

1.5. Contiguous and homologous subspaces of a configuration. Two subspaces A_i and A_j of a configuration $A = \{A_1, \ldots, A_m\}$ are said to be *contiguous* if either they coincide, or there exists a degeneration $A(t) = \{A_1(t), \ldots, A_m(t)\}$ of A such that:

1) the restrictions of $A(t)$ to the following subconfigurations are rigid isotopies:

$$\{A_1, \ldots, A_{i-1}, A_{i+1}, \ldots, A_m\} \quad \text{and} \quad \{A_1, \ldots, A_{j-1}, A_{j+1}, \ldots, A_m\};$$

2) $A_i(1) = A_j(1)$.

Subspaces C_i, C_j of a nonsingular $(2k+1;k)$-configuration $C = \{C_1, \ldots, C_m\}$ are called *homologous* if $\text{lk}(C_i, C_l, C_n) = \text{lk}(C_j, C_l, C_n)$ (or, what is equivalent, $\text{lk}(C_i, C_j, C_l) = \text{lk}(C_i, C_j, C_n)$) for any $l, n \in \{1, \ldots, m\} \setminus \{i, j\}$. Two subspaces of a 1-singular $(2k+1;k)$-configuration are called *homologous* if they are homologous subspaces of the nonsingular configurations obtained from the 1-singular configuration by perturbations. The contiguity of two subspaces of a 1-singular (nonsingular) $(2k+1;k)$-configuration obviously implies that they are homologous. The converse is generally not true (see [7, 2]).

The following lemmas are obvious.

1.5.1. LEMMA. *Let subspaces A_i and A_j of a configuration $A = \{A_1, \ldots, A_m\}$ be contiguous. Then the subconfigurations $A^i = \{A_1, \ldots, A_{i-1}, A_{i+1}, \ldots, A_m\}$ and $A^j = \{A_1, \ldots, A_{j-1}, A_{j+1}, \ldots, A_m\}$ of A are rigidly isotopic.*

1.5.2. LEMMA. *Let $S = \{S_1, \ldots, S_m, S_{m+1}, S_{m+2}\}$ be a 1-singular (nonsingular) $(2k+1;k)$-configuration. The subconfigurations $\{S_1, \ldots, S_m, S_{m+1}\}$ and $\{S_1, \ldots, S_m, S_{m+2}\}$ of S are homology equivalent if and only if the subspaces S_{m+1} and S_{m+2} of S are homologous.*

1.5.3. LEMMA. *Let disjoint subspaces S_i and S_j of a 1-singular (nonsingular) $(2k+1;k)$-configuration $S = \{S_1, \ldots, S_m\}$ be homologous. Then there exist a rigid isotopy $S_i(t)$ of S_i in $\mathbb{R}P^{2k+1}$ and a subdivision $0 = t_0 < t_1 < t_2 < \cdots < t_{q-1} < t_q = 1$ of $[0, 1]$ such that:*

1) $S_i(t) \cap S_j = \varnothing$ *for any* $t \in [0, 1)$;
2) *if* $t' \in (t_{p-1}, t_p)$ *and* $t'' \in (t_p, t_{p+1})$, *where* $p \in \{1, \ldots, q-1\}$, *then the configuration* $S'' = \{S_1, \ldots, S_{i-1}, S_i(t''), S_{i+1}, \ldots, S_m\}$ *is obtained from* $S' = \{S_1, \ldots, S_{i-1}, S_i(t'), S_{i+1}, \ldots, S_m\}$ *by passing across the subspaces $S_i(t')$ and S_{l_p}, where $l_p \in \{1, \ldots, m\} \setminus \{i, j\}$;*
3) $S_i(1) = S_j$;
4) *in the process of the rigid isotopy $S_i(t)$, the subspace S_i intersects every subspace S_l, $l \in \{1, \ldots, m\} \setminus \{i, j\}$, an even number of times.*

1.6. Join sum and suspension. Let $A = \{A_1, \ldots, A_m\}$ be an ordered configuration of k-dimensional subspaces of $\mathbb{R}P^n$, and let $B = \{B_1, \ldots, B_m\}$ be an ordered configuration of l-dimensional subspaces of $\mathbb{R}P^s$. Suppose that $\mathbb{R}P^n$ and $\mathbb{R}P^s$ are imbedded into $\mathbb{R}P^{n+s+1}$ as disjoint linear subspaces. If n and s are odd, suppose, in addition, that $\mathbb{R}P^n$, $\mathbb{R}P^s$, and $\mathbb{R}P^{n+s+1}$ are oriented, and the linking number of the images of $\mathbb{R}P^n$ and $\mathbb{R}P^s$ in $\mathbb{R}P^{n+s+1}$ equals $+1$. Let C_i be the projective hull of the images of A_i and B_i in $\mathbb{R}P^{n+s+1}$, $i = 1, \ldots, m$. It is clear that $C = \{C_1, \ldots, C_m\}$ is an ordered $(n+s+1; k+l+1)$-configuration of degree m. The configuration C is called the *join* of A and B. We denote this join by $A \vee B$. A configuration is called an *isotopy join* if it is rigidly isotopic to the join of some two configurations.

In the same manner, one can define the join sum of two unordered configurations, but in this case it depends on the choice of the order of their elements.

A nonsingular ordered (unordered) $(3; 1)$-configuration consisting of the generatrices of a quadric in $\mathbb{R}P^3$ is called an ordered (respectively unordered) *trivial configuration* of lines of $\mathbb{R}P^3$. A trivial configuration of lines of oriented space $\mathbb{R}P^3$ with positive linking numbers of triples of the lines is called the *Hopf configuration*. The join of a configuration of k-dimensional subspaces of $\mathbb{R}P^n$ and the trivial configuration (the Hopf configuration, if n is odd) is called the *suspension* of this $(n; k)$-configuration.

It is easy to see that any two lines of a trivial configuration can be transposed by a rigid autoisotopy of the configuration that keeps other lines of the configuration fixed. From this fact it follows that one can find a rigid autoisotopy of this configuration that permutes its lines in an arbitrary way. Hence, up to rigid isotopy, the join of an unordered configuration of k-dimensional subspaces of $\mathbb{R}P^n$ and the trivial configuration does not depend on the orders of the elements of these configurations.

1.6.1. LEMMA. *Let a nonsingular ordered configuration $S = \{S_1, \ldots, S_m\}$ of $(p + q + 1)$-dimensional subspaces of oriented space $\mathbb{R}P^{2(p+q)+3}$ be the join of a nonsingular ordered configuration $T = \{T_1, \ldots, T_m\}$ of p-dimensional subspaces of oriented space $\mathbb{R}P^{2p+1}$ and a nonsingular ordered configuration $R = \{R_1, \ldots, R_m\}$ of q-dimensional subspaces of oriented space $\mathbb{R}P^{2q+1}$. Then*

$$\mathrm{lk}_{\mathbb{R}P^{2(p+q)+3}}(S_i, S_j, S_l) = \mathrm{lk}_{\mathbb{R}P^{2p+1}}(T_i, T_j, T_l)\,\mathrm{lk}_{\mathbb{R}P^{2q+1}}(R_i, R_j, R_l)$$

for any $i, j, l = 1, \ldots, m$, $i < j < l$.

1.6.2. COROLLARY. *Let a nonsingular ordered configuration $S = \{S_1, \ldots, S_m\}$ of $(k + 2)$-dimensional subspaces of oriented space $\mathbb{R}P^{2k+5}$ be the suspension of a nonsingular ordered configuration $T = \{T_1, \ldots, T_m\}$ of k-dimensional subspaces of oriented space $\mathbb{R}P^{2k+1}$. Then $\mathrm{lk}_{\mathbb{R}P^{2k+5}}(S_i, S_j, S_l) = \mathrm{lk}_{\mathbb{R}P^{2k+1}}(T_i, T_j, T_l)$ for any $i, j, l = 1, \ldots, m$, $i < j < l$.*

1.6.3. THEOREM. *If $m \leqslant k + 5$ ($k > 0$), then any 1-singular (nonsingular) $(2k + 5; k + 2)$-configuration of degree m is rigidly isotopic to the suspension of a 1-singular (nonsingular) $(2k + 1; k)$-configuration; if $m \leqslant k + 2$, the suspensions of two $(2k + 1; k)$-configurations are rigidly isotopic iff the original configurations are rigidly isotopic.*

The proof of this theorem for nonsingular configurations was announced in [3]; the proof of the whole theorem will be published separately.

1.7. Proof of the main result. The following lemma is the key result for proving the main theorem.

1.7.1. LEMMA. *Let a 1-singular (nonsingular) $(2k + 5; k + 2)$-configuration $S = \{S_1, \ldots, S_m\}$ be the join of a 1-singular (nonsingular) $(2k + 1; k)$-configuration $T = \{T_1, \ldots, T_m\}$ and a nonsingular $(3; 1)$-configuration $K = \{K_1, \ldots, K_m\}$. Assume also that the subspaces T_i and T_j of T and the lines K_i and K_j of K can be moved up to intersection at a point. Then S is rigidly isotopic to the join of the 1-singular (nonsingular) $(2k + 1; k)$-configuration T' and the nonsingular $(3; 1)$-configuration K' which are obtained from T and K by passing across the subspaces T_i, T_j and K_i, K_j respectively.*

PROOF. By the assumptions of the theorem, one can suppose that the configuration $T = \{T_1, \ldots, T_m\}$ is a 1-singular (nonsingular) configuration of k-dimensional subspaces in a $(2k+1)$-dimensional linear subspace Γ of $\mathbb{R}P^{2k+5}$, that $K = \{K_1, \ldots, K_m\}$ is a nonsingular configuration of lines in a three-dimensional linear subspace Δ of $\mathbb{R}P^{2k+5}$, and that S_i is the projective hull of T_i and K_i for any $i = 1, \ldots, m$.

Since the lines K_i and K_j of K can be moved up to intersection at a point, there is a degeneration $K(t) = \{K_1(t), \ldots, K_m(t)\}$ of K in Δ such that:
a) the restrictions of $K(t)$ to the subconfigurations

$$\{K_1, \ldots, K_{i-1}, K_{i+1}, \ldots, K_m\} \quad \text{and} \quad \{K_1, \ldots, K_{j-1}, K_{j+1}, \ldots, K_m\}$$

of K are rigid isotopies;
b) the lines $K_i(1)$ and $K_j(1)$ intersect at a point.

Denote the subconfiguration $\{K_i(1), K_j(1)\}$ of $K(1)$ by $K^{ij}(1)$.

Let ε and δ be so small that the union of the ε-neighborhoods of $K_i(1)$ and $K_j(1)$ in Δ is a regular neighborhood of $K_i(1) \cup K_j(1)$, and the perturbation

$$[K^{ij}(1)]^\delta(t) = \{K_i(1 - \delta t), K_j(1 - \delta t)\}, \qquad t \in [0, 1],$$

of $K^{ij}(1)$ is an ε-perturbation. Denote the point of intersection of $K_i(1)$ and $K_j(1)$ by A and fix a point B of $K_i(1)$ different from A. Let C be a point of Δ close to A that does not belong to the projective hull of B and $K_j(1)$. Let σ be a segment joining A and C, and let $A(t)$ be a rectilinear isotopy of A such that $A(1) = C$ and $A(t) \in \sigma$ for any $t \in [0, 1]$. In addition we choose C and σ in such a way that:
1) the line passing through B and $A(t)$ does not intersect K_l for any $t \in [0, 1]$ and $l \in \{1, \ldots, m\} \setminus \{i, j\}$;
2) the perturbation

$$[K^{ij}(1)]^\sigma(t) = \{[K_i(1)]^\sigma(t), K_j(1)\}$$

of $K^{ij}(1)$, where $[K_i(1)]^\sigma(t)$ is the line passing through B and $A(t)$ for any $t \in [0, 1]$, is an ε-perturbation;
3) the ε-perturbations $[K^{ij}(1)]^\sigma(t)$ and $[K^{ij}(1)]^\delta(1)$ of $K^{ij}(1)$ belong to the same type (see Lemma 1.2.1).

Let $\bar{K} = \{\bar{K}_1, \ldots, \bar{K}_m\}$ be a nonsingular configuration of lines of Δ for which $\bar{K}_l = K_l(1)$ for any $l = 1, \ldots, m$, $l \neq i$, and $\bar{K}_i$ is the line passing through B and C. By conditions 1)–3), K and $\bar{K}$ are rigidly isotopic.

Since the subspaces T_i and T_j of T can be moved up to intersection at a point, there is a degeneration $T(t) = \{T_1(t), \ldots, T_m(t)\}$ of T in Γ such that:
a) the restrictions of $T(t)$ to the subconfigurations $\{T_1, \ldots, T_{i-1}, T_{i+1}, \ldots, T_m\}$ and $\{T_1, \ldots, T_{j-1}, T_{j+1}, \ldots, T_m\}$ of T are rigid isotopies;
b) the subspaces $T_i(1)$ and $T_j(1)$ intersect at a point.

Denote the point of intersection of $T_i(1)$ and $T_j(1)$ by O.

Let $\bar{S} = \{\bar{S}_1, \ldots, \bar{S}_m\}$ be the configuration of $(k+2)$-dimensional subspaces of $\mathbb{R}P^{2k+5}$ which is the join of $T(1)$ and $\bar{K}$. Let Q_1 be a hyperplane of $\mathbb{R}P^{2k+5}$ such that $\Delta \subset Q_1$ and $O \notin Q_1$. Denote the projection of $\mathbb{R}P^{2k+5}$ onto Q_1 from O by Pr_{O,Q_1}. Let $\bar{S}_l^1 = \mathrm{Pr}_{O,Q_1}(\bar{S}_l)$, where $l = 1, \ldots, m$. It is clear that $\bar{S}_i^1$ and $\bar{S}_j^1$ are $(k+1)$-dimensional subspaces of Q_1, and $\bar{S}_l^1$ is a $(k+2)$-dimensional subspace of Q_1 for any $l \in \{1, \ldots, m\} \setminus \{i, j\}$. Let P be the plane passing through C and $K_j(1)$. Consider a $(2k+3)$-dimensional linear subspace Q_2 of Q_1 such that $P \subset Q_2$ and $B \notin Q_2$. Denote the projection of Q_1 onto Q_2 from B by Pr_{B,Q_2}. Let $\bar{S}_l^2 = \mathrm{Pr}_{B,Q_2}(\bar{S}_l^1)$, where $l = 1, \ldots, m$. It is clear that $\bar{S}_i^2$ is a k-dimensional subspace of Q_2, $\bar{S}_j^2$ is a $(k+1)$-dimensional subspace of Q_2, and $\bar{S}_l^2$ is a $(k+2)$-dimensional subspace of Q_2 for any $l \in \{1, \ldots, m\} \setminus \{i, j\}$.

It is convenient further to distinguish the following two cases.

Case 1: $k = 0$. In this case $\bar{S}_i^2$ is the point C, $\bar{S}_j^2$ is the line $K_j(1)$, and $\bar{S}_l^2$ is a plane of the three-dimensional subspace Q_2 for any $l \in \{1, \ldots, m\} \setminus \{i, j\}$.

Since, by condition 1), the planes $\bar{S}_l^2$, $l = 1, \ldots, m$, $l \neq i$, $l \neq j$, do not intersect the segment σ, there exists an isotopy $C(t)$ of C in the three-dimensional subspace Q_2 such that:
1') $C(t) \cap \bar{S}_l^2 = \varnothing$ for any $t \in [0, 1]$ and $l = 1, \ldots, m$, $l \neq i$;
2') $C(1)$ belongs to the line passing through A and C and does not belong to σ;
3') $C(1)$ is in the ε-neighborhood of A.

Let $K' = \{K'_1, \ldots, K'_m\}$ be the nonsingular configuration of lines of Δ for which $K'_l = K_l(1)$ for any $l = 1, \ldots, m$, $l \neq i$, and K'_i is the line passing through B and $C(1)$. Notice that the nonsingular $(3;1)$-configuration K' is obtained from $\overline{K}$ by passing across the lines $\overline{K}_i$ and $\overline{K}_j$, and therefore is obtained from K by passing across the lines K_i and K_j. By conditions $1')$–$3')$, the configuration $\overline{S}$ is rigidly isotopic to the join of $T(1)$ and K'. Then, by Lemmas 1.2.2 and 1.6.1, $T \vee \overline{K}$ is rigidly isotopic to $T' \vee K'$, where T' is obtained from T by passing across the subspaces T_i and T_j. Since K and $\overline{K}$ are rigidly isotopic, S is rigidly isotopic to $T' \vee K'$.

Case 2: $k > 0$. Let Q_3 be a $(2k+2)$-dimensional linear subspace of Q_2 and O_3 be a point of $\overline{S}_i^2$ such that $P \subset Q_3$ and $O_3 \notin Q_3$. Denote the projection of Q_2 onto Q_3 from O_3 by Pr_{O_3, Q_3}. Let

$$\overline{S}_l^3 = \mathrm{Pr}_{O_3, Q_3}(\overline{S}_l^2), \qquad l = 1, \ldots, m.$$

It is clear that $\overline{S}_i^3$ is a $(k-1)$-dimensional subspace of Q_3 and $\overline{S}_j^3$ is a $(k+1)$-dimensional subspace of Q_3, while $\overline{S}_l^3$ is a $(k+2)$-dimensional subspace of Q_3 for any $l \in \{1, \ldots, m\} \setminus \{i, j\}$. Projecting in this way from points of $\overline{S}_i^2$, finally we come to a $(k+3)$-dimensional linear subspace Q_{k+2} of a $(k+4)$-dimensional linear subspace Q_{k+1} and a point O_{k+2} of a line $\overline{S}_i^{k+1}$ such that $P \subset Q_{k+2}$ and $O_{k+2} \notin Q_{k+2}$. Denote the projection of Q_{k+1} onto Q_{k+2} from O_{k+2} by $\mathrm{Pr}_{O_{k+2}, Q_{k+2}}$. Let

$$\overline{S}_l^{k+2} = \mathrm{Pr}_{O_{k+2}, Q_{k+2}}(\overline{S}_l^{k+1}),$$

where $l = 1, \ldots, m$. It is clear that $\overline{S}_i^{k+2}$ is the point C, $\overline{S}_j^{k+2}$ is a $(k+1)$-dimensional linear subspace of Q_{k+2}, and $\overline{S}_l^{k+2}$ is a $(k+2)$-dimensional linear subspace of Q_{k+2} for any $l \in \{1, \ldots, m\} \setminus \{i, j\}$. Since, due to condition 1), the $(k+2)$-dimensional subspaces $\overline{S}_l^{k+2}$, $l = 1, \ldots, m$, $l \neq i$, $l \neq j$, do not intersect the segment σ, there exists an isotopy $C(t)$ of C in the $(k+3)$-dimensional subspace Q_{k+2} such that:

$1'')$ $C(t) \cap \overline{S}_l^{k+2} = \emptyset$ for any $t \in [0, 1]$ and $l = 1, \ldots, m$, $l \neq i$;
$2'')$ $C(1)$ belongs to the line passing through A and C and does not belong to σ;
$3'')$ $C(1)$ is in the ε-neighborhood of A.

Reasoning the same way as in Case 1, we can further show that S is rigidly isotopic to $T' \vee K'$, where T' and K' are obtained from T and K by passing across the subspaces T_i, T_j and K_i, K_j respectively. □

1.7.2. LEMMA. *Let $K = \{K_1, \ldots, K_{m-1}, K_m\}$ be a trivial configuration of lines of $\mathbb{R}P^3$. Then for any $j = 1, \ldots, m-1$ there exists a rigid isotopy $K_m^j(t)$ of K_m such that*:

1) $K_m^j(t) \cap K_i = \emptyset$ *for any* $t \in [0; 1)$ *and* $i = 1, \ldots, m-1$;
2) $K_m^j(1) \cap K_i = \emptyset$ *for any* $i = 1, \ldots, m-1$, $i \neq j$;
3) $K_m^j(1)$ *and* K_j *intersect at a point.*

PROOF. Trivial. □

1.7.3. LEMMA. *Let $A = \{A_1, \ldots, A_{m-1}, A_m\}$ be a 1-singular (nonsingular) $(2k+1; k)$-configuration, where $m \leq k+2$ and $A_m \cap A_i = \emptyset$ for any $i = 1, \ldots, m-1$. Then for any $j = 1, \ldots, m-1$ there exists a rigid isotopy $A_m^j(t)$ of A_m such that*:

1) $A_m^j(t) \cap A_i = \emptyset$ *for any* $t \in [0, 1)$ *and* $i = 1, \ldots, m-1$;
2) $A_m^j(1) \cap A_i = \emptyset$ *for any* $i = 1, \ldots, m-1$, $i \neq j$;
3) $A_m^j(1)$ *and* A_j *intersect at a point.*

PROOF. Trivial. □

1.7.4. LEMMA. *Let $B = \{B_1, \ldots, B_{m-1}, B_m\}$ be the $(m-1)$-fold suspension of a 1-singular (nonsingular) $(2k+1; k)$-configuration $A = \{A_1, \ldots, A_{m-1}, A_m\}$, where $m \leq k + 2$ and $A_m \cap A_i = \emptyset$ for any $i = 1, \ldots, m-1$. Let*

$$A'(t) = \{A'_1(t), \ldots, A'_{m-1}(t)\}$$

be a rigid isotopy of the subconfiguration $A' = \{A_1, \ldots, A_{m-1}\}$ of A, and

$$B'(t) = \{B'_1(t), \ldots, B'_{m-1}(t)\}$$

be the corresponding rigid isotopy of the subconfiguration $B' = \{B_1, \ldots, B_{m-1}\}$ of B. Then there exists a rigid isotopy

$$B(t) = \{B_1(t), \ldots, B_{m-1}(t), B_m(t)\}$$

of B such that $B_i(1) = B'_i(1)$ for any $i = 1, \ldots, m-1$.

PROOF. It is easy to see that there exist a rigid isotopy $A_m(t)$ of A_m in $\mathbb{R}P^{2k+1}$ and a subdivision $0 = t_0 < t_1 < t_2 < \cdots < t_{q-1} < t_q = 1$ of $[0, 1]$ such that: if $t' \in (t_{p-1}, t_p)$ and $t'' \in (t_p, t_{p+1})$, $p \in \{1, \ldots, q-1\}$, then the configuration $\{A'_1(t''), \ldots, A'_{m-1}(t''), A_m(t'')\}$ is obtained from $\{A'_1(t'), \ldots, A'_{m-1}(t'), A_m(t')\}$ by passing across the subspaces $A_m(t')$ and $A'_{l_p}(t')$, where $l_p \in \{1, \ldots, m-1\}$. By Lemma 1.7.3, one can also assume that in the process of the rigid isotopy $A_m(t)$ the subspace A_m intersects every subspace A_l, $l \in \{1, \ldots, m-1\}$, an even number of times.

By the assumptions of the theorem, $B = A \vee (\bigvee_{s=1}^{m-1} K^s)$, where K^s is a trivial configuration for any $s = 1, \ldots, m-1$. Let $\overline{K}^s = \{K_1^s, \ldots, K_{m-1}^s, \overline{K}_m^s\}$ be a configuration of lines of $\mathbb{R}P^3$ which is obtained from K^s by passing across the lines K_m^s and K_s^s, $s = 1, \ldots, m-1$. By Lemma 1.7.2, there exists a rigid isotopy $K_m^s(t)$ of the line K_m^s in $\mathbb{R}P^3$ such that:
  a) $K_m^s(t) = K_m^s$ for any $t \in [0, t_1/2]$;
  b) if $s \neq l_p$, then

$$K_m^s(t) = K_m^s(t_{p-1} + (t_p - t_{p-1})/2)$$

  for any $t \in [t_{p-1} + (t_p - t_{p-1})/2; t_p + (t_{p+1} - t_p)/2]$;
  c) if $s = l_p$, then for any $t' \in (t_{p-1}, t_p)$ and $t'' \in (t_p, t_{p+1})$ the configuration $\{K_1^s, \ldots, K_{m-1}^s, K_m^s(t'')\}$ is obtained from $\{K_1^s, \ldots, K_{m-1}^s, K_m^s(t')\}$ by passing across $K_m^s(t')$ and $K_{l_p}^s$. And if, in the process of the restriction of $A_m(t)$ to the interval $[0, t_{p+1})$, the subspace A_m intersects A_{l_p} an odd number of times, then $K_m^s(t_p + (t_{p+1} - t_p)/2) = \overline{K}_m^s$, otherwise $K_m^s(t_p + (t_{p+1} - t_p)/2) = K_m^s$;
  d) $K_m^s(t) = K_m^s(t_{q-1} + (t_q - t_{q-1})/2)$ for any $t \in [t_{q-1} + (t_q - t_{q-1})/2, 1]$.

Since in the process of the rigid isotopy $A_m(t)$ the subspace A_m intersects every subspace A_l, $l \in \{1, \ldots, m-1\}$, an even number of times, we have $K_m^s(1) = K_m^s$. By

Lemma 1.7.1, the configurations

$$\left\{ A'_1\left(t_{p-1} + \frac{t_p - t_{p-1}}{2}\right), \ldots, A'_{m-1}\left(t_{p-1} + \frac{t_p - t_{p-1}}{2}\right), A_m\left(t_{p-1} + \frac{t_p - t_{p-1}}{2}\right)\right\}$$

$$\vee \left(\bigvee_{s=1}^{m-1} \left\{K_1^s, \ldots, K_{m-1}^s, K_m^s\left(t_{p-1} + \frac{t_p - t_{p-1}}{2}\right)\right\}\right),$$

$$\left\{A'_1\left(t_p + \frac{t_{p+1} - t_p}{2}\right), \ldots, A'_{m-1}\left(t_p + \frac{t_{p+1} - t_p}{2}\right), A_m\left(t_p + \frac{t_{p+1} - t_p}{2}\right)\right\}$$

$$\vee \left(\bigvee_{s=1}^{m-1} \left\{K_1^s, \ldots, K_{m-1}^s, K_m^s\left(t_p + \frac{t_{p+1} - t_p}{2}\right)\right\}\right)$$

are rigidly isotopic for any $p = 1, \ldots, q - 1$. Therefore

$$B(t) = \{B_1(t), \ldots, B_{m-1}(t), B_m(t)\}$$

$$= \{A'_1(t), \ldots, A'_{m-1}(t), A_m(t)\} \vee \left(\bigvee_{s=1}^{m-1} \{K_1^s, \ldots, K_{m-1}^s, K_m^s(t)\}\right)$$

is a rigid isotopy of B. It is also easy to see that $B_i(t) = B'_i(t)$ for any $i = 1, \ldots, m - 1$. □

1.7.5. LEMMA. *Let $C = \{C_1, \ldots, C_m, C_{m+1}\}$ be the $(m-1)$-fold suspension of a 1-singular (nonsingular) $(2k+1; k)$-configuration $A = \{A_1, \ldots, A_m, A_{m+1}\}$. If A_m and A_{m+1} are homologous subspaces of A, then C_m and C_{m+1} are contiguous subspaces of C.*

PROOF. By Lemma 1.5.3, there exists a rigid isotopy $A_{m+1}(t)$ of A_{m+1} in $\mathbb{R}P^{2k+1}$ and a subdivision $0 = t_0 < t_1 < t_2 < \cdots < t_{q-1} < t_q = 1$ of $[0, 1]$ such that:
1) $A_{m+1}(t) \cap A_m = \varnothing$ for any $t \in [0, 1)$;
2) if $t' \in (t_{p-1}, t_p)$ and $t'' \in (t_p, t_{p+1})$, where $p \in \{1, \ldots, q-1\}$, then the configuration $A'' = \{A_1, \ldots, A_m, A_{m+1}(t'')\}$ is obtained from $A' = \{A_1, \ldots, A_m, A_{m+1}(t')\}$ by passing across the subspaces $A_{m+1}(t')$ and A_{l_p}, where $l_p \in \{1, \ldots, m-1\}$;
3) $A_{m+1}(1) = A_m$;
4) in the process of the rigid isotopy $A_{m+1}(t)$ the subspace A_{m+1} intersects every subspace A_l, $l \in \{1, \ldots, m-1\}$, an even number of times.

By the assumptions of the theorem, $C = A \vee (\bigvee_{s=1}^{m-1} K^s)$, where K^s is a trivial configuration for any $s = 1, \ldots, m-1$. Let $\overline{K}^s = \{K_1^s, \ldots, K_m^s, \overline{K}_{m+1}^s\}$ be a configuration of lines of $\mathbb{R}P^3$ which is obtained from K^s by passing across the lines K_m^s and K_s^s, $s = 1, \ldots, m-1$. By Lemma 1.7.2, there exists a rigid isotopy $K_{m+1}^s(t)$ of the line K_{m+1}^s in $\mathbb{R}P^3$ such that:
a) $K_{m+1}^s(t) = K_{m+1}^s$ for any $t \in [0, t_1/2]$;
b) if $s \neq l_p$, then

$$K_{m+1}^s(t) = K_{m+1}^s(t_{p-1} + (t_p - t_{p-1})/2)$$

for any $t \in [t_{p-1} + (t_p - t_{p-1})/2; t_p + (t_{p+1} - t_p)/2]$;
c) if $s = l_p$, then for any $t' \in (t_{p-1}, t_p)$ and $t'' \in (t_p, t_{p+1})$ the configuration $\{K_1^s, \ldots, K_m^s, K_{m+1}^s(t'')\}$ is obtained from $\{K_1^s, \ldots, K_m^s, K_{m+1}^s(t')\}$ by passing across $K_{m+1}^s(t')$ and $K_{l_p}^s$. And, if in the process of the restriction of $A_{m+1}(t)$ to

the interval $[0, t_{p+1})$, the subspace A_{m+1} intersects A_{l_p} an odd number of times, then $K^s_{m+1}(t_p + (t_{p+1} - t_p)/2) = \overline{K}^s_{m+1}$, otherwise $K^s_{m+1}(t_p + (t_{p+1} - t_p)/2) = K^s_{m+1}$;

d) $K^s_{m+1}(t) = K^s_{m+1}(t_{q-1} + (t_q - t_{q-1})/2)$ for any $t \in [t_{q-1} + (t_q - t_{q-1})/2, 1]$.

Condition 4) implies that $K^s_{m+1}(1) = K^s_{m+1}$. By Lemma 1.7.1, the configurations

$$\left\{ A_1, \ldots, A_m, A_{m+1}\left(t_{p-1} + \frac{t_p - t_{p-1}}{2}\right)\right\}$$
$$\vee \left(\bigvee_{s=1}^{m-1} \left\{ K_1^s, \ldots, K_m^s, K_{m+1}^s\left(t_{p-1} + \frac{t_p - t_{p-1}}{2}\right)\right\}\right),$$
$$\left\{ A_1, \ldots, A_m, A_{m+1}\left(t_p + \frac{t_{p+1} - t_p}{2}\right)\right\}$$
$$\vee \left(\bigvee_{s=1}^{m-1} \left\{ K_1^s, \ldots, K_m^s, K_{m+1}^s\left(t_p + \frac{t_{p+1} - t_p}{2}\right)\right\}\right)$$

are rigidly isotopic for any $p = 1, \ldots, q-1$. Therefore the degeneration

$$C(t) = \{A_1, \ldots, A_m, A_{m+1}(t)\} \vee \left(\bigvee_{s=1}^{m-1} \{K_1^s, \ldots, K_m^s, K_{m+1}^s(t)\}\right)$$

of C satisfies the conditions imposed in the definition of contiguity of the subspaces C_m and C_{m+1}. □

We shall say that two $(n; k)$-configurations of degree m are *stably equivalent* if their s-fold suspensions are rigidly isotopic for some s.

1.7.6. THEOREM. *Two ordered (unordered) 1-singular (nonsingular) $(2k + 1; k)$-configurations of degree m are stably equivalent if and only if they are homology equivalent.*

PROOF. By Corollary 1.6.2, the stable equivalence of two 1-singular (nonsingular) $(2k + 1; k)$-configurations implies their homology equivalence. Now let us prove the converse statement, i.e., that the stable equivalence of two 1-singular (nonsingular) $(2k + 1; k)$-configurations follows from their homology equivalence.

We proceed by induction on the number m of subspaces of the configuration. Suppose that the theorem is true for all 1-singular (nonsingular) $(2k + 1; k)$-configurations of degree $< m$; let us show that in this case it remains true for all 1-singular (nonsingular) $(2k + 1; k)$-configurations of degree m.

Let $A = \{A_1, \ldots, A_{m-1}, A_m\}$ and $\overline{A} = \{\overline{A}_1, \ldots, \overline{A}_{m-1}, \overline{A}_m\}$ be some 1-singular (nonsingular) $(2k + 1; k)$-configurations of degree m that are homology equivalent. Assume, without loss of generality, that $A_m \cap A_i = \emptyset$ for any $i = 1, \ldots, m-1$. Then $A' = \{A_1, \ldots, A_{m-1}\}$ and $\overline{A}' = \{\overline{A}_1, \ldots, \overline{A}_{m-1}\}$ are 1-singular (nonsingular) homology equivalent $(2k + 1; k)$-configurations of degree $m - 1$. By the induction hypothesis, A' and $\overline{A}'$ are stably equivalent. Consider a number s satisfying the following conditions:

1) the s-fold suspensions of A' and $\overline{A}'$ are rigidly isotopic;
2) $m \leqslant k + 2s + 2$.

Consider the $(s + m - 1)$-fold suspensions $B = \{B_1, \ldots, B_{m-1}, B_m\}$ and $\overline{B} = \{\overline{B}_1, \ldots, \overline{B}_{m-1}, \overline{B}_m\}$ of A and $\overline{A}$ respectively. By conditions 1), 2) and Lemma 1.7.4, there exists a rigid isotopy $B(t) = \{B_1(t), \ldots, B_{m-1}(t), B_m(t)\}$ of B such that

$B_i(1) = \overline{B}_i$ for any $i = 1, \ldots, m-1$. By Corollary 1.6.2, B and $\overline{B}$ are homology equivalent. Hence, by Lemma 1.5.2, the subspaces $\overline{B}_m$ and $B_m(1)$ of the configuration $\{\overline{B}_1, \ldots, \overline{B}_{m-1}, \overline{B}_m, B_m(1)\}$ are homologous. Then, by Lemmas 1.7.5 and 1.5.1, the $(m-1)$-fold suspensions of $B(1)$ and $\overline{B}$ are rigidly isotopic. By definition, this means that the initial $(2k+1;k)$-configurations A and $\overline{A}$ are stably equivalent. □

The next corollaries follow from Theorems 1.6.3 and 1.7.6.

1.7.7. COROLLARY. *If $m \leq k+2$, then two 1-singular (nonsingular) $(2k+1;k)$-configurations of degree m are rigidly isotopic if and only if they are homology equivalent.*

1.7.8. COROLLARY. *If $m \leq k+2$, then the adjacency graphs of the spaces of $(2k+1;k)$-configurations of degree m and $(2k+5;k+2)$-configurations of degree m are isomorphic. This isomorphism is induced by the suspension operator.*

§2. Algebraic proof of the main result

Note that a nonsingular real projective $(2k+1;k)$-configuration can be regarded as a collection of pairwise transversal $(k+1)$-dimensional vector subspaces of $\mathbb{R}^{2k+2}$, and a rigid isotopy of the configuration can be regarded as a motion of these subspaces in the process of which they remain pairwise transversal $(k+1)$-dimensional vector subspaces. Therefore, instead of considering the rigid isotopies of nonsingular projective configurations of the middle dimension, we can investigate the rigid isotopies of the corresponding configurations of pairwise transversal vector subspaces.

2.1. DEFINITION. An *ordered (unordered) vector $(2k;k)$-configuration of degree m* is an ordered (respectively unordered) collection of m vector subspaces $\{A_1, \ldots, A_m \subset V\}$ of dimension k in a space $V \cong \mathbb{R}^{2k}$. The number k is called the *dimension* of the configuration. A vector $(2k;k)$-configuration is said to be *nonsingular* if $A_i \cap A_j = \{0\}$ for any $i, j = 1, \ldots, m, i \neq j$. Two nonsingular configurations are called *rigidly isotopic* if they belong to the same connected component of the space of the nonsingular configurations. We denote by $R(k;m)$ the set of all rigid isotopy classes of ordered nonsingular vector $(2k;k)$-configurations of degree m. It is clear that the permutation group S_m acts naturally on $R(k;m)$. In what follows $[A]$ denotes the rigid isotopy class of the configuration A. In the sequel all configurations are assumed ordered.

Let $A = \{A_1, \ldots, A_m \subset V\}$ be a nonsingular vector $(2k;k)$-configuration. Suppose that V and A_i ($i = 1, \ldots, m$) are oriented. Then to any two subspaces A_i and A_j of A ($i, j = 1, \ldots, m, i \neq j$) one can assign their *intersection number* $i(A_i, A_j)$, which is equal to $+1$ if the orientation of V coincides with that of $A_i \oplus A_j$, and equals -1 otherwise. Then the *intersection number* of a triple A_i, A_j, A_l of subspaces of A ($i, j, l = 1, \ldots, m, i \neq j, i \neq l, j \neq l$)

$$i(A_i, A_j, A_l) = i(A_i, A_j) \cdot i(A_j, A_l) \cdot i(A_l, A_i)$$

does not depend on the choice of the orientations of the subspaces of A. Two nonsingular configurations are said to be *homology equivalent* if for any pair of corresponding triples of subspaces of these configurations their intersection numbers coincide.

Thus, if H is a vector space over the field $\mathbb{F}_2$ with basis $a_1, \ldots, a_m$, then the set $R_h(m)$ of all possible homology equivalence classes of nonsingular $(2k;k)$-configurations of degree m coincides with $\wedge^2 H^*/H_0$, where H_0 is the image of H^* under the map $\varphi: H^* \to \wedge^2 H^*$, determined by the following formula:

$$(\varphi l)(a_i, a_j) = l(a_i) - l(a_j) \quad \text{for all } l \in H^*.$$

It is easy to see that the dimension of $R_h(m)$ over $\mathbb{F}_2$ is equal to $(m-1)(m-2)/2$. Since H_0 is invariant under the action of the permutation group S_m, the action of S_m on $R_h(m)$ is also well defined. Therefore the set of all possible homology equivalence classes of unordered nonsingular $(2k; k)$-configurations of degree m coincides with the quotient of $R_h(m)$ under the action of the permutation group S_m.

2.2. DEFINITION. Let $A = \{A_1, \ldots, A_m \subset V\}$ be a $(2k; k)$-configuration and $B = \{B_1, \ldots, B_m \subset W\}$ be a $(2l; l)$-configuration. Denote by $A \oplus B$ the $(2(k+l); k+l)$-configuration $\{A_1 \oplus B_1, \ldots, A_m \oplus B_m \subset V \oplus W\}$. Note that if k or l is even, then $A \oplus B$ is rigidly isotopic to $B \oplus A$, and if A and B are nonsingular, then $A \oplus B$ is also nonsingular and

$$i(A_i \oplus B_i, A_j \oplus B_j) = i(A_i, A_j) \cdot i(B_i, B_j).$$

2.3. DEFINITION. A nonsingular configuration $A = \{A_1, \ldots, A_m \subset V\}$ is called *trivial* if $\dim A_i = (\dim V)/2$ is an even integer and there exists a quadratic form $q \in S^2 V^*$ such that its restriction to any A_i is equal to zero, i.e., A_i is the linear generatrix of the quadric $q(v) = 0$ for any $i = 1, \ldots, m$.

It is well known that a quadric $q(v) = 0$ in a $2k$-dimensional vector space V has k-dimensional linear generatrices iff the signature of the quadratic form $q \in S^2 V^*$ is equal to $(k; k)$. If, in addition, k is even, then there are two connected families of k-dimensional linear generatrices of the quadric. For a fixed orientation of V, the intersection number of any three pairwise transversal subspaces from one family equals $+1$, and the intersection number of any three pairwise transversal subspaces from the other family equals -1. Denote the rigid isotopy class of a configuration of m pairwise transversal subspaces from the first family by $E_+(k; m)$ and the rigid isotopy class of a configuration of m pairwise transversal subspaces from the second family by $E_-(k; m)$.

A configuration rigidly isotopic to a trivial one is also called *trivial*.

2.4. DEFINITION. Let $A = \{A_1, \ldots, A_m \subset V\}$ be a $(2k; k)$-configuration. By A^* we shall denote the configuration $\{\text{Ann}(A_1), \ldots, \text{Ann}(A_m) \subset V^*\}$ of the dual vector space. It also is a $(2k; k)$-configuration of degree m.

2.5. LEMMA. *If A is nonsingular, then A^* is also nonsingular.*

PROOF. Trivial. □

2.6. PROPOSITION. *If A is a nonsingular vector $(2k; k)$-configuration of degree m, then $[A \oplus A^*] \in E_+(2k; m)$.*

PROOF. Consider the bilinear form q_0 on $V \oplus V^*$ defined as follows:

$$q_0((v_1, w_1), (v_2, w_2)) = v_1 \cdot w_2 + v_2 \cdot w_1 \quad \text{for all } v_1, v_2 \in V \text{ and all } w_1, w_2 \in V^*.$$

Then the restriction of the corresponding quadratic form to every subspace of the space $A \oplus A^*$ is obviously equal to zero. Since $i(A_i, A_j) = i(\text{Ann}(A_i), \text{Ann}(A_j))$, $i(A_i \oplus \text{Ann}(A_i), A_j \oplus \text{Ann}(A_j)) = (i(A_i, A_j))^2 = +1$. Therefore $[A \oplus A^*] \in E_+(2k; m)$, and the statement follows. □

2.7. DEFINITION. Let A be a $(2k; k)$-configuration of degree m and B be a $(2l; l)$-configuration of degree m. The configurations A and B are said to be *stably equivalent* if there exists a positive integer s such that the rigid isotopy types $[A] \oplus E_+(s - k; m)$ and $[B] \oplus E_+(s - l; m)$ coincide. This relation is an equivalence on $\bigcup_{k=1}^{\infty} R(k; m)$.

2.8. PROPOSITION. *Let A and B be some configurations, C a nonsingular configuration, and let $A \oplus C$ be rigidly isotopic to $B \oplus C$. Then A is stably equivalent to B.*

PROOF. From the assumptions it follows that $A \oplus C \oplus C^*$ is rigidly isotopic to $B \oplus C \oplus C^*$. But, by Proposition 2.6, $C \oplus C^*$ is a trivial configuration. $\square$

By $R(m)$ we shall denote the set of the stable equivalence classes of elements of $\bigcup_{k=1}^{\infty} R(k; m)$. It is easy to see that the subset of the stable equivalence classes of even-dimensional configurations is closed under addition. This subset is denoted by $R^e(m)$.

The next theorem follows from the properties of stable equivalence proved above.

2.9. THEOREM. *The set $R^e(m)$ is an Abelian group with respect to the operation $\oplus$.*

Let V be an oriented k-dimensional vector space and φ be a bilinear form on V. It is obvious that the graph of φ regarded as a homomorphism from V into V^* is a k-dimensional vector subspace in $V \oplus V^*$. This k-dimensional vector subspace with the orientation induced by the orientation of V with the help of the projection $V \oplus V^* \to V$ is denoted by A_φ.

2.10. LEMMA. a) *Consider the bilinear form q_0 on $V \oplus V^*$ that was defined in Proposition 2.6. The subspace A_φ belongs to the quadric $q_0((v, w), (v, w)) = 0$ if and only if the bilinear form φ is antisymmetric.*

b) *Let φ and ψ be bilinear forms on V. Then the subspaces A_φ and A_ψ have a nonzero intersection if and only if the difference $\varphi - \psi$ is degenerate.*

c) *A k-dimensional subspace A of $V \oplus V^*$ can be represented as A_φ for some bilinear form φ if and only if $A \cap V^* = \{0\}$.*

d) *Let φ, ψ, ξ be bilinear forms on V for which the differences $\psi - \varphi$, $\xi - \psi$, and $\varphi - \xi$ are nondegenerate. Then $i(A_\varphi, A_\psi) = \operatorname{sign} \det(\psi - \varphi)$ and $i(A_\varphi, A_\psi, A_\xi) = \operatorname{sign}(\det(\psi - \varphi) \det(\xi - \psi) \det(\varphi - \xi))$.*

e) $i(A_\varphi, V^*) = +1$.

PROOF. a) A point $(v, w) \in V \oplus V^*$ belongs to the quadric $q_0((v, w), (v, w)) = 0$ if and only if $v \cdot w = 0$. Therefore, A_φ belongs to the quadric if and only if $v \cdot \varphi(v) = \varphi(v, v) = 0$ for any $v \in V$.

b) Let $(v, w) \in A_\varphi \cap A_\psi$. Then

$$(v, w) = (v, \varphi(v)) = (v, \psi(v)), \quad \text{or} \quad (\varphi - \psi)(v) = 0 \in V^*.$$

This means that the bilinear form $\varphi - \psi$ is degenerate.

c) By the assumptions, A projects onto V isomorphically. Therefore $\varphi \colon V \to V^*$ can be given by $\varphi(v) = p_2(p_1^{-1}(v))$, where p_1 and p_2 are the projections of A onto V and V^* respectively.

d) Choose some basis $e = \{e_1, \ldots, e_k\}$ in V. Let $e^* = \{e_1^*, \ldots, e_k^*\}$ be the dual basis in V^*. Denote by M_φ and M_ψ the matrices of the maps $\varphi \colon V \to V^*$ and $\psi \colon V \to V^*$ in these bases. Then the vectors $\{(e_i, \varphi(e_i)), (e_i, \psi(e_i))\}$ form a basis in

$V \oplus V^*$, its matrix in the basis $e \oplus e^*$ being the following

$$M = \begin{pmatrix} E_k & M_\varphi \\ E_k & M_\psi \end{pmatrix},$$

where E_k is the unit matrix of size $k \times k$. It is easy to see that $i(A_\varphi, A_\psi)$ coincides with the sign of the determinant of M, which is equal to $\det(\psi - \varphi)$.

e) The proof is analogous to the proof of d). □

2.11. COROLLARY. *Let φ be a nondegenerate antisymmetric bilinear form on a $2k$-dimensional vector space W, $\lambda_1, \ldots, \lambda_{m-1}$ be some pairwise different real numbers, and $A = \{A_{\lambda_1\varphi}, A_{\lambda_2\varphi}, \ldots, A_{\lambda_{m-1}\varphi}, W^* \subset W \oplus W^*\}$. Then $[A] \in E_+(2k; m)$.*

Lemma 2.10 allows us to reduce all the results obtained before to matrix form.

2.12. DEFINITION. An *ordered nonsingular matrix configuration of type $(k; m)$* is an ordered collection of m matrices $\{B_1, \ldots, B_m \in M_k\}$ such that $\det(B_i - B_j) \neq 0$ for any $i, j = 1, \ldots, m$, $i \neq j$. Two nonsingular matrix configurations are said to be *rigidly isotopic* if they belong to the same connected component of the space of nonsingular configurations. Two configurations $B = \{B_1, \ldots, B_m \in M_k\}$ and $B' = \{B'_1, \ldots, B'_m \in M_k\}$ are called *homology equivalent* if

$$\operatorname{sign} \det(B_i - B_j) = \operatorname{sign} \det(B'_i - B'_j)$$

for any $i, j = 1, \ldots, m$, $i \neq j$. The sum $B \oplus B''$ of B and $B'' = \{B''_1, \ldots, B''_m \in M_l\}$ is the configuration $\{B_1 \oplus B''_1, \ldots, B_m \oplus B''_m \in M_{k+l}\}$, where $B_i \oplus B''_i$ is the block sum of matrices B_i and B''_i for any $i = 1, \ldots, m$.

2.13. THEOREM. *The set of rigid isotopy classes of nonsingular matrix configurations of type $(k; m)$ is in one-to-one correspondence with $R(k; m+1)$. Besides, the sum of the matrix configurations corresponds to the sum of vector configurations, and $\{B_1^t, \ldots, B_m^t \in M_k\}$, where B_i^t is the transpose of B_i, is dual to $\{B_1, \ldots, B_m \in M_k\}$. Two nonsingular vector $(2k; k)$-configurations are homology equivalent if and only if the corresponding matrix configurations are homology equivalent.*

PROOF. Any nonsingular vector $(2k; k)$-configuration of $V \oplus V^*$ can be reduced by a linear transformation which preserves the orientation of $V \oplus V^*$ to a form such that the last subspace of the configuration coincides with V^*. Then all the other subspaces of the configuration can be represented as A_{φ_i} for some bilinear forms φ_i, $i = 1, \ldots, m$. □

2.14. THEOREM. *Let $B \in M_k$, where k is an even integer, $\det(B) > 0$, and let $\lambda_1, \ldots, \lambda_m$ be pairwise different real numbers. Then $\{\lambda_1 B, \ldots, \lambda_m B \in M_k\}$ corresponds to a nonsingular vector $(2k; k)$-configuration that belongs to $E_+(k; m)$.*

PROOF. Note that if A is a nondegenerate antisymmetric matrix, then $\det(A) > 0$. Let $C = AB^{-1}$. Then the configuration $\{\lambda_1 CB, \ldots, \lambda_m CB\}$ is rigidly isotopic to the original one, and the statement follows from Corollary 2.11. □

The following theorem is the crucial result of this section.

2.15. Theorem. *If $m \leq k$, then:*

a) *two nonsingular vector $(2k; k)$-configurations of degree m are rigidly isotopic if and only if they are homology equivalent;*

b) *there exists a nonsingular vector $(2k; k)$-configuration of degree m realizing any given collection of intersection numbers.*

Proof. First, let us replace all the nonsingular vector $(2k; k)$-configurations of degree m by the corresponding nonsingular matrix configurations of type $(k; m-1)$. In what follows we consider only matrix configurations.

Two 1-dimensional vector subspaces in general position divide $\mathbb{R}^2$ into 2^2 parts, three 2-dimensional vector subspaces in general position divide $\mathbb{R}^3$ into 2^3 parts. The following obvious lemma is a generalization of this fact.

2.16. Lemma. *Let $V \cong \mathbb{R}^k$ and $w_1, \ldots, w_s$ be some linearly independent elements of V^*. Set $V' = \{l \in V \mid \forall i = 1, \ldots, s : w_i(l) \neq 0\}$. Then the set of connected components of V' is in one-to-one correspondence with the set of collections $(\varepsilon_1, \ldots, \varepsilon_s)$, where $\varepsilon_i = \pm 1$.*

2.17. Lemma. *Let $V \cong \mathbb{R}^k$ and s be a positive integer. Denote*

$$M = \{(w_1, \ldots, w_s) \in (V^*)^s :$$
$$w_1, \ldots, w_s \text{ are linearly independent elements of } V^*\},$$
$$N = \{(l, w_1, \ldots, w_s) \in V \times M : w_i(l) \neq 0 \text{ for any } i = 1, \ldots, s\}.$$

Then the projection $\pi \colon N \to M$ satisfies the path lifting property.

Proof. Let $\{(w_1(t), \ldots, w_s(t))\}$, $t \in [0, 1]$, be a path in M, and l_0 be an element of V for which $w_i(0) \cdot l_0 \neq 0$ for any $i = 1, \ldots, s$. We must find a path $l(t)$ in V such that:

1) $l(0) = l_0$;
2) $w_i(t)l(t) \neq 0$ for any $t \in [0, 1]$ and $i = 1, \ldots, s$.

Such a path can be constructed by the following way. Let $W(t) = \{l \in V : w_i(t)l = w_i(0)l_0 \text{ for any } i = 1, \ldots, s\}$. It is easy to see that $W(t)$ is a $(k-s)$-dimensional affine subspace of $\mathbb{R}^k$. For example, we can define $l(t)$ as the nearest point to l_0 from W. □

Let $k > 1$, $V \cong \mathbb{R}^k$, $q \colon V^{k-1} \to V^*$, be the map determined by the formula

$$q(v_1, \ldots, v_{k-1}) = v_1 \wedge \cdots \wedge v_{k-1}.$$

2.18. Lemma. *For any $w \in V^*$*

$$\operatorname{codim} q^{-1}(w) = \begin{cases} k & \text{if } w \neq 0, \\ 2 & \text{if } w = 0. \end{cases}$$

Proof. Trivial. □

2.19. Lemma. *If X is a linear codimension > 1 subspace of V^*, then*

$$\operatorname{codim} q^{-1}(X) = 2.$$

PROOF. Represent the algebraic variety $q^{-1}(X)$ as the union $q^{-1}(X) = Y_0 \cup Y_1$, where $Y_0 = q^{-1}(0)$ and $Y_1 = q^{-1}(X \setminus \{0\})$. By Lemma 2.18, codim $Y_0 = 2$. Consider the map $q: Y_1 \to X \setminus \{0\}$. Any fiber of this map has the same dimension $(\dim V^{k-1}) - k$. Therefore

$$\dim Y_1 = (\dim V^{k-1}) - k + \dim(X \setminus \{0\}) \leqslant \dim V^{k-1} - 2. \qquad \square$$

Let $1 < m < k$, $V \cong \mathbb{R}^k$, $M \cong V^{(m-1)(k-1)}$. Any element of M can be represented as a matrix

$$b = \begin{pmatrix} v_{11} & \cdots & v_{m-1,1} \\ \vdots & \ddots & \vdots \\ v_{1,k-1} & \cdots & v_{m-1,k-1} \end{pmatrix},$$

where $v_{ij} \in V$ for any $i = 1, \ldots, m-1$ and $j = 1, \ldots, k-1$. Denote by $q_i(b)$ the vector $v_{i1} \wedge \cdots \wedge v_{i,k-1} \in \wedge^{k-1} V \cong V^*$. Set

$$Y = \{b \in M : q_1(b), \ldots, q_{m-1}(b) \text{ are linearly dependent}\}.$$

2.20. LEMMA. $\operatorname{codim}_M(Y) > 1$.

PROOF. Consider the projection π of M into $V^{(m-2)(k-1)}$ that for any matrix $b \in M$ forgets its first column. It is easy to see that if the intersection of the algebraic variety Y with any fiber of π is a codimension > 1 subset in the fiber, then $\operatorname{codim}_M(Y) > 1$.

Suppose $b \in Y$, i.e., $q_1(b), \ldots, q_{m-1}(b)$ are linearly dependent. Without loss of generality, one can assume that $q_1(b)$ is linearly expressed in $q_2(b), \ldots, q_{m-1}(b)$. Then $q_1(b)$ belongs to the subspace X of dimension $\leqslant (m-2)$ generated in V^* by $q_2(b), \ldots, q_{m-1}(b)$. Since $m - 2 < k - 1$, then, due to Lemma 2.19, the set of the vectors

$$\begin{pmatrix} v_{11} \\ \vdots \\ v_{1,k-1} \end{pmatrix} \text{ such that } q(v_{11}, \ldots, v_{1,k-1}) \in X$$

has codimension > 1. Therefore the intersection of Y with any fiber of π is a codimension > 1 subset in the fiber, and the statement follows. $\square$

Represent any matrix

$$B = \begin{pmatrix} b_{11} & \cdots & b_{1k} \\ \vdots & \ddots & \vdots \\ b_{k1} & \cdots & b_{kk} \end{pmatrix}$$

as a collection of k vectors

$$v_1 = (b_{11}, \ldots, b_{1k}), \ \ldots, \ v_k = (b_{k1}, \ldots, b_{kk}), \quad \text{where } v_l \in V \cong \mathbb{R}^k,\ l = 1, \ldots, k.$$

We suppose that V is oriented, i.e., we choose an isomorphism $\wedge^k V \cong \mathbb{R}$ or, what is equivalent, an isomorphism $\wedge^{k-1} V \cong V^*$. Then $\det(B) = v_1 \wedge \cdots \wedge v_k$. Taking into account this agreement, we shall represent any configuration $B = \{B_1, \ldots, B_m \in M_k\}$ as the collection of vectors

$$\begin{pmatrix} v_{11} & \cdots & v_{m1} \\ \vdots & \ddots & \vdots \\ v_{1k} & \cdots & v_{mk} \end{pmatrix}.$$

Then the conditions $\det(B_i - B_j) \neq 0$ for the configuration B to be nonsingular can be written as
$$(v_{i1} - v_{j1}) \wedge \cdots \wedge (v_{ik} - v_{jk}) \neq 0.$$
Without loss of generality, we can suppose that
$$\sum_{i=1}^m B_i = 0 \quad \text{or} \quad \sum_{i=1}^m v_{il} = 0 \quad \text{for any fixed } l = 1, \ldots, k.$$

Let us choose the vectors v_{il}, $i = 1, \ldots, m$, $l = 1, \ldots, k-1$, and investigate the condition for the configuration B to be nonsingular. Denote by Q_{ij} the product
$$(v_{i1} - v_{j1}) \wedge \cdots \wedge (v_{i,k-1} - v_{j,k-1}) \in \wedge^{k-1} V \cong V^*, \qquad i, j = 1, \ldots, m.$$
Note that $Q_{ji} = (-1)^{k-1} Q_{ij}$. Now the condition under consideration can be written as
$$Q_{ij} \wedge (v_{ik} - v_{jk}) \neq 0.$$
Denote by W the space V^m. Let
$$W' = \{w = (v_1, \ldots, v_m) \in W \mid v_1 + \cdots + v_m = 0\}.$$
Define linear forms $P_{ij} \in W^*$ as
$$P_{ij} = (0, \ldots, Q_{ij}, \ldots, -Q_{ij}, \ldots, 0),$$
where Q_{ij} is at the ith position, and $-Q_{ij}$ is at the jth position. It is easy to check that $P_{ij} \in (W')^*$. Now the condition for B to be nonsingular, expressed in terms of $w = (v_{1k}, \ldots, v_{mk}) \in W$, is the following: $P_{ij}(w) \neq 0$ for any i, j, where $1 \leq i < j \leq m$. By Lemma 2.16, if all the $m(m-1)/2$ vectors P_{ij} of W^* ($1 \leq i < j \leq m$) are linearly independent, then the conditions $P_{ij}(w) \neq 0$ divide W exactly into $2^{m(m-1)/2}$ connected components, the components being in one-to-one correspondence with the set of collections $\{\varepsilon_{ij} \mid 1 \leq i < j \leq m\}$ of the intersection numbers, i.e., in one-to-one correspondence with the set of the homology equivalence classes.

2.21. DEFINITION. A collection of vectors
$$\{v_{il} \in V \mid i = 1, \ldots, m, \ l = 1, \ldots, k-1\}$$
is called *general* if the corresponding vectors P_{ij} ($1 \leq i < j \leq m$) are linearly independent, and is called *special* otherwise.

Fix the following notation.

Let $U = V^{(k-1)m}$ be the set of all the collections of vectors
$$\begin{pmatrix} v_{11} & \cdots & v_{m1} \\ \vdots & \ddots & \vdots \\ v_{1,k-1} & \cdots & v_{m,k-1} \end{pmatrix},$$
where $v_{ij} \in V$ for any $i = 1, \ldots, m$, $j = 1, \ldots, k-1$, and let U' be the subset of all the general vectors from U. For a given matrix configuration
$$B = \{B_1, \ldots, B_m\} = \begin{pmatrix} v_{11} & \cdots & v_{m1} \\ \vdots & \ddots & \vdots \\ v_{1k} & \cdots & v_{mk} \end{pmatrix},$$

where $B_i \in M_k$, $v_{ij} \in V$ for any $i = 1,\ldots,m$, $j = 1,\ldots,k$, denote by $u(B)$ the following element from U:

$$\begin{pmatrix} v_{11} & \cdots & v_{m1} \\ \vdots & \ddots & \vdots \\ v_{1,k-1} & \cdots & v_{m,k-1} \end{pmatrix},$$

i.e., the operator u forgets the lower line of the collection above.

2.22. LEMMA. *If the special collections of vectors*

$$\{v_{il} \in V \mid i = 1,\ldots,m,\ l = 1,\ldots,k-1\}$$

form a codimension $\geqslant 2$ subset in U for some k and m, then
  a) *two nonsingular matrix configurations of type $(k;m)$ are rigidly isotopic if and only if they are homology equivalent;*
  b) *there exists a nonsingular matrix configuration of type $(k;m)$ realizing any given collection of intersection numbers.*

PROOF. a) Let B' and B'' be some nonsingular homology equivalent matrix configurations for which $u(B')$ and $u(B'')$ belong to U'. Since $\operatorname{codim}_U(U \setminus U') > 1$, $u(B')$ and $u(B'')$ can be joined by a path in U', and by Lemma 2.17 this path can be lifted to a path which joins B' and B'' in the space of all the nonsingular matrix configurations.

b) This immediately follows from Lemma 2.16. □

Now the statement of Theorem 2.15 follows from the next lemma.

2.23. LEMMA. *If $m < k$, then the special collections of vectors*

$$\{v_{il} \in V \mid i = 1,\ldots,m,\ l = 1,\ldots,k-1\}$$

form a codimension $\geqslant 2$ subset in U.

PROOF. Suppose that the vectors

$$P_{12} = (Q_{12}, -Q_{12}, 0, 0, \ldots, 0),$$
$$P_{13} = (Q_{13}, 0, -Q_{13}, 0, \ldots, 0),$$
$$\ldots\ldots\ldots\ldots\ldots\ldots\ldots\ldots$$
$$P_{m-1,m} = (0, 0, \ldots, 0, Q_{m-1,m}, -Q_{m-1,m})$$

are linearly dependent for some $\{v_{il}\}$. Then for some $r \in \{1,\ldots,m\}$ the vectors $Q_{r1},\ldots,Q_{rm}$ are linearly dependent. But, by Lemma 2.20, for $k > m$ the set of vectors

$$\{v_{il} : i = 1,\ldots,r-1, r+1,\ldots,m;\ l = 1,\ldots,k-1\},$$

for which $Q_{r1},\ldots,Q_{rm}$ are linearly dependent, is a codimension $\geqslant 2$ subset in $V^{(m-1)(k-1)}$. Therefore the special collections of vectors

$$\{v_{il} \in V \mid i = 1,\ldots,m,\ l = 1,\ldots,k-1\}$$

form a codimension $\geqslant 2$ subset in U. □

Theorem 2.15 is proved. □

The condition $m < k$ of Lemma 2.23 is obviously too rough. There are grounds to believe that Lemma 2.23 remains valid if the number of vectors P_{ij} (which equals $m(m-1)/2$) is less than the dimension of the ambient space $(W')^*$ (which equals $k(m-1)$).

2.24. CONJECTURE. *In Theorem 2.15, the inequality $m \leqslant k$ can be replaced by $m < 2k$.*

As a consequence of Theorem 2.15, we obtain the following theorems.

2.25. THEOREM. *Homology equivalence of configurations coincides with stable equivalence.*

2.26. THEOREM. *The group $R^e(m)$ is isomorphic to $\mathbb{Z}^{(m-1)(m-2)/2}$.*

In this section we considered only ordered configurations. It is easy to see that the set of unordered configurations up to stable equivalence coincides with the set of the orbits of the natural action of S_m on $R^e(m)$.

§3. Configurations of lines of $\mathbb{R}P^3$

Recall that a real projective $(3; 1)$-configuration of degree m can be regarded as a vector $(4; 2)$-configuration of degree m.

3.1. THEOREM. *A nonsingular homology trivial vector $(4; 2)$-configuration of degree m is trivial and belongs to the rigid isotopy class $E_+(2; m)$.*

PROOF. Let a matrix configuration $A = \{A_1, \ldots, A_m \in M_2\}$ be homology trivial, i.e., $\det(A_i - A_j) > 0$ for any $1 \leqslant i < j \leqslant m$. Consider the basis in M_2 consisting of the following matrices:

$$e_1 = \begin{pmatrix} 1 & 0 \\ 0 & 1 \end{pmatrix}, \quad e_2 = \begin{pmatrix} 0 & -1 \\ 1 & 0 \end{pmatrix}, \quad e_3 = \begin{pmatrix} 1 & 0 \\ 0 & -1 \end{pmatrix}, \quad e_4 = \begin{pmatrix} 0 & 1 \\ 1 & 0 \end{pmatrix}.$$

Then $\det(\alpha e_1 + \beta e_2 + \gamma e_3 + \delta e_4) = \alpha^2 + \beta^2 - \gamma^2 - \delta^2$. Let $A_i = \alpha_i e_1 + \beta_i e_2 + \gamma_i e_3 + \delta_i e_4$. Since A is homology trivial,

$$(\alpha_i - \alpha_j)^2 + (\beta_i - \beta_j)^2 > (\gamma_i - \gamma_j)^2 + (\delta_i - \delta_j)^2$$

for any i, j. Consider the rigid isotopy $A(t)$ consisting of the matrices $A_i(t) = \alpha_i e_1 + \beta_i e_2 + t(\gamma_i e_3 + \delta_i e_4)$. This rigid isotopy joins the initial configuration A with the configuration $A(0) = \{\alpha_i e_1 + \beta_i e_2\}$. Without loss of generality, we can suppose that $\beta_i \neq \beta_j$ for any $i, j = 1, \ldots, m$, $i \neq j$. Therefore, the rigid isotopy $B(t) = \{t\alpha_i e_1 + \beta_i e_2\}$ moves the configuration $\{\alpha_i e_1 + \beta_i e_2\}$ to $B(0) = \{\beta_i e_2\}$, which is obviously trivial. $\square$

In [7] Viro proved that all the nonsingular configurations of $\leqslant 5$ lines of $\mathbb{R}P^3$ are isotopy joins and are determined up to rigid isotopy by the linking numbers of the lines of a configuration. In [2, 5] it was shown that there exist nonsingular configurations of six lines of $\mathbb{R}P^3$ which are neither isotopy joins, nor are determined up to rigid isotopy by their linking numbers. On the other hand, in these articles it was proved that nonsingular isotopy join configurations of six lines are determined by their linking numbers. Komarov and Mazurovskiĭ showed that nonsingular isotopy join $(3; 1)$-configurations of degree $\leqslant 8$ are determined by their linking numbers too (see [4]). Recently Penne proved that nonsingular isotopy join $(3; 1)$-configurations with the same linking numbers are flexibly isotopic (see [6]). We prove a stronger result.

3.2. THEOREM. *Two ordered (unordered) 1-singular (nonsingular) isotopy join (3; 1)-configurations are rigidly isotopic if and only if they are homology equivalent.*

We shall prove this theorem only for nonsingular configurations. The proof for 1-singular configurations is similar.

First of all, we consider the corresponding matrix configurations.

3.3. LEMMA. *The nonsingular matrix configurations*

$$\left\{ \begin{pmatrix} c_1 & 0 \\ 0 & d_1 \end{pmatrix}, \ldots, \begin{pmatrix} c_m & 0 \\ 0 & d_m \end{pmatrix} \right\} \text{ and } \left\{ \begin{pmatrix} 0 & c_1 \\ -d_1 & 0 \end{pmatrix}, \ldots, \begin{pmatrix} 0 & c_m \\ -d_m & 0 \end{pmatrix} \right\}$$

are rigidly isotopic.

PROOF. These configurations can be joined by the rigid isotopy

$$\left\{ (1-t) \cdot \begin{pmatrix} c_i & 0 \\ 0 & d_i \end{pmatrix} + t \cdot \begin{pmatrix} 0 & c_i \\ -d_i & 0 \end{pmatrix} \right\}. \quad \square$$

3.4. LEMMA. *Let two nonsingular matrix configurations*

$$\left\{ \begin{pmatrix} a_1 & 0 \\ 0 & b_1 \end{pmatrix}, \ldots, \begin{pmatrix} a_m & 0 \\ 0 & b_m \end{pmatrix} \right\} \text{ and } \left\{ \begin{pmatrix} c_1 & 0 \\ 0 & d_1 \end{pmatrix}, \ldots, \begin{pmatrix} c_m & 0 \\ 0 & d_m \end{pmatrix} \right\}$$

be homology equivalent. Then these configurations are rigidly isotopic.

PROOF. By Lemma 3.3, the configuration

$$\left\{ \begin{pmatrix} c_1 & 0 \\ 0 & d_1 \end{pmatrix}, \ldots, \begin{pmatrix} c_m & 0 \\ 0 & d_m \end{pmatrix} \right\}$$

is rigidly isotopic to

$$\left\{ \begin{pmatrix} 0 & c_1 \\ -d_1 & 0 \end{pmatrix}, \ldots, \begin{pmatrix} 0 & c_m \\ -d_m & 0 \end{pmatrix} \right\}.$$

But the configurations

$$\left\{ \begin{pmatrix} a_1 & 0 \\ 0 & b_1 \end{pmatrix}, \ldots, \begin{pmatrix} a_m & 0 \\ 0 & b_m \end{pmatrix} \right\} \text{ and } \left\{ \begin{pmatrix} 0 & c_1 \\ -d_1 & 0 \end{pmatrix}, \ldots, \begin{pmatrix} 0 & c_m \\ -d_m & 0 \end{pmatrix} \right\}$$

can be joined by the rigid isotopy

$$\left\{ (1-t) \cdot \begin{pmatrix} a_i & 0 \\ 0 & b_i \end{pmatrix} + t \cdot \begin{pmatrix} 0 & c_i \\ -d_i & 0 \end{pmatrix} \right\}. \quad \square$$

PROOF OF THEOREM 3.2. Let $A = \{A_1, \ldots, A_m\}$ and $C = \{C_1, \ldots, C_m\}$ be two nonsingular join (3; 1)-configurations that are homology equivalent. Then there exist lines L^1, L^2 and K^1, K^2 in $\mathbb{R}P^3$ such that

$$L^1 \cap L^2 = \varnothing, \quad K^1 \cap K^2 = \varnothing, \quad A_i \cap L^j \neq \varnothing, \quad C_i \cap K^j \neq \varnothing$$

for any $i = 1, \ldots, m$, $j = 1, 2$. It is easy to see that there exists a projective transformation of $\mathbb{R}P^3$ which preserves the orientation of $\mathbb{R}P^3$ and puts L^j onto K^j ($j = 1, 2$) and A_m onto C_m. Let $A' = \{A'_1, \ldots, A'_m\}$ be the image of A under this projective transformation. By Theorem 2.13, A' and C correspond to the homology equivalent matrix configurations

$$\left\{ \begin{pmatrix} a_1 & 0 \\ 0 & b_1 \end{pmatrix}, \ldots, \begin{pmatrix} a_{m-1} & 0 \\ 0 & b_{m-1} \end{pmatrix} \right\} \text{ and } \left\{ \begin{pmatrix} c_1 & 0 \\ 0 & d_1 \end{pmatrix}, \ldots, \begin{pmatrix} c_{m-1} & 0 \\ 0 & d_{m-1} \end{pmatrix} \right\}$$

which are rigidly isotopic due to Lemma 3.4. Therefore, by Theorem 2.13, A and C are also rigidly isotopic. □

Acknowledgements

The authors are deeply grateful to Professors O. Ya. Viro and A. I. Degtyarev for their attention to this work.

References

1. S. M. Finashin, *Configurations of seven points in* $\mathbb{R}P^3$, Topology and Geometry—Rokhlin seminar (O. Ya. Viro, ed.), Lecture Notes in Math., vol. 1346, Springer-Verlag, Berlin and Heidelberg, 1988, pp. 501–526.
2. V. F. Mazurovskiĭ, *Configurations of six skew lines*, Zap. Nauchn. Sem. Leningrad. Otdel. Mat. Inst. Steklov. (LOMI) **167** (1988), 121–134; English transl., J. Soviet Math. **52** (1990), no. 1, 2825–2832.
3. _____, *Non-singular configurations of k-dimensional subspaces of $(2k+1)$-dimensional real projective space*, Vestnik Leningrad. Univ. Mat. Mekh. Astronom. **1990**, no. 3, 21–26; English transl., Vestnik Leningrad Univ. Math. **1990**, no. 3, 26–32.
4. _____, *Rigid isotopies of the real projective configurations*, Ph. D. Thesis, Leningrad Univ., 1990. (Russian)
5. _____, *Configurations of at most six lines of* $\mathbb{R}P^3$, Real Algebraic Geometry, Proceedings of the conference held in Rennes, France, June 24–28, 1991, Lecture Notes in Math. (M. Coste, L. Mahe, and M.-F. Roy, eds.), vol. 1524, Springer-Verlag, Berlin and Heidelberg, 1992, pp. 354–371.
6. R. Penne, *Moves on pseudoline diagrams*, Research Report 91-42 (1991), Universiteit Instelling Antwerp.
7. O. Ya. Viro, *Topological problems concerning lines and points of three-dimensional space*, Dokl. Akad. Nauk SSSR **284** (1985), no. 5, 1049–1052; English transl., Soviet Math. Dokl. **32** (1985), no. 2, 528–531.
8. O. Ya. Viro and Yu. V. Drobotukhina, *Configurations of skew lines*, Algebra i Analiz **1** (1989), no. 4, 222–246; English transl., Leningrad Math. J. **1** (1990), no. 4, 1027–1050.

Translated by V. F. MAZUROVSKIĬ

Smoothing of 6-Fold Singular Points and Constructions of 9th Degree M-Curves

A. B. Korchagin

ABSTRACT. Theorems on the possibility of smoothing nondegenerate 6-fold points by using charts of 6th degree M-curves for some special cases are proved. Series of smoothings of 3-fold points of types J_{2k}, $k \geqslant 6$, are described. Eight new M-curves of degree 9 are constructed.

Introduction

This paper continues the series of papers [P], [K1–K6] on constructions of new M-curves of degree 9. The main tool of the constructions in these papers is Viro's method of gluing charts of polynomials. But, as a rule, each paper contains some new specific construction tricks. The first idea in this paper has to do with smoothings of 6-fold singular points of degree 9 curves. We had constructed some curves having 6-fold points before, but did not obtain new M-curves of degree 9. So the second idea is to find special displacements of branches of the curves that actually give us new M-curves. These special curves having 6-fold singular points are constructed in §3. The possibility of gluing a 9th degree curve possessing a 6-fold point with a 6th degree nonsingular curve is the central point of the construction. It is provided by the theorems of §2.

Now the number of constructed M-curves of degree 9 has reached 404. To find all the information about realizable and prohibited schemes of M-curves of degree 9 it is sufficient to supplement [K6] by Theorem 3.5 of the present paper.

§1. Curves of degree 3

In this section we construct special 3rd degree curves in the projective coordinate system $(x_0 : x_1 : x_2)$; these curves are needed for the construction of certain curves of degrees 6 and 9 in the later sections.

LEMMA 1.1 ([V]). *For any four integers $0 < a_1 < a_2 < a_3$ and $b < 0$ there exists a neighborhood $U \subset \mathbb{R}$ of the integer b such that for any distinct $b_1, b_2, b_3 \in U$ there exists a curve of degree 3 passing through the points $A_i = (0 : 1 : a_i)$ and $B_i = (1 : b_i : 0)$, $i = 1, 2, 3$, and having the chart shown in Figure* 1.1.

1991 *Mathematics Subject Classification.* Primary 14P25, 14H99.

©1996, American Mathematical Society

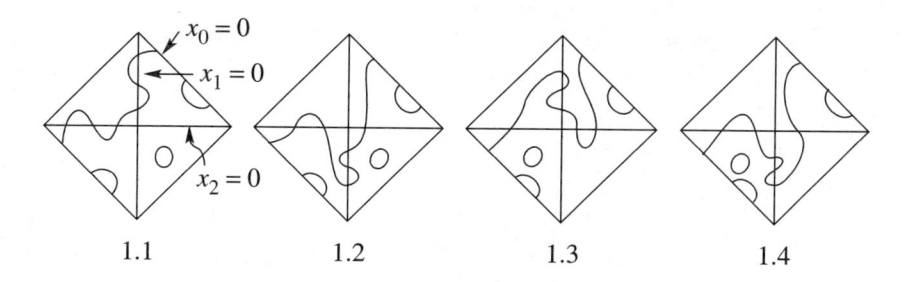

FIGURE 1

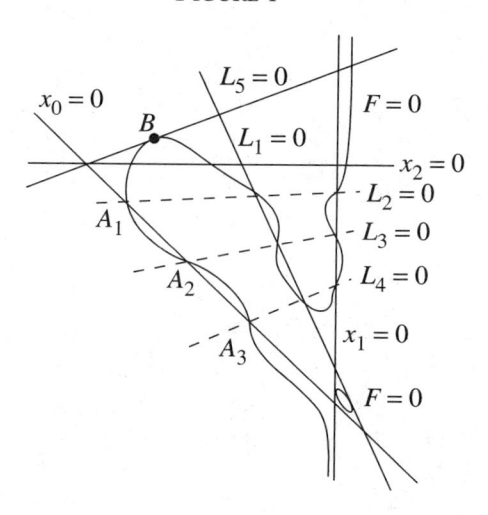

FIGURE 2

LEMMA 1.2. *For any three integers $0 < a_1 < a_2 < a_3$ there exist two integers $b < 0 < b_3$ and a neighborhood $U \subset \mathbb{R}$ of the integer b such that for any distinct $b_1, b_2 \in U$ there exists a curve of degree 3 passing through the points $A_i = (0:1:a_i)$ and $B_i = (1:b_i:0)$, $i = 1, 2, 3$, and having the chart shown in Figure 1.2.*

PROOF. Let us draw the lines $L_i = 0$, $i = 1, 2, 3, 4$, as shown in Figure 2. Then the 3rd degree curve $F \equiv x_0 x_1 L_1 + t L_2 L_3 L_4 = 0$ for sufficient small $|t|$ and a suitable sign of t is located as shown in Figure 2. Obviously we can draw the tangent $L_5 = 0$ to the curve $F = 0$ at the point $(0:1:0)$. Now we choose a new system of coordinates. Let the new axes $x_0 = 0$ and $x_1 = 0$ be the same and the new axis $x_2 = 0$ be the tangent $L_5 = 0$. Let $B = (1:b:0)$ and $B_3 = (1:b_3:0)$ be the points of intersection of the 3rd degree curve and axis $x_2 = 0$ in new coordinates (B is the point of tangency). Note that the coordinates of the points $A_i = (0:1:a_i)$ are not changed. Keeping points A_i, $i = 1, 2, 3$, fixed, we can perturb the curve of degree 3 and obtain any two points $B_i = (1:b_i:0)$, $i = 1, 2$, of the curve in the small neighborhood W of B, where $W \cap \{x_2 = 0\} = U$.

LEMMA 1.3. *For any three integers $0 < a_1 < a_2 < a_3$ there exist two integers $b_1 < 0 < b$ and a neighborhood $U \subset \mathbb{R}$ of the integer b such that for any distinct $b_2, b_3 \in U$ there exists a curve of degree 3 passing through the points $A_i = (0:1:a_i)$ and $B_i = (1:b_i:0)$, $i = 1, 2, 3$, and having the chart shown in Figure 1.3.*

LEMMA 1.4. *For any three integers $0 < a_1 < a_2 < a_3$ there exist two integers $b < 0 < b_3$ and a neighborhood $U \subset \mathbb{R}$ of the integer b such that for any distinct $b_1, b_2 \in U$ there exists a curve of degree 3 passing through the points $A_i = (0:1:a_i)$ and $B_i = (1:b_i:0)$, $i = 1, 2, 3$, and having the chart shown in Figure 1.4.*

The proofs of Lemmas 1.3 and 1.4 are analogous to the proof of Lemma 1.2.

§2. Smoothing of 6-fold nondegenerate singular points and curves of degree 6

We denote 6-fold nondegenerate singularities by M_{25}. Smoothings of singularities of type M_{25} are interesting for constructing nonsingular curves. The corresponding problem of classifying affine curves of degree 6 is solved only partially. In this problem the most interesting is the case when the curve is an M-curve and its points of intersection with the infinite line are distinct, real and belong to the same oval of the curve. In this section we study only this case. This investigation was begun in [K7] and [K-S], where the next two theorems were proved.

THEOREM 2.1. *There exist M-curves of degree 6 having the 33 types shown in Figures 3.1–3.9.*

We denote by $\langle \alpha \rangle$ the fragment of the curve consisting of the α empty ovals. In Figures 3.1–3.9 one can also see the codes of the charts.

THEOREM 2.2. *There does not exist any 6th degree M-curves of types different from those listed in Theorem 2.1 and from the following 11 types: $A_3(1,4,5)$, $A_3(3,2,5)$, $A_4(1,6,3)$, $A_4(5,2,3)$, $B_2(1,8,1)$, $B_2(1,4,5)$, $B_2(3,2,5)$, $C_1(9,0,1)$, $C_1(5,0,5)$, $C_1(1,0,9)$, $D(1,3,3,3)$.*

About the notation $D(1,3,3,3)$, see [K-S]. The existence of M-curves of these 11 types is neither proved nor disproved.

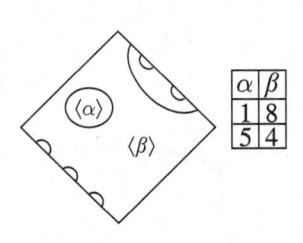

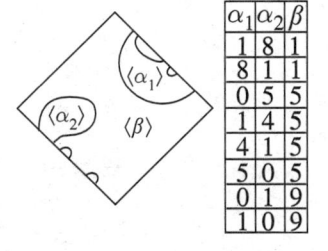

FIGURE 3.1. Code $A_1(\alpha, \beta)$

α	β
1	8
5	4

FIGURE 3.2. Code $A_2(\alpha_1, \alpha_2, \beta)$

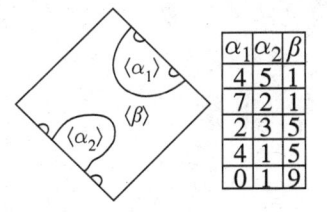

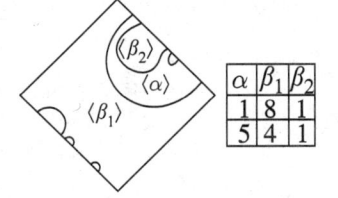

α_1	α_2	β
1	8	1
8	1	1
0	5	5
1	4	5
4	1	5
5	0	5
0	1	9
1	0	9

FIGURE 3.3. Code $A_3(\alpha_1, \alpha_2, \beta)$

α_1	α_2	β
4	5	1
7	2	1
2	3	5
4	1	5
0	1	9

FIGURE 3.4. Code $A_4(\alpha, \beta_1, \beta_2)$

α	β_1	β_2
1	8	1
5	4	1

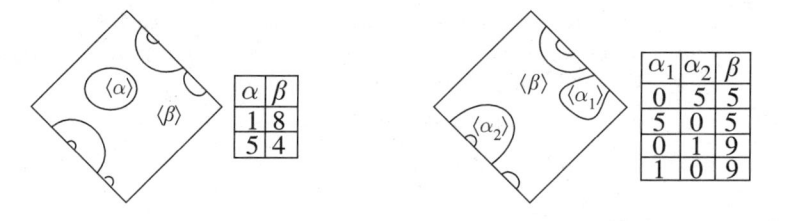

FIGURE 3.5. Code $B_1(\alpha, \beta)$

FIGURE 3.6. Code $B_2(\alpha_1, \alpha_2, \beta)$

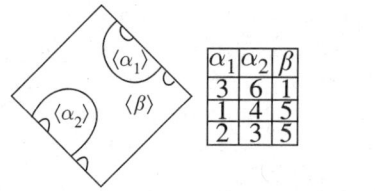

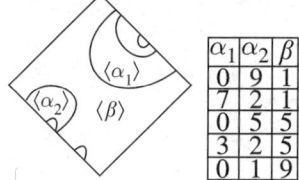

FIGURE 3.7. Code $B_3(\alpha_1, \alpha_2, \beta)$

FIGURE 3.8. Code $C_1(\alpha_1, \alpha_2, \beta)$

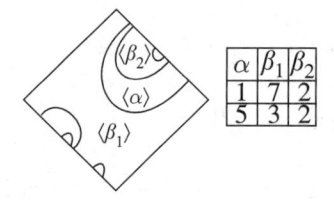

FIGURE 3.9. Code $C_2(\alpha, \beta_1, \beta_2)$

It is also unknown how the points of an M-curve of degree 6 can be placed on the infinite line. Only the realizability of arbitrary positions of points in some special cases is known. Namely, we have the following theorem.

THEOREM 2.3 ([**H, G**]). *Any gem of a plane curve belonging to type M_{25} and having the chart*

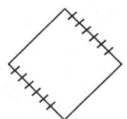

can be smoothed by the charts $A_1(1, 8)$, $A_1(5, 4)$, $A_2(1, 0, 9)$, $A_2(5, 0, 5)$. In other words, for any 6 integers $0 < a_1 < \cdots < a_6$ there exists an M-curve of degree 6 passing through the points $A_i = (0:1:a_i)$, $i = 1, \ldots, 6$, and having the charts $A_1(1, 8)$, $A_1(5, 4)$, $A_2(1, 0, 9)$, $A_2(5, 0, 5)$ (see Figures 3.1, 3.2).

But the point is that for the construction of the required nonsingular curves we do not need the complete classification of arrangements of the 6 points on the infinite line. In all the intersecting cases, the partial results proved below suffice.

THEOREM 2.4. *For any two integers $0 < a < a'$ there exist disjoint neighborhoods $U, U' \subset \mathbb{R}_+$ ($a \in U$, $a' \in U'$) such that for any distinct $a_1, a_2, a_3 \in U$ and for any distinct $a_4, a_5, a_6 \in U'$ one can find 6th degree M-curves passing through the points*

$A_i = (0:1:a_i)$, $i = 1,\ldots,6$, *and having the charts* $A_3(4,5,1)$, $A_3(7,2,1)$, $A_3(2,3,5)$, $A_3(4,1,5)$, $A_3(0,1,9)$, $B_2(0,5,5)$, $B_2(0,1,9)$, $B_3(3,6,1)$, $B_3(1,4,5)$, $B_3(2,3,5)$ *shown in Figures* 3.3, 3.6, *and* 3.7.

PROOF. Denote the 3rd degree curves shown in Figure 1.1 by F_i. Under quadratic transformation $(x_0:x_1:x_2) \mapsto (x_1x_2:x_0x_2:x_0x_1)$ the curve F_i maps into the 6th degree curve G_i, $i = 1,2,3,4$. The curve G_i has three 3-fold points at $(1:0:0)$, $(0:1:0)$, $(0:0:1)$. Gluing the curves F_i and G_j by Viro's method, we obtain the 6th degree curves H'_{ij} having two 3-fold points. Under the quadratic transformation $(x_0:x_1:x_2) \mapsto (x_0x_1:x_0^2:x_1x_2)$, the curve H'_{ij} maps into the 6th degree curve H_{ij}, $i,j = 1,2,3,4$, having two 3-fold points. Some of the curves H_{ij} are shown in Figures 4–6.

By using Viro's Lemma [V] about smoothing 3-fold singular points, we see that there exist neighborhoods $W, W' \subset \mathbb{R}P^1 = \{x_0 = 0\}$ of the points $(0:0:1)$, $(0:1:0)$ respectively such that for any distinct $B_i = (0:b_i:1) \in W$, $i = 1,2,3$, and for any distinct $B_i = (0:1:b_i) \in W'$, $i = 4,5,6$, the M-curves of degree 6 shown in Figures 4–6 exist. Obviously, for any two points $A = (0:1:a)$ and $A' = (0:1:a')$ there exists a linear projective transformation $l: \mathbb{R}P^2 \to \mathbb{R}P^2$ such that $l(0:0:1) = A$, $l(0:1:0) = A'$, $l(W) = U$, $l(W') = U'$, $l(B_i) = A_i$, $i = 1,\ldots,6$.

Similarly we can construct an M-curve having the chart $B_2(0,5,5)$, where the infinite line is $x_2 = 0$.

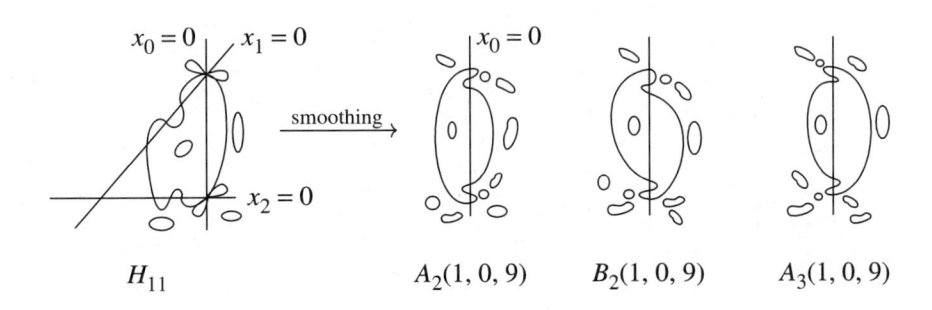

FIGURE 4

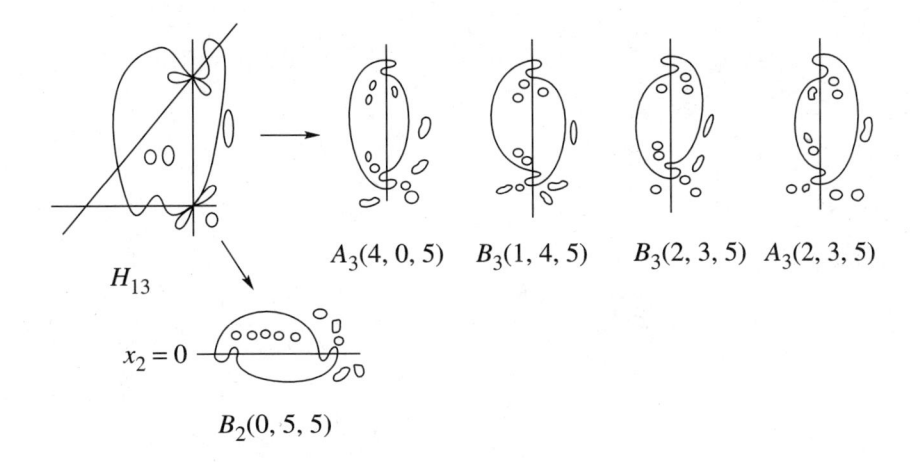

FIGURE 5

H_{33} $A_3(7, 2, 1)$ $B_3(3, 6, 1)$ $A_3(4, 5, 1)$

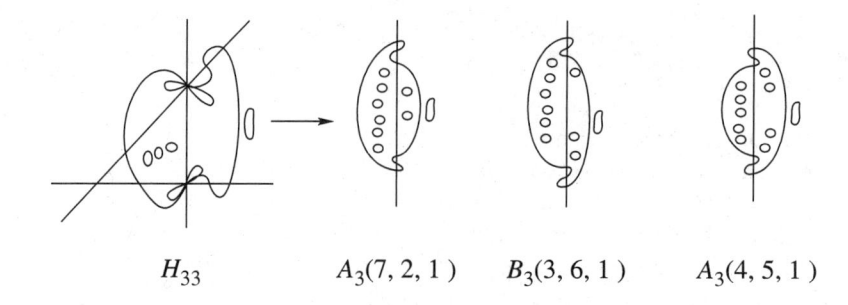

FIGURE 6

THEOREM 2.5. *For any three integers $0 < a_1 < a < a'$ there exist disjoint neighborhoods $U, U' \subset \mathbb{R}$ ($a \in U$, $a' \in U'$) such that for any distinct $a_2, a_3, a_4 \in U$ and for any distinct $a_5, a_6 \in U'$ the 6th degree M-curves passing through the points $A_i = (0 : 1 : a_i)$, $i = 1, \ldots, 6$, and having the charts $A_2(8, 1, 1)$, $A_2(4, 1, 5)$, $A_4(1, 8, 1)$, $A_4(5, 4, 1)$, $C_1(7, 2, 1)$, $C_1(3, 2, 5)$, $C_2(1, 7, 2)$, $C_2(5, 3, 2)$ shown in Figures* 3.2, 3.4, 3.8, *and* 3.9 *exist.*

PROOF. In the proof of Theorem 2.4, we constructed 6th degree curves H_{ij}, $i, j = 1, 2, 3, 4$. The curves H_{12}, H_{13}, H_{32}, and H_{33} are shown in Figures 7–10.

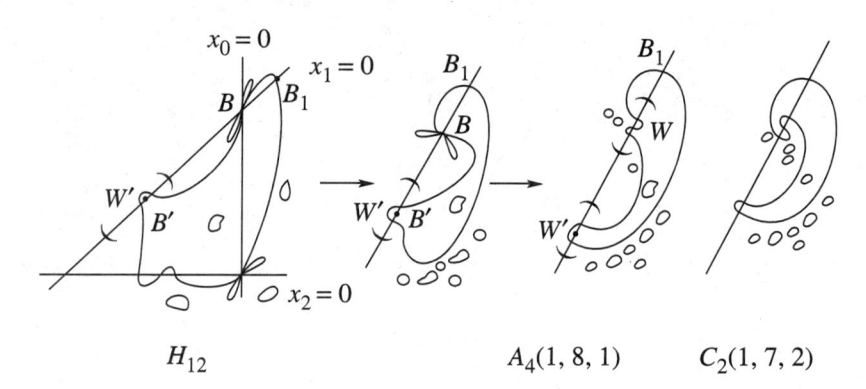

H_{12} $A_4(1, 8, 1)$ $C_2(1, 7, 2)$

FIGURE 7

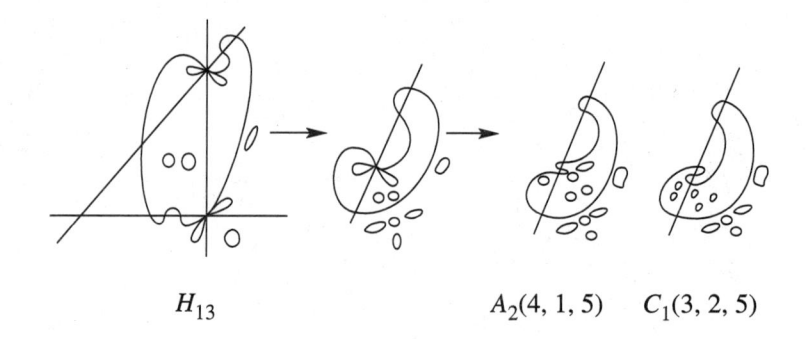

H_{13} $A_2(4, 1, 5)$ $C_1(3, 2, 5)$

FIGURE 8

H_{32} $A_4(5, 4, 1)$ $C_2(5, 3, 2)$

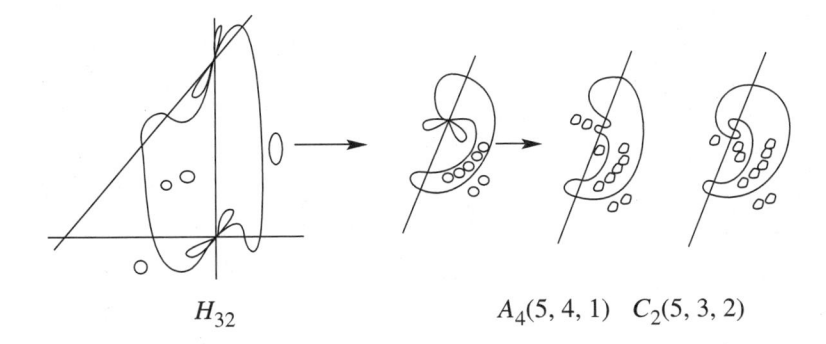

FIGURE 9

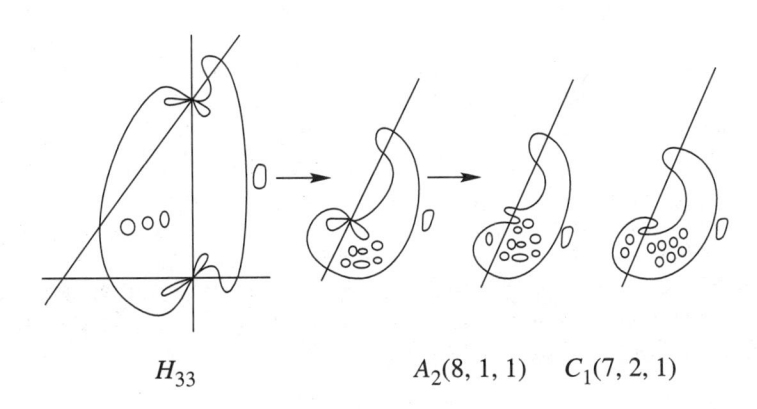

H_{33} $A_2(8, 1, 1)$ $C_1(7, 2, 1)$

FIGURE 10

The curve H_{12} has four points of intersection with the line $x_1 = 0$; one of them is a 3-fold point and three points are simple. According to Lemma 1.2, two simple points belong to some neighborhood W' of the point B' and may be chosen arbitrarily in W'. Using Viro's Lemma [V] about smoothing 3-fold points, we see that there exists a neighborhood $W \subset \mathbb{R}P^1 = \{x = 0\}$ of the point $B = (0:0:1)$ such that for any distinct $B_i = (b_i:0:1) \in W$, $i = 2, 3, 4$, and for any distinct $B_i = (b_i:0:1) \in W'$, $i = 5, 6$, one can find M-curves having charts $A_4(1, 8, 1)$ and $C_2(1, 7, 2)$. Obviously, for any three points $A_1 = (0:1:a_1)$, $A = (0:1:a)$ and $A' = (0:1:a')$ there exists a linear projective transformation $l: \mathbb{R}P^2 \to \mathbb{R}P^2$ such that $l(0:0:1) = A$, $l(B') = A'$, $l(W) = U$, $l(W') = U'$, $l(B_i) = A_i$, $i = 1, \ldots, 6$.

Similarly we may construct the M-curves shown in Figures 8–10.

THEOREM 2.6. *For any two integers $0 < a_1 < a$ there exists a neighborhood $U \subset \mathbb{R}_+$ of a ($a \in U$, $a_1 \notin U$) such that for any distinct a_i, $i = 2, \ldots, 6$, one can find 6th degree M-curves passing through the points $A_i = (0:1:a_i)$, $i = 1, \ldots, 6$, and having the charts $A_2(1, 8, 1)$, $A_2(0, 5, 5)$, $A_2(1, 4, 5)$, $A_2(0, 1, 9)$.*

PROOF. There exists a 4th degree curve which is placed in a projective system of coordinates as shown in Figure 11.1. This curve can be constructed by using Viro's gluing method as shown in Figure 11.2. Under the transformation $(x_0:x_1:x_2) \mapsto (x_0^3:x_0x_1x_2:x_1x_2^2)$ the 4th degree curve maps into the 5th degree curve $F = 0$ which has a 3-fold point of type D_6 at $(1:0:0)$ and a 5-fold point of tangency with

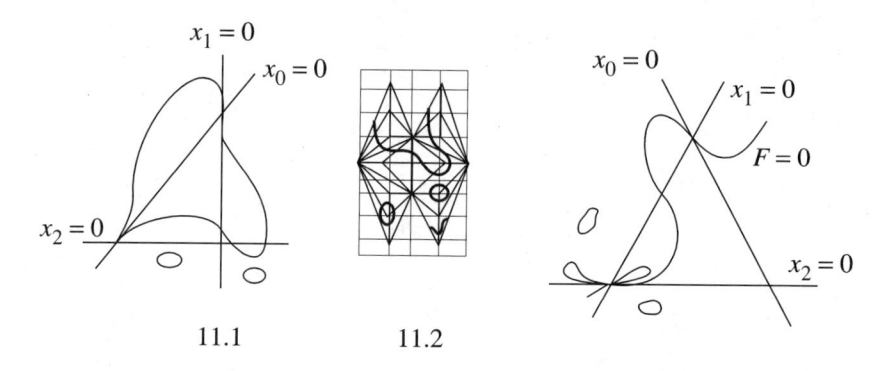

11.1   11.2

FIGURE 11   FIGURE 12

the axis $x_0 = 0$ at $(0:0:1)$; see Figure 12. This transformation is the composition of two birational quadratic transformations $(x_0:x_1:x_2) \mapsto (x_0^2:x_1x_2:x_0x_2)$ and $(x_0:x_1:x_2) \mapsto (x_0^2:x_0x_1:x_1x_2)$.

Let us consider the 6th degree curve $x_2F = 0$. This curve has a singular point of type Z_{15} at $(1:0:0)$ and a 5-fold point of tangency with the axis $x_0 = 0$ at $(0:0:1)$. We can smooth these points independently. The point of type Z_{15} can be smoothed using Viro's gluing method of the charts shown in Figure 13. These charts can be constructed by methods of [**K8**]. After smoothing the point Z_{15}, there appears a point $B_1 = (0:1:b_1)$ of intersection of the 6th degree curve with the axis $x_0 = 0$ in a small neighborhood of $(0:1:0)$.

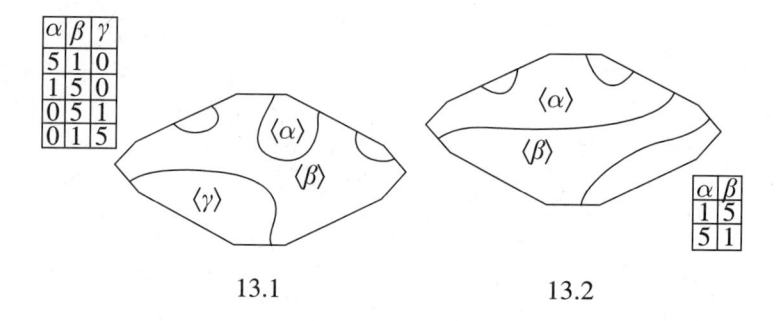

13.1   13.2

FIGURE 13

The 5-fold point can be smoothed so that any 5 distinct points $B_i = (0:b_i:1)$, $i = 2, \ldots, 6$, of intersection in a small neighborhood V of $(0:0:1)$ arise. Obviously for any two points $A_1 = (0:1:a_1)$ and $A = (0:1:a)$ there exists a linear projective transformation $l: \mathbb{R}P^2 \to \mathbb{R}P^2$ such that $l(B_1) = A_1$, $l(0:0:1) = A$, $l(V) = U$ and $U \ni l(B_i) = A_i$, $i = 2, \ldots, 6$.

THEOREM 2.7. *For any three integers $0 < a_1 < a < a'$ there exist disjoint neighborhoods $U, U' \subset \mathbb{R}_+$ of a, a' respectively such that for any distinct $a_2, a_3 \in U$ and for any distinct $a_4, a_5, a_6 \in U'$ one can find 6th degree M-curves passing through the points $A_i = (0:1:a_i)$, $i = 1, \ldots, 6$, and having the charts $B_1(1,8)$, $B_1(5,4)$, $B_2(1,0,9)$, $B_2(5,0,5)$ shown in Figures 3.5 and 3.6.*

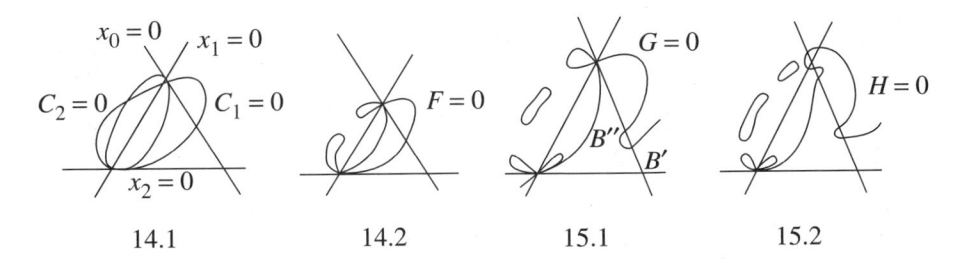

FIGURE 14

FIGURE 15

PROOF. Let us consider two conics
$$C_1 \equiv x_0 x_2 - k_1 x_1^2 + \varepsilon^{-2} x_1 x_2 = 0, \qquad C_2 \equiv x_0 x_2 - k_2 x_1^2 = 0,$$
where $0 < k_1 < k_2$ and $\varepsilon > 0$ is sufficiently small; see Figure 14.1. Let us perturb the union of the conics according to the formula $F \equiv C_1 C_2 - t x_1^3 (x_2 - k_1 \varepsilon x_1) = 0$, where $t > 0$ is sufficiently small. The 4th degree curve $F = 0$ is shown in Figure 14.2.

According to Theorem 3.B from [K7], for a sufficiently small $\tau > 0$ the 5th degree curve $G(\tau) \equiv x_2 F - \tau x_1 (x_0 x_2 - k_3 x_1^2)(x_0 x_2 - k_4 x_1^2) = 0$, where $k_2 < k_3 < k_4$, is situated as shown in Figure 15.1. This curve has a point of type A_1 at $(0:0:1)$ and a point of type D_6 at $(1:0:0)$.

If the value of τ increases, then, in the first place, the points B' and B'' of the curve $G(\tau) = 0$ approach each other and at some moment $\tau = \tau_1$ the curve is tangent to the axis $x_0 = 0$ at some point $B = (0:1:b)$, and, second, the oval of the curve collapses and at some moment $\tau = \tau_2$ the oval becomes a point. It is not difficult to check that if εk_1 is small enough, then $\tau_1 < \tau_2$.

Let us consider the curve $G(\tau_1) = 0$. Keeping the point of the type D_6, we can perturb the tangency at the point $B = (0:1:b)$ and at the singular point of type A_1 so that in some neighborhood V of B there arise any two distinct points $B_i = (0:1:b_i) \in V$, $i = 2, 3$, of intersection and in some neighborhood V' of $(0:0:1)$ there arise any three points $B_i = (0:b_i:1) \in V'$, $i = 4, 5, 6$. Let $H = 0$ be the equation of the perturbed 5th degree curve. The curve $H = 0$ is shown in Figure 15.2.

The end of the proof is similar to the preceding one. We smooth the singular point of type Z_{15} of the 6th degree curve $x_2 H = 0$ by Viro's gluing method for the charts shown in Figure 13.1. After a suitable linear transformation, we obtain the required M-curves of degree 6.

THEOREM 2.8. *For any three integers $0 < a_1 < a < a'$ there exist disjoint neighborhoods $U, U \subset \mathbb{R}_+$ of a, a' respectively such that for any distinct $a_2, a_3, a_4 \in U$ and for any distinct $a_5, a_6 \in U'$ one can find 6th degree M-curves passing through the points $A_i = (0:1:a_i)$, $i = 1, \ldots, 6$, and having the charts $C_1(0, 9, 1)$ and $C_1(0, 5, 5)$ shown in Figure 3.8.*

PROOF. Take the curve $x_2 H = 0$ from the preceding proof and smooth its singular point of type Z_{15} using Viro's gluing method for the charts shown in Figure 13.2.

To prove Theorem 2.10 below, we must smooth a singular point that has the normal form $y^3 - ax^{11} = 0$. But now we prove a more general result about smoothing of singular points which have normal form $y^3 - ax^n = 0$, $n \in \mathbb{N}$. The type of such singular points is denoted by J_μ, where $\mu = 2n - 2$ is Milnor's number. In particular

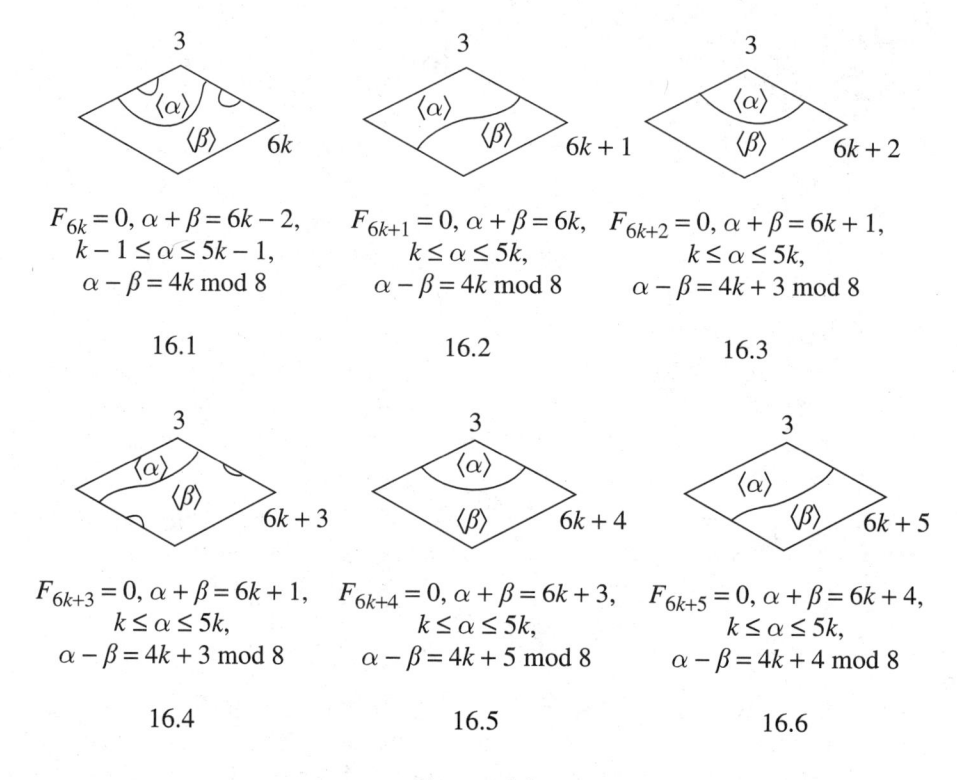

$F_{6k} = 0$, $\alpha + \beta = 6k - 2$,
$k - 1 \leq \alpha \leq 5k - 1$,
$\alpha - \beta = 4k \bmod 8$

16.1

$F_{6k+1} = 0$, $\alpha + \beta = 6k$,
$k \leq \alpha \leq 5k$,
$\alpha - \beta = 4k \bmod 8$

16.2

$F_{6k+2} = 0$, $\alpha + \beta = 6k + 1$,
$k \leq \alpha \leq 5k$,
$\alpha - \beta = 4k + 3 \bmod 8$

16.3

$F_{6k+3} = 0$, $\alpha + \beta = 6k + 1$,
$k \leq \alpha \leq 5k$,
$\alpha - \beta = 4k + 3 \bmod 8$

16.4

$F_{6k+4} = 0$, $\alpha + \beta = 6k + 3$,
$k \leq \alpha \leq 5k$,
$\alpha - \beta = 4k + 5 \bmod 8$

16.5

$F_{6k+5} = 0$, $\alpha + \beta = 6k + 4$,
$k \leq \alpha \leq 5k$,
$\alpha - \beta = 4k + 4 \bmod 8$

16.6

FIGURE 16

J_0 is the type of the flex point, $J_2 = A_2$ is the type of the cusp, $J_4 = D_4$ is the type of the nondegenerate 3-fold point, $J_6 = E_6$, $J_8 = E_8$. Note that the smoothings of singular points of types J_μ for $\mu = 2, 4, 6, 8, 10$ were studied in [**V**]. If $\mu \equiv 0$ or $2 \pmod 8$, then there is only one branch passing through the point. If $\mu \equiv 4 \pmod 6$, then there are three branches passing through the point. We consider the case when all the branches are real and when the maximal number of ovals is generated by smoothing.

LEMMA 2.9. *Each singular point of type $J_{12k+2i-2}$ admits smoothings to all charts of the curve $F_{6k+i} = 0$ shown in Figure* 16.2, *where $i = 0, 1, \ldots, 5$ and $k = 0, 1, 2, \ldots$*.

PROOF. The curves $F_i = 0$, $i = 1, \ldots, 6$, whose charts are shown in Figure 17, were constructed in [**V**] and [**K5**].

Denote $hy(x_0 : x_1 : x_2) = (x_0 x_1 : x_1^2 : x_0 x_2)$, $hy^k = hy \circ \cdots \circ hy$. Following Newton, the birational quadratic transformation hy is called a *hyperbolism*. Gluing in a suitable fashion all the charts of the polynomials of the sequence

$$F_6^{p_1},\ F_6^{p_2} \circ hy,\ F_6^{p_3} \circ hy^2, \ldots, F_6^{p_k} \circ hy^{k-1},\ F_i \circ hy^k,$$

where $p_1, \ldots, p_k = 1, 2, 3, 4$, $k = 0, 1, 2, \ldots$, $i = 0, 1, \ldots, 5$ $((k, i) \neq (0, 0))$, we obtain all the polynomials F_{6k+1} (when $i = 0$ we exclude the term $F_i \circ hy^k$, and when $k = 0$ the series consists only of this term). By the words "suitable fashion", we mean the reduction of $F_6^p \circ hy^q$ by x_1^r, where $r = 6 \cdot 2^q - (3q + 6)$, and the symmetric transformations with repsect to the axes $x_1 = 0$ and $x_2 = 0$ when they are necessary.

THEOREM 2.10. *For any three integers $0 < a_1 < a < a'$ there exist disjoint neighborhoods $U, U' \subset \mathbb{R}_+$ of a, a' respectively such that for any distinct $a_2, a_3, a_4 \in U$ and*

SMOOTHING OF 6-FOLD POINTS AND CONSTRUCTIONS OF M-CURVES

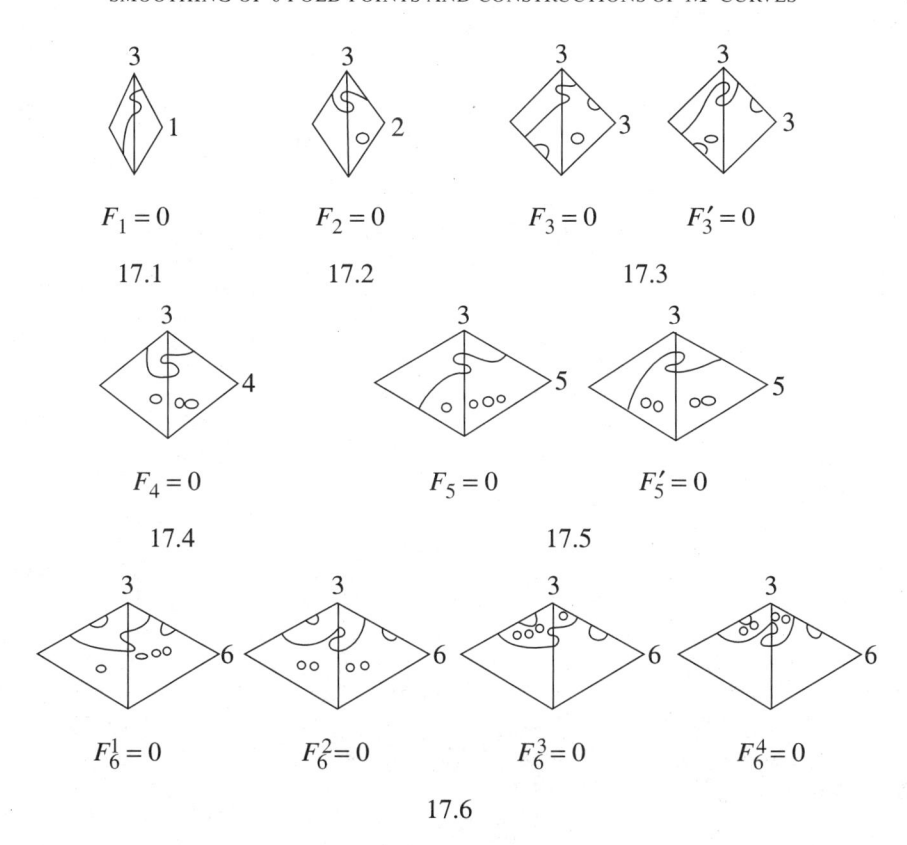

$F_1 = 0$ $F_2 = 0$ $F_3 = 0$ $F_3' = 0$

17.1   17.2   17.3

$F_4 = 0$ $F_5 = 0$ $F_5' = 0$

17.4   17.5

$F_6^1 = 0$ $F_6^2 = 0$ $F_6^3 = 0$ $F_6^4 = 0$

17.6

FIGURE 17

for any distinct $a_5, a_6 \in U'$ one can find a 6th degree M-curve passing through the points $A_i = (0:1:a_i)$, $i = 1, \ldots, 6$, and having the chart $C_1(0, 1, 9)$ shown in Figure 3.8.

PROOF. Let us consider three conics $C_i \equiv y - x^2 - i^{-1}y^2 = 0$, $i = 1, 2, 3$. The conics are tangent to each other with 4-fold tangency, so the curve $C_1 C_2 C_3 = 0$ has a singular point of type J_{22}. It is not difficult to check that the curve $F(t) \equiv C_1 C_2 C_3 + txy^5 = 0$ for $t \neq 0$ has only one singular point of type J_{20} at $(0,0)$. For $t > 0$ the curve $F(t)$ is shown in Figure 18. Obviously, it has a flex point B on the arc PQ. Let $L(t) = 0$ be the tangent to the curve $F(t) = 0$ at the flex point. It is clear that when t increases, the tangent $L(t) = 0$ moves so as to become tangent to the curve at one more point B' on the arc QR. Let $t = t_0$ be the value when $L(t_0)$ is this double tangent to the curve. The curve $F(t) = 0$ satisfies Shustin's Proposition 1 from [S] for every value of $t \neq 0$. So, by using the chart shown in Figure 16.6 ($k = 1$, $\alpha = 1$, $\beta = 9$), we can smooth the singular point of the type J_{20} so that the smoothed curve will have the same points of tangency B and B'; see Figure 18.2. We denote these points by the same letters.

After a suitable linear transformation that maps B_1, B, B' onto A_1, A, A', we perturb the points A, A' and obtain the required curve.

REMARK 2.11. Theorems 2.4–2.8 and 2.10 are in fact true for any points $A_1, A, A' \in \mathbb{R}P^1 = \{x_0 = 0\}$. To formulate similar theorems in detail in this case, we would have

18.1   18.2

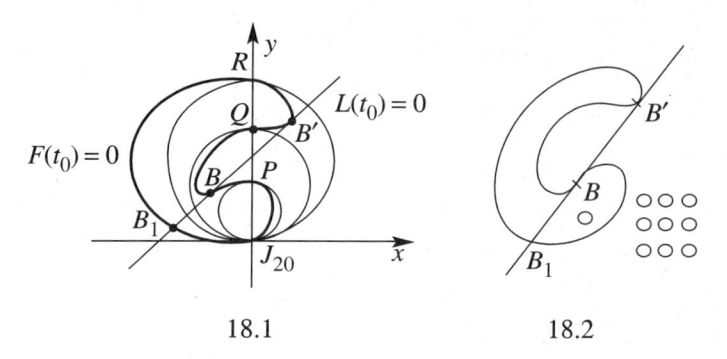

FIGURE 18

to draw a lot of charts. Theorem 2.3 is true also for any points $A_i \in \mathbb{R}P^1 = \{x_0 = 0\}$, $i = 1, \ldots, 6$.

§3. The 9th degree curves

In this section we construct 9th degree curves with two singularities of types J_{10} and M_{25}. By smoothing these singularities, we obtain 8 new M-curves.

We begin with Viro's wonderful curve constructed in [V]. This curve of degree 5 is shown in Figure 19. It consists of an odd branch, 5 ovals and one isolated singular point of type A_1. Let $B_i = (1:0:b_i)$, $i = 1, \ldots, 5$, be the points of intersection of Viro's curve and the axis $x_1 = 0$, and $b_1 < \cdots < b_5$; see Figure 19. It is easy to improve Viro's construction, obtaining the following theorem.

THEOREM 3.1 [V]. *For any disjoint neighborhoods $U, U' \subset \mathbb{R}P^1 = \{x_1 = 0\}$ of points $(1:0:0)$, $(0:0:1)$ respectively and for any distinct points in each of following cases*:
  (1) $B_1, B_2 \in U'$ and $B_3, B_4, B_5 \in U$;
  (2) $B_1, B_2, B_3 \in U$ and $B_4, B_5 \in U'$;
  (3) $B_1 \notin U, U'$, $B_2, B_3 \in U$, and $B_4, B_5 \in U'$
Viro's curve shown in Figure 19 *exists.*

THEOREM 3.2. *There exist two 9th degree curves with two singular points of types J_{10} and M_{25} and shown in Figure 20 such that the tangents to the branches at the singular point of type M_{25} of the curve in Figure 20.1 (20.2) intersects the axis $x_0 = 0$ at the points $A_1, \ldots, A_6$ provided by Theorem 2.4 (by Theorems 2.4 and 2.5 respectively).*

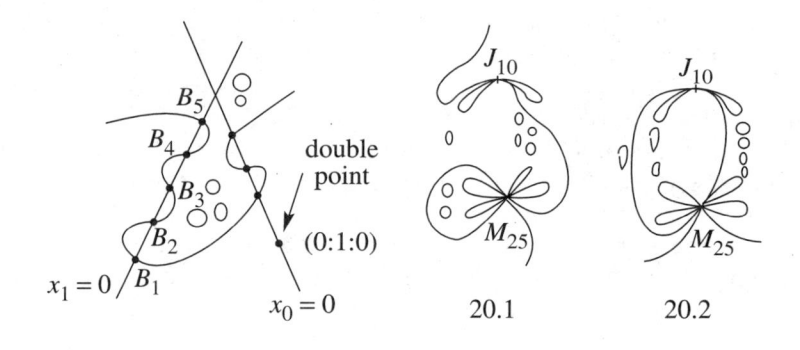

20.1   20.2

FIGURE 19   FIGURE 20

21.1 21.2 21.3

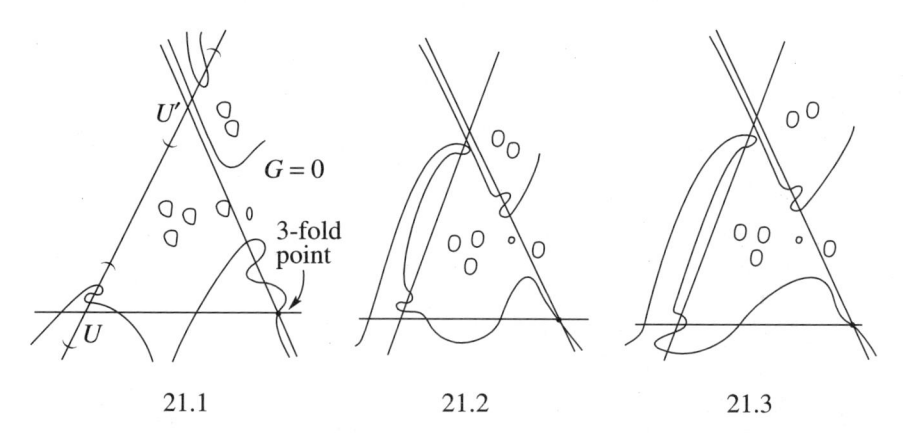

FIGURE 21

PROOF. 1) Let $F = 0$ be the equation of Viro's curve provided by Theorem 3.1, in case (1). Keeping the 3-fold singular point and the points $B_1, \ldots, B_5$ of the curve $x_0 F = 0$, let us perturb it so as to obtain the 6th degree curve $G = 0$ shown in Figure 21.1. Under the quadratic transformation $(x_0 : x_1 : x_2) \mapsto (x_0^2 : x_0 x_1 : x_1 x_2)$, the curve $G = 0$ maps onto the 9th degree curve with singular points of types J_{10} and M_{25} shown in Figure 20.1. It is easy to make the tangents to the branches at the singular point of type M_{25} to intersect the axis $x_0 = 0$ at the points $A_1, \ldots, A_6$ provided by Theorem 2.4.

2) Similarly, starting from the curves of Theorem 3.1 in cases (2) and (3), one can construct the 6th degree curves shown in Figure 21.2 and 21.3 and then obtain the curves shown in Figure 20.2.

The next curve of degree 5 shown in Figure 22 comes from another curve constructed by Viro in [V].

THEOREM 3.3. *For any two integers $b < b'$ there exist disjoint neighborhoods $U, U' \subset \mathbb{R}$ of b, b' respectively such that for any distinct $b_1, b_2 \in U$ and $b_3, b_4, b_5 \in U'$ one can find a 5th degree curve passing through the points $B_i = (1 : 0 : b_i)$, $i = 1, \ldots, 5$, and having one isolated singular point of type A_1 at $(0 : 1 : 0)$; see Figure 22.*

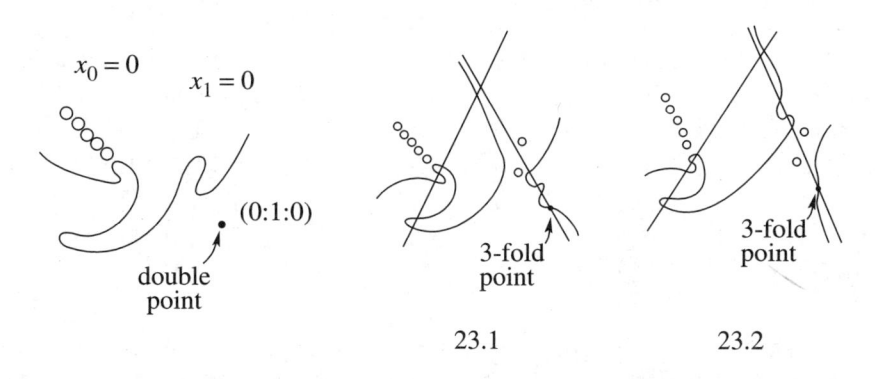

23.1 23.2

FIGURE 22   FIGURE 23

154   A. B. KORCHAGIN

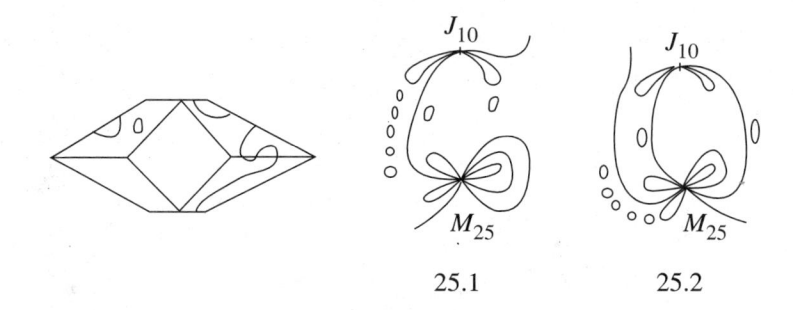

25.1   25.2

FIGURE 24   FIGURE 25

PROOF. Let us consider the curve $H = 0$ constructed in the proof of Theorem 2.7; see Figure 15.2. Its points belonging to the axis $x_0 = 0$ satisfy the conditions imposed on the points $B_1, \ldots, B_6$ in Theorem 3.3. Let us perturb the singular point of type D_6 of the curve $H = 0$ by using the chart shown in Figure 24; after the mapping $(x_0 : x_1 : x_2) \mapsto (x_2 : x_0 : x_1)$, we obtain the curve shown in Figure 22.

THEOREM 3.4. *There exist two 9th degree curves with two singular points of types J_{10} and M_{25} shown in Figure 25 such that the tangents to the branches at the singular point of type M_{25} intersect the axis $x_0 = 0$ at the points $A_1, \ldots, A_6$ provided by Theorems 2.5 and 2.10.*

PROOF. This proof is similar to the proof of Theorem 3.2. We construct the 6th degree curves shown in Figure 23 from the 5th degree curve shown in Figure 22; after the transformation $(x_0 : x_1 : x_2) \mapsto (x_0^2 : x_0 x_1 : x_1 x_2)$, we obtain the 9th degree curves shown in Figure 25.

THEOREM 3.5. *There exist 8 M-curves of degree 9 having the schemes which are contained in the last column of Table 1.*

TABLE 1

#	Singular curve Figure	Smoothing point J_{10} α, β	point M_{25}	Schemes of M-curve
1	20.1	0, 4	$A_3(7, 2, 1)$	$\langle 1\langle 2\rangle \amalg 1\langle 11\rangle \amalg 13 \amalg J\rangle$
2	20.2	0, 4	$A_3(7, 2, 1)$	$\langle 1\langle 2\rangle \amalg 1\langle 12\rangle \amalg 12 \amalg J\rangle$
3	20.2	4, 0	$A_2(8, 1, 1)$	$\langle 1\langle 17\rangle \amalg 10 \amalg J\rangle$
4	20.2	4, 0	$A_3(4, 5, 1)$	$\langle 1\langle 5\rangle \amalg 1\langle 13\rangle \amalg 8 \amalg J\rangle$
5	20.2	4, 0	$A_3(7, 2, 1)$	$\langle 1\langle 2\rangle \amalg 1\langle 16\rangle \amalg 8 \amalg J\rangle$
6	25.1	4, 0	$C_2(1, 7, 2)$	$\langle 22 \amalg 1\langle 1 \amalg 1\langle 3\rangle\rangle \amalg J\rangle$
7	25.1	4, 0	$C_2(5, 3, 2)$	$\langle 18 \amalg 1\langle 5 \amalg 1\langle 3\rangle\rangle \amalg J\rangle$
8	25.2	0, 4	$C_1(0, 9, 1)$	$\langle 24 \amalg 1\langle 1 \amalg 1\langle 1\rangle\rangle \amalg J\rangle$

PROOF. Let us smooth the singular points of types J_{10} and M_{25} of the 9th degree curves shown in Figures 20 and 25. To smooth the points of type J_{10}, we use the charts shown in Figure 16.1 for $k = 0$. To smooth the points of type M_{25}, we use the charts provided by Theorems 2.4, 2.5, and 2.10. The results of gluing are incorporated in Table 1.

References

[G] D. A. Gudkov, *Construction of a new series of M-curves*, Dokl. Akad. Nauk SSSR **200** (1971), no. 6, 1269–1272; English transl., Soviet Math. Dokl. **12** (1971), no. 5, 1559–1563.

[H] A. Harnack, *Über die Vieltheiligkeit der ebenen algebraischen Kurven*, Math. Ann. **10** (1876), 189–199.

[K1] A. B. Korchagin, *New possibilities of the Brusotti method for the construction of M-curves of degree 8*, Methods of Qualitative Theory of Differential Equations, Gor'ky State Univ., 1978, pp. 149–159. (Russian)

[K2] _____, *M-curves of 9th degree: construction of 141 curves*, Manuscript No. 7459-B86, deposited at VINITI (1986), 1–73. (Russian)

[K3] _____, *M-curves of 9th degree: realizability of 32 types*, Manuscript No. 2566-B87, deposited at VINITI (1987), 1–17. (Russian)

[K4] _____, *M-curves of 9th degree: realizability of 24 types*, Manuscript No. 3049-B87, deposited at VINITI (1987), 1–17. (Russian)

[K5] _____, *M-curves of 9th degree: realizability of 167 types*, Manuscript No. 7884-B87, deposited at VINITI (1987), 1–69. (Russian)

[K6] _____, *Construction of new M-curves of 9th degree*, Proc. Intern. Conf. Real Algebraic Geometry, Rennes, June 24–29, 1991, Lecture Notes in Math., vol. 1524, Springer-Verlag, Berlin, Heidelberg, and New York, 1991, pp. 296–307.

[K7] _____, *On the reduction of singularities and the classification of nonsingular affine curves of degree 6*, Manuscript No. 1107-B86, deposited at VINITI, 1986, pp. 1–16. (Russian)

[K8] _____, *Isotopy classification of plane seventh degree curves with only singular point Z_{15}*, Lecture Notes in Math., vol. 1346, Springer-Verlag, Berlin, Heidelberg, and New York, 1988, pp. 407–426.

[K-S] A. B. Korchagin and E. I. Shustin, *Affine curves of degree 6 and smoothings of a nondegenerate sixth order singular point*, Izv. Akad. Nauk SSSR Ser. Mat. **52** (1988), no. 6, 1181–1199; English transl., Math. USSR-Izv. **33** (1989), no. 3, 501–520.

[P] G. M. Polotovskiĭ, *To the problem of the topological classification of arrangements of ovals of a nonsingular algebraic curve in projective plane*, Methods of Qualitative Theory of Differential Equations, Gor'ky State Univ., 1975, pp. 101–128. (Russian)

[S] E. I. Shustin, *Topology of real plane algebraic curves*, Proc. Intern. Conf. Real Algebraic Geometry, Rennes, June 24–29, 1991, Lecture Notes in Math., vol. 1524, Springer-Verlag, Berlin, Heidelberg, and New York, 1992, pp. 97–109.

[V] O. Ya. Viro, *Real algebraic varieties with prescribed topological properties*, D. Sc. Thesis, Leningrad State Univ., 1983. (Russian)

Translated by THE AUTHOR

Automaton Model of Relations Between Two Countries

Mark Kushelman

ABSTRACT. In the model of international relations between two countries [2], *leaders* are objects of influence of *influence groups* and their positions are presented by attractors of a dynamical system on the plane. If we admit that leaders also can influence their influence groups, we get a situation in which parameters of the dynamical system are connected with the state point *only* in cases in which the state point is located in the vicinity of attractors of the dynamical system. An automaton is used for the description of this situation and a computer experiment is presented.

§1. Introduction

We consider the situation when a modeled object is represented by attractors of a dynamical system and the parameters of the dynamical system can be influenced by the state point that presents the modeled object only when the state point is located in the vicinity of the attractors of the dynamical system. A nondeterministic automaton is used for the description of this situation. Then an automaton connected with the model of international relations between two countries [2] is considered and a computer experiment is presented.

§2. Construction of the formal tool

Below we use the following notation. If $Y = \{y_1, \ldots, y_I\}$ is a finite set, we choose a copy YR in R^I, representing $y_i \in Y$ by the vector $(0, \ldots, 1_i, \ldots, 0) \in R^I$, and we write $y_i = (0, \ldots, 1_i, \ldots, 0)$. If $y \in YR$, then the ith coordinate is denoted by $[y]_i$. For the unit $(I-1)$-dimensional simplex in YR we use the notation Y_U, i.e.,

$$Y_U = \{y \in YR : \sum [y]_i = 1, [y]_i \geq 0\}.$$

The set Y_U is the set of distributions on the set Y.

Let the rectangular part of Figure 1 be a finite nondeterministic automaton. If there is a linear map $F_{AD} : A_U \to D_U$, then the following Markov process is defined in Figure 1:

(1) $$w(k+1) = F_{WW}(F_{AD}(F_{WA}(w(k))), w(k)).$$

1991 *Mathematics Subject Classification*. Primary 68Q68; Secondary 92K10.

©1996, American Mathematical Society

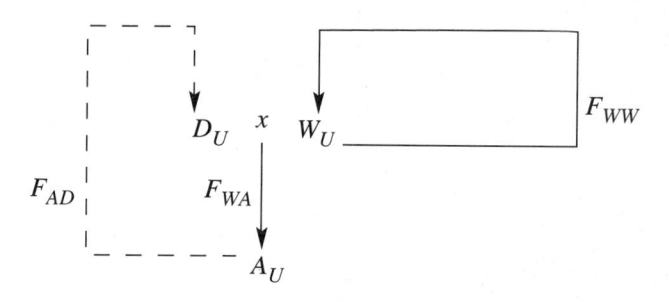

FIGURE 1

THEOREM 1. *Let the maps F_{WA}, $F_{WW}(d_i)$, and F_{AD} be presented by matrices H_{WA}, $H_{WW}(d_i)$, $d_i \in D$, and H_{AD}, operating from the right. Then*

$$w(k+1) = w(k) \cdot \sum_i [w(k) \cdot H_{WA} \cdot H_{AD}]_i H_{WW}(d_i).$$

CONJECTURE 1. *For the dynamical system*

(2) $$\dot{x} = f(x, p),$$

where $x \in R^M$, $p \in R^N$, and $f(x, p)$ is a sufficiently smooth function, there is a finite set of regions $S = \{s_1, \ldots, s_{\dim S}\}$ in R^N, such that $s_i \cap s_j = \varnothing$ and every structural stability region is the union of some $\mathrm{Cl}(s_i)$'s.

Let A be the set of attractors of (2) and $W \subset S \times A$ be the set of (s_i, a_j) such that if $p \in s_i$, then the qualitative picture of (2) includes the attractor a_j. Let $\mathrm{Pr}: W \to A$ be the projection and and $F_{WA}: W_U \to A_U$ be linear, so that $F_{WA}|_W = \mathrm{Pr}$.

Let S^{N-1} be the unit $(N-1)$-sphere in the parameter space of (2) and $D = \{d_1, \ldots, d_{\dim}\}$ be a finite covering of S^{N-1}, so that $d_i \cap d_j = \varnothing$. The set of points from d_i will be called the *direction* of d_i.

CONJECTURE 2. *For every $d_i \in D$, there is a linear map $L_{WW}(d_i): W_U \to W_{U'}$.*

We define the linear map $F_{WW}: D_U \times W_U \to W_U$, so that $F_{WW}|_D = L_{WW}$.

CONJECTURE 3. *There is a linear map $F_{AD}: A_U \to D_U$.*

Conjectures 1–3 claim that the process (1) is connected with (2), and the rest of this paper is devoted to the construction of an example.

§3. Automaton model of relations between two countries

In [2], a model of international relations between two countries was described by the dynamical system

(3) $$\dot{x} = bx + ay - x^3, \qquad \dot{y} = cx + dy - y^3,$$

where x (y) is the position of the leader of the first (second) country ST_1 (ST_2) and everybody from the influence group of ST_1 (ST_2) relating to ST_2 (ST_1), b (d) is the expected damage in ST_1 (ST_2) from war, with ST_2 (ST_1), a (c) the inverse coefficient of the influence group of ST_1 (ST_2).

We consider the following example: *The armed forces of ST_1 and ST_2 have means of attack so powerful that a further arms race does not influence the sufficiently large*

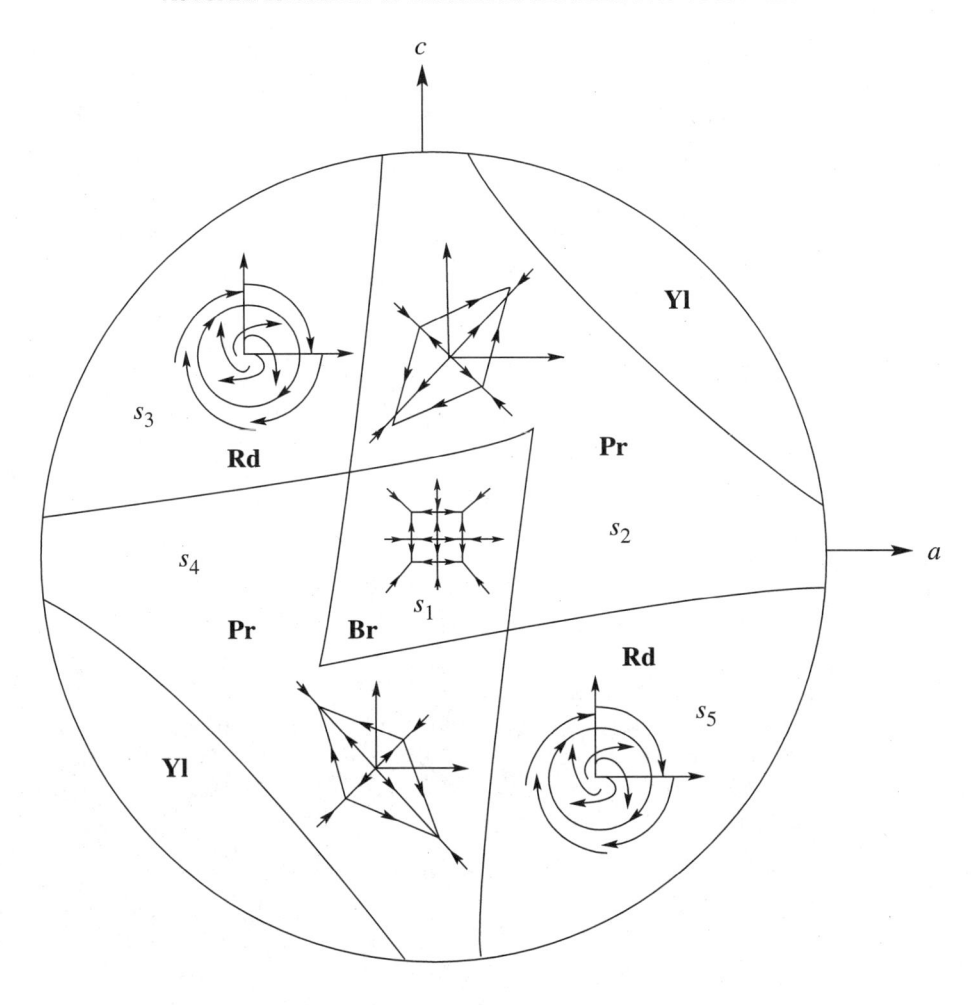

FIGURE 2

value of the damage. These assumptions imply that b and d are positive and constant. Structural stability regions of (3) are defined by the following theorem (we adopt the notation for structurally stable regions from [1]).

THEOREM 2. *Intersections of structurally stable regions of* (3) *by the planes* $b =$ constant *and* $d =$ constant $(b, d > 0)$ *are the regions* **Br** ("BRown"), **Yl** ("YeLlow"), **Pr** ("PuRple"), **Rd** ("ReD") *in Figure* 2, *where*
  a) *the boundary of* **Yl** *is the hyperbola* $ac - bd = 0$;
  b) *the remaining part of the lines in Figure* 2 *is defined by* $F(a, c) = 0$, *where*

$$F(p) = 256a^3c^3 - 192bda^2c^2 - 6b^2d^2ac + 27d^4a^2 + 27b^4c^2 - 4b^3d^3.$$

Let D^2 be the 2-disk in the plane (a, c) with center at $(0, 0)$ that does not intersect the region **Yl**. We define (see Figure 2)

$$s_1 = \mathbf{Br}, \quad s_2 = (\text{northeastern } \mathbf{Pr}) \cup D^2, \quad s_3 = (\text{northwestern } \mathbf{Rd}) \cup D^2,$$
$$s_4 = (\text{southwestern } \mathbf{Pr}) \cup D^2, \quad s_5 = (\text{southeastern } \mathbf{Rd}) \cup D^2.$$

THEOREM 3. *Attractors of* (3) *include* (*see Figure* 2):
a) *four nodes* a_1, a_2, a_3, a_4, *where the node* a_i *is located in the ith quadrant of the* (x, y)-*plane, if* $(a, c) \in s_1$;
b) *two nodes* a_1, a_3, *if* $(a, c) \in s_2$ *and* a_2, a_4, *if* $(a, c) \in s_4$;
c) *the limit cycle* a_5 *surrounding* $(0, 0)$, *if* $(a, c) \in s_3$ *or* $(a, c) \in s_5$.

So $S = \{s_1, s_2, s_3, s_4, s_5\}$ and $A = \{a_1, a_2, a_3, a_4, a_5\}$. For $w_i \in W$ we use the notation

$$w_1 = (s_1, a_1), \quad w_2 = (s_1, a_2), \quad w_3 = (s_1, a_3), \quad w_4 = (s_1, a_4), \quad w_5 = (s_2, a_1),$$

$$w_6 = (s_2, a_3), \quad w_7 = (s_3, a_5), \quad w_8 = (s_4, a_2), \quad w_9 = (s_4, a_4), \quad w_{10} = (s_5, a_5).$$

Let

$$d_1 = \{(a, c) \in S^1 : a > 0, c > 0\}, \quad d_2 = \{(a, c) \in S^1 : a < 0, c > 0\},$$

$$d_3 = \{(a, c) \in S^1 : a < 0, c < 0\}, \quad d_4 = \{(a, c) \in S^1 : a > 0, c < 0\}.$$

Here d_i is the set of directions of increase or decrease of the parameters a or c.

The transition graphs of (1) are shown in Figure 3. We suppose that nonzero transitions between automaton states occur with equal probability.

Various types of the dynamic process (1) and hence various types of long-term relations between the two countries are obtained by using various matrices H_{AD}. We show the results of computer experiments with some of them.

3.1. Let the leader have the possibility to influence only his/her influence group. In addition, let the leader be able to both a) intensify or b) slacken the inversion tendency. In other words, let $e = a$ or c. If $e > 0$ ($e < 0$), then in case a) the corresponding leader influences his/her group so that e increases (decreases) and in case b) e decreases (increases). There are only three cases for H_{AD}:

$$H1 = \begin{Vmatrix} 1 & 0 & 0 & 0 \\ 0 & 1 & 0 & 0 \\ 0 & 0 & 1 & 0 \\ 0 & 0 & 0 & 1 \\ 0.25 & 0.25 & 0.25 & 0.25 \end{Vmatrix},$$

$$H2 = \begin{Vmatrix} 0 & 0 & 1 & 0 \\ 0 & 0 & 0 & 1 \\ 1 & 0 & 0 & 0 \\ 0 & 1 & 0 & 0 \\ 0.25 & 0.25 & 0.25 & 0.25 \end{Vmatrix},$$

$$H3 = \begin{Vmatrix} 0 & 0 & 0 & 1 \\ 0 & 0 & 1 & 0 \\ 0 & 1 & 0 & 0 \\ 1 & 0 & 0 & 0 \\ 0.25 & 0.25 & 0.25 & 0.25 \end{Vmatrix},$$

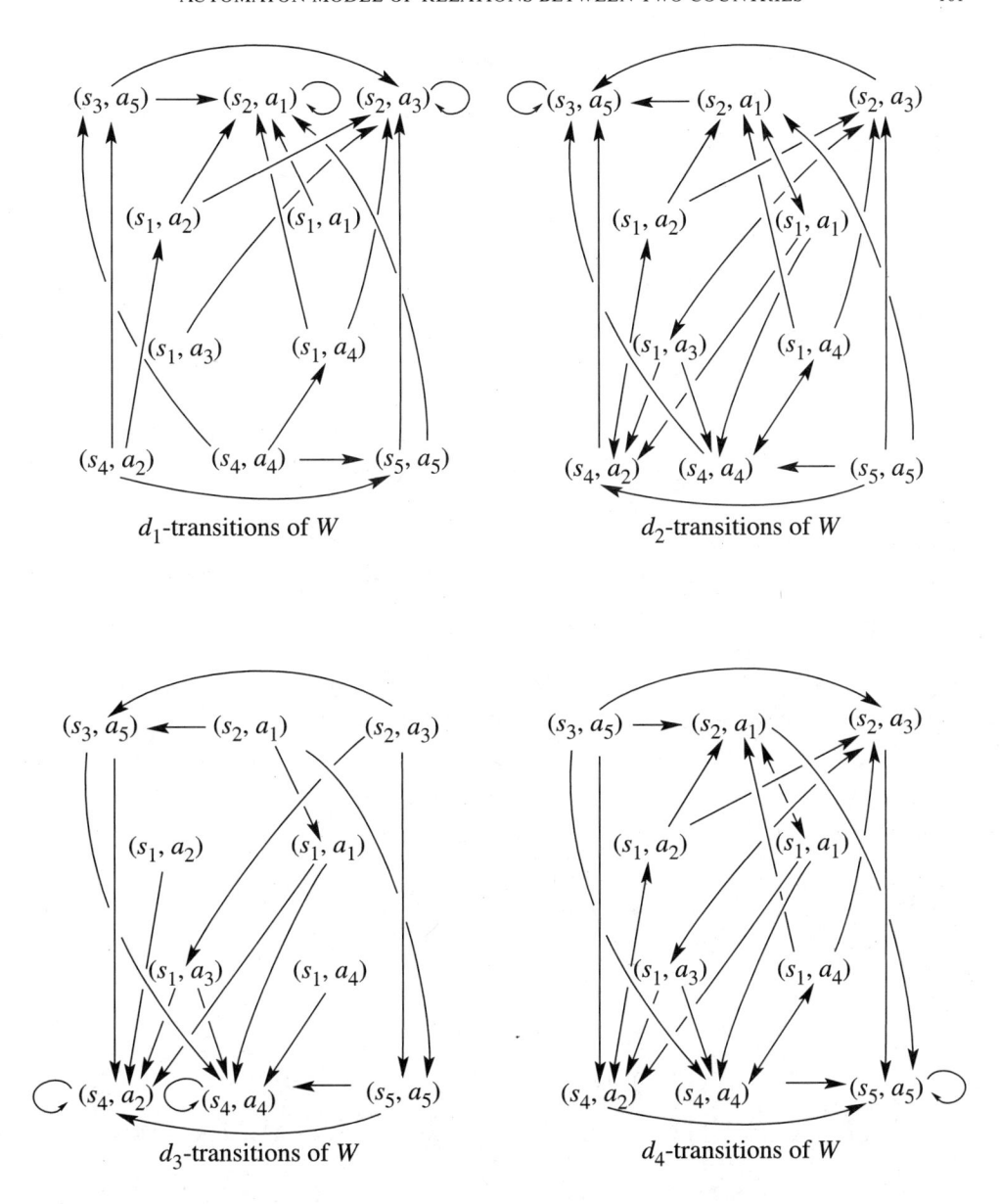

FIGURE 3

where $H1$ ($H3$) is chosen if both leaders intensify (slacken) the inversion tendency, $H2$ is chosen if one of the leaders intensifies and the other one slackens the inversion tendency.

NOTE. If the state point of (3) moves along the limit cycle a_5, this means that the positions of the leaders are uncertain: in the long-term both leaders implement in succession hard and soft politics in the contrary phase. We suppose that this is equivalent to every leader implementing hard and soft politics with equal probability. This is the reason for our choice of the last row in the matrices H_{AD}.

Calculations for $H1$ with $w(0) = w_2$ yield

$$a = \begin{Vmatrix} 0 & .333 & .204 & .207 & .168 & .182 & .172 & .179 & .175 & .177 & .176 \\ 1 & .333 & .204 & .233 & .193 & .197 & .18 & .183 & .177 & .179 & .176 \\ 0 & .333 & .204 & .207 & .168 & .182 & .172 & .179 & .175 & .177 & .176 \\ 0 & .0 & .0 & .126 & .135 & .168 & .165 & .175 & .173 & .176 & .175 \\ 0 & .0 & .389 & .228 & .337 & .271 & .311 & .284 & .3 & .29 & .297 \end{Vmatrix},$$

$$d = \begin{Vmatrix} 0 & .333 & .204 & .207 & .168 & .182 & .172 & .179 & .175 & .177 & .176 \\ 0 & .333 & .301 & .264 & .252 & .25 & .25 & .25 & .25 & .25 & .25 \\ 1 & .333 & .301 & .29 & .277 & .265 & .258 & .254 & .252 & .251 & .251 \\ 0 & .333 & .301 & .264 & .252 & .25 & .25 & .25 & .25 & .25 & .25 \\ 0 & .0 & .097 & .183 & .22 & .236 & .242 & .246 & .248 & .249 & .249 \\ 0 & .0 & .097 & .183 & .22 & .236 & .242 & .246 & .248 & .249 & .249 \end{Vmatrix}.$$

Thus for $k = 0$ both influence groups are direct ($a > 0$, $c > 0$), the first leader maintains a soft position and the second leader maintains a hard one. At the 10th step the behavior of the leaders is chaotic: with probability 0.175–0.176 it is presented by the attractors a_1, a_2, a_3, a_4, and with probability 0.297 by the attractor a_5.

3.2. "Political straightforwardness". Independently from politics, the leader influences the influence group only in one direction, i.e., either only increases, or only decreases its inversion. This group behavior is described by the matrices H_{AD} with one column of ones, the other columns consisting of zeros:

$$H1 = \begin{Vmatrix} 1 & 0 & 0 & 0 \\ 1 & 0 & 0 & 0 \\ 1 & 0 & 0 & 0 \\ 1 & 0 & 0 & 0 \\ 1 & 0 & 0 & 0 \end{Vmatrix}, \quad H2 = \begin{Vmatrix} 0 & 1 & 0 & 0 \\ 0 & 1 & 0 & 0 \\ 0 & 1 & 0 & 0 \\ 0 & 1 & 0 & 0 \\ 0 & 1 & 0 & 0 \end{Vmatrix},$$

$$H3 = \begin{Vmatrix} 0 & 0 & 1 & 0 \\ 0 & 0 & 1 & 0 \\ 0 & 0 & 1 & 0 \\ 0 & 0 & 1 & 0 \\ 0 & 0 & 1 & 0 \end{Vmatrix}, \quad H4 = \begin{Vmatrix} 0 & 0 & 0 & 1 \\ 0 & 0 & 0 & 1 \\ 0 & 0 & 0 & 1 \\ 0 & 0 & 0 & 1 \\ 0 & 0 & 0 & 1 \end{Vmatrix}.$$

In case $H1$ (in case $H3$) both leaders decrease (increase) the inversion of their groups independently of their positions, in case $H2$ (in case $H4$) the first leader decreases (increases) the inversion of the first group, the second leader increases (decreases) the inversion of the second group. For any $w(0)$ a) in the case $H_{AD} = H1$, we have $a(k) \to (0.5; 0; 0.5; 0; 0)$; b) in the cases $H_{AD} = H2$ or $H4$, we have $a(k) \to (0; 0; 0; 0; 1)$; c) in the case $H_{AD} = H3$, we have $a(k) \to (0; 0.5; 0; 0.5; 0)$. Thus in the long run a) both leaders will maintain both hard or soft politics with probability 0.5; b) both leaders successively maintain hard and soft politics in the contrary phase; c) one of leaders will maintain hard politics and the other, soft politics.

Acknowledgements

The author would like to thank Professor Valentine Afraimovich (School of Mathematics, Georgia [U.S.A.] Institute of Technology) for fruitful discussions, and Vadim Tsoglin (Detroit) for help in preparing the English version of the manuscript.

References

1. R. H. Abraham, A. Keith, M. Koebbe, and G. Mayer-Kress, *Computational unfolding of double-cusp models of opinion formation*, Internat. J. Bifurcation Chaos **1** (1991), no. 2, 417–430.
2. M. Kadyrov, *A mathematical model of the relations between two states*, Global Development Processes **3** (1984), Institute for System Studies. (Russian)

Translated by THE AUTHOR

225 Franklin Rd., Apt.44–"O", Atlanta, Georgia 30342

Classification of Curves of Degree 6 Decomposing into a Product of M-Curves in General Position

T. V. Kuzmenko and G. M. Polotovskiĭ

In memory of Professor D. A. Gudkov

ABSTRACT. We consider arrangements of real branches of real 6th degree curves that can be decomposed in a product of more than two M-curves in general position with the maximal number of points of intersection. A detailed description of their isotopy classification, consisting of 163 types, is given in the case of "line×conic×cubic."

§1. Statement of the problem and statistics of the results

The isotopy classification of plane real projective curves of degree 6 decomposed into a product of two transversally intersecting nonsingular curves was obtained by 1979 in [1–3]. Different applications of this classification to other questions relating to the Hilbert 16th problem were found by O. Ya. Viro, D. A. Gudkov, E. I. Shustin, G. M. Polotovskiĭ and others; see details and references in the survey [4]. In the present paper we continue the classification of decomposed curves of degree 6. Namely, we consider the situations when the number of factors is *more than two*. To make the problem more visual, we shall consider only the most important cases determined by general position conditions and by some "maximality conditions."

1.1. Let C_m be a plane projective real curve of degree m (below we call it a curve of degree m). Let $\mathbb{R}C_m$ be the set of its real points and $C_{m_1}, \ldots, C_{m_k}$ be the irreducible (over $\mathbb{R}$) factors of C_m. We assume that the following conditions are fulfilled:
  (i) *every curve C_{m_i} is an M-curve (i.e., $\mathbb{R}C_{m_i}$ consists of $\frac{1}{2}(m_i - 1)(m_i - 2) + 1$ connected components);*
  (ii) *$\mathbb{R}C_{m_i}$ intersects $\mathbb{R}C_{m_j}$ transversally when $i \neq j$;*
  (iii) *every point of the real projective plane $\mathbb{R}P^2$ belongs to at most two sets $\mathbb{R}C_{m_i}$;*
  (iv) *$\mathrm{card}(\mathbb{R}C_{m_i} \cap \mathbb{R}C_{m_j}) = m_i m_j$ for any $i \neq j$.*

1.2. Let S_l be a collection of $\frac{1}{2}(l - 1)(l - 2) + 1$ pairwise disjoint circles, and if l is even (odd), then each circle is two-sided embedded in $\mathbb{R}P^2$ and is called an *oval*

1991 *Mathematics Subject Classification.* Primary 14P25, 14H99.
This work was supported in part by the AMS fSU Aid Fund.

©1996, American Mathematical Society

(resp., except one of them, which is called the *odd branch*). The union of k collections $S_{m_1}, \ldots, S_{m_k}$ will be called a *scheme* $S_{m_1,\ldots,m_k}$ of degree $m = \sum_{i=1}^{k} m_i$ if the conditions (ii)–(iv) that were stated in 1.1 for $\mathbb{R}C_{m_1}, \ldots, \mathbb{R}C_{m_k}$, are fulfilled for $S_{m_1}, \ldots, S_{m_k}$.

Below, the oval (the odd branch) of the scheme S_l is denoted by $O(S_l)$ (resp., $J(S_l)$); similar notation will be used to indicate the type of embedding of branches of a nonsingular curve C_m in $\mathbb{R}P^2$. If $m = 1$ and $m = 2$, then the symbols J and O (resp.) will be omitted.

We shall say that the schemes $S_{m_1,\ldots,m_k}$ and $\widetilde{S}_{m_1,\ldots,m_k}$ are *equivalent* (or that they are of the same type) if there is a homeomorphism of $\mathbb{R}P^2$ transforming $S_{m_1,\ldots,m_k}$ into $\widetilde{S}_{m_1,\ldots,m_k}$ which takes S_{m_i} to $\widetilde{S}_{m_j}$ with $m_i = m_j$.

1.3. The goal of our investigations is to obtain the list of all pairwise different types of schemes $S_{m_1,\ldots,m_k}$, $k > 2$, realizable by curves of degree 6. The statistics of the answers is given in Table 1. For comparison, the cases $k = 2$ (see [1]) and $k = 1$ (i.e., the fragment of the classical result of D. A. Gudkov [5] on nonsingular curves of degree 6) are included in this Table.

TABLE 1

# of the row	k	\multicolumn{6}{c	}{Degrees of factors}	The number of realizable
		m_1	m_2	m_3	m_4	m_5	m_6	types of $S_{m_1,m_2,m_3,m_4,m_5,m_6}$
1	6	1	1	1	1	1	1	4
2	5	1	1	1	1	2	–	23
3	4	1	1	1	3	–	–	19
4	4	1	1	2	2	–	–	109
5	3	1	1	4	–	–	–	26
6	3	1	2	3	–	–	–	163
7	3	2	2	2	–	–	–	105
8	2	1	5	–	–	–	–	15
9	2	2	4	–	–	–	–	42
10	2	3	3	–	–	–	–	46
11	1	6	–	–	–	–	–	3

In the present article we shall describe only the simplest case (six straight lines, row 1 of Table 1) and the most complicated case (row #6).

§2. Arrangements of six straight lines in $\mathbb{R}P^2$

Since the realizability of the schemes $S_{1,1,1,1,1,1}$ is evident, the problem is the enumeration of all pairwise nonequivalent arrangements of six straight lines in $\mathbb{R}P^2$. The answer is the four pictures in Figure 1. They are pairwise different, since they give distinct decompositions of $\mathbb{R}P^2$ into polygons (the label r_s in brackets near a picture means that this decomposition includes r s-gons).

REMARK. The arrangement of n straight lines in general position in the plane was studied earlier (this should be expected, because the problem seems natural), in particular in the context of real algebraic geometry. We found it (after having obtained the answer for $n = 6$) in [6], and then in [7–10]; it turned out that one of our approaches to the solution is practically the same as that in [10]. It is known from [6] and [10] that

$(1_6, 0_5, 9_4, 6_3)$ $(0_6, 6_5, 0_4, 10_3)$

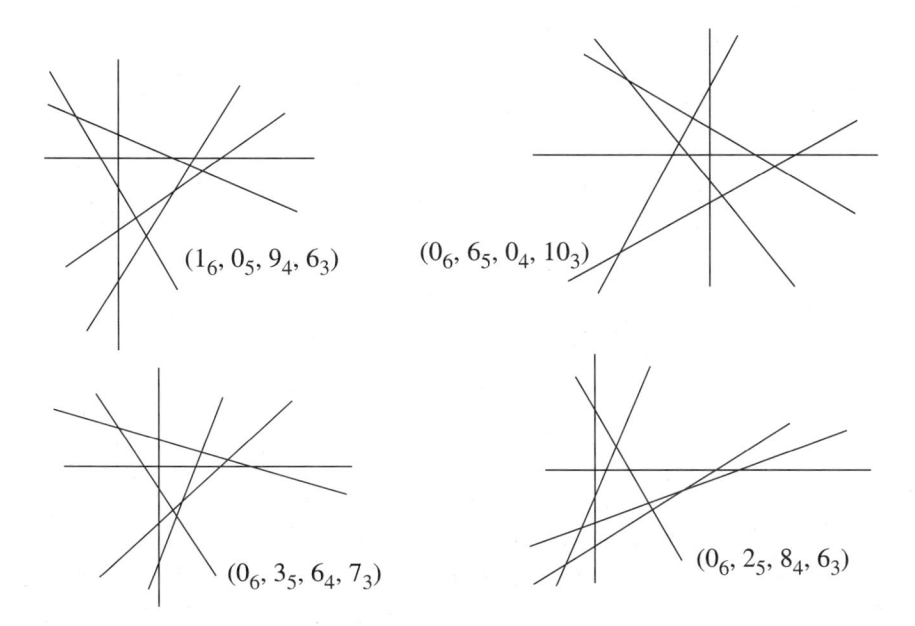

$(0_6, 3_5, 6_4, 7_3)$ $(0_6, 2_5, 8_4, 6_3)$

FIGURE 1

the number of types of arrangements of seven straight lines is equal to 11, and of eight lines is more than 135. We were not able to obtain [11].

§3. The mutual arrangements of straight line, conic and cubic

Below, as the model of $\mathbb{R}P^2$ we use the *Poincaré disk* (i.e., a disk whose diametrically opposite boundary points are identified, so that the boundary becomes a straight line $l \subset \mathbb{R}P^2$).

3.1. Enumeration of logically possible schemes $S_{1,2,3}$ for curves. One can assume without loss of generality that the oval S_2 does not intersect l, and S_3 intersects l transversally. Then, either $\text{card}(l \cap S_3) = 1$, i.e.,

$$\text{card}(l \cap J(S_3)) = 1 \quad \text{and} \quad \text{card}(l \cap O(S_3)) = 0,$$

or $\text{card}(l \cap S_3) = 3$. In the last case, $\text{card}(l \cap O(S_3))$ is equal to 0 or 2. Since $O(C_3)$ is convex, we obtain seven pairwise different types for $S_{2,3}$ (see Figure 2), where in cases A, B, G the oval $O(C_3)$ does not intersect S_2 and it is not shown.

The justification of Figure 2 is not difficult (compare with 3.4.1 below): mark six points $1, 2, \ldots, 6$ on the ellipse C_2, consider all 6! substitutions on the set $\{1, 2, \ldots, 6\}$ (actually it is sufficient to consider 5! substitutions by fixing the point "1") and for each substitution try to draw branches of the nonsingular curve C_3. In many cases this is impossible by the Jordan curve theorem; in other cases it is not difficult to find pairwise nonequivalent pictures.[1] In cases A, B, G, the domains of $\mathbb{R}P^2$ where the oval $O(S_3)$ of a realizable scheme $S_{2,3}$ may be situated may be determined by the following two lemmas.

---

[1] The picture G in Figure 2 was omitted in Figure 1 of [4] (by mistake) but it was only used as an illustration in [4] and was not used further in [4].

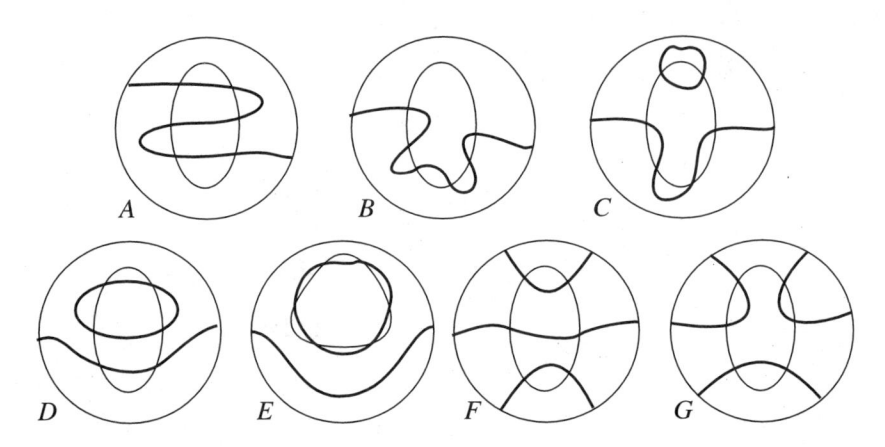

FIGURE 2

3.1.A. LEMMA. *If* $\text{card}(\mathbb{R}C_2 \cap J(C_3)) = 6$ *and the disk bounded by* C_2 *is divided by arcs of* $J(C_3)$ *into three two-gons and a six-gon, then* $O(C_3)$ *lies in the six-gon.*

The conditions of this lemma are fulfilled in situations B and G. In case G, if $O(C_3)$ lies outside the disk bounded by C_2, then a conic intersecting C_3 in eight points may be obtained "by inflation" of C_2, but this contradicts the Bézout theorem; if $O(C_3)$ lies inside a two-gon, then double points of $\mathbb{R}C_2 \cap \mathbb{R}C_3$ may be removed by a small perturbation of degree 5 so that each of the three two-gons turn into an oval (such a perturbation exists by the Brusotti theorem; see for example [**12**, p. 14]). Then we obtain a curve of degree 5 containing four ovals, which lies within one of the others. This also contradicts the Bézout theorem (consider an intersection with a straight line passing through the interior oval).

In case B, the disposition of $O(C_3)$ inside the two-gon is prohibited for the same reasons, and outside C_2 it is prohibited by complex orientation theory (about this theory, see [**13**]), or by quadratic transformations (this is done in [**3**, pp. 147–148]).

3.1.B. LEMMA. *If* $\text{card}(\mathbb{R}C_2 \cap J(C_3)) = 6$ *and the disk bounded by* C_2 *is divided by arcs of* $J(C_3)$ *into two two-gons and two four-gons, then* $O(C_3)$ *lies outside this disk and outside the two-gons bounded by arcs of* C_2 *and* $J(C_3)$.

The conditions of this lemma are fulfilled in situation A in Figure 2. The proof is quite similar to that in the case of two-gons in the previous lemma. In the same way one can obtain the following statement.

3.1.C. LEMMA. *If* $\text{card}(\mathbb{R}C_1 \cap J(C_3)) = 3$, *then* $O(C_3)$ *cannot lie inside a two-gon bounded by a segment of straight line* $\mathbb{R}C_1$ *and an arc of* $J(C_3)$.

Now we must add the scheme S_1 (i.e., a straight line) to each scheme in Figure 2. For this purpose we note that S_2 is divided by its points of intersection with S_3 into six arcs (for example, into the arcs 1–6 in case A in Figure 2). Therefore we must consider all possible pairs (i, j), $i \leqslant j$, which are formed by numbers of arcs on which common points of S_1 and S_2 lie. Since $1 \leqslant i \leqslant j \leqslant 6$, there exist 21 such pairs. But in fact it is sufficient to consider a smaller number of pairs (i, j). For example, in situation A the arcs 1 and 4, as well as 2 and 5, 3 and 6, are equivalent by virtue of symmetry; therefore it is sufficient to consider only 12 pairs: $(1, j)$ with $1 \leqslant j \leqslant 6$ and $(2, 2)$, $(2, 3)$, $(2, 5)$, $(2, 6)$, $(3, 3)$, $(3, 6)$.

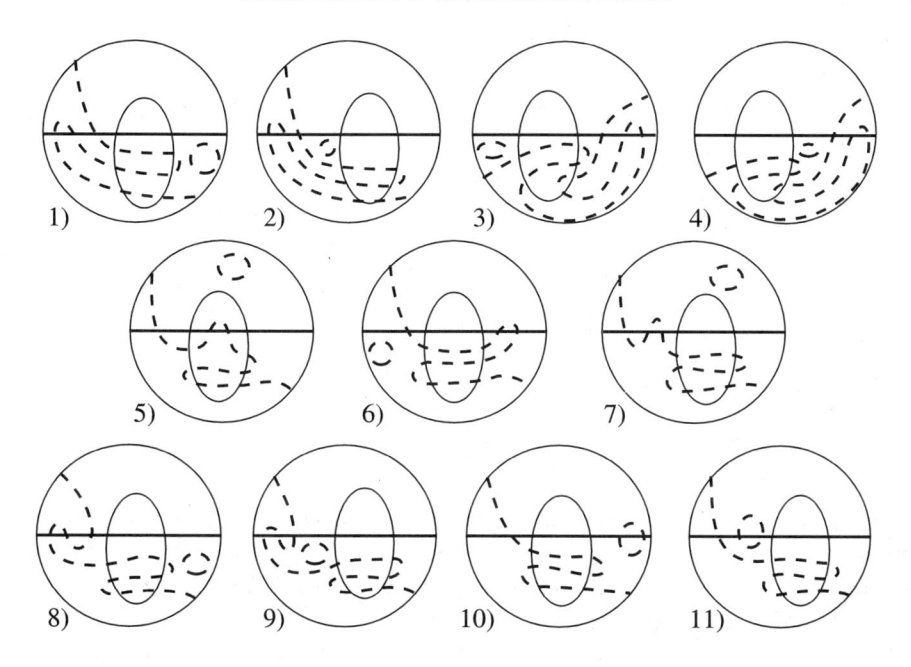

FIGURE 3

Now note that branches of S_3 are divided by points of intersection with S_2 into six arcs. Therefore, for each (i, j) under consideration, we must enumerate all the different possibilities of intersection of S_1 with these arcs under the condition card $(S_1 \cap S_3) = 3$ (see the condition (iv) in 1.1). Note that two cases card $(S_1 \cap S_3) = 0$ or 2 are possible.

Taking into account everything stated above, including Lemmas 3.1.B and 3.1.C, we obtain the list (see Figure 3) of logically possible types of schemes $S_{1,2,3}$ for situation A in Figure 2 in the case $(i, j) = (1, 1)$.

Similar lists were obtained for every pair (i, j) for all situations in Figure 2. In all there are more than 200 logically possible types of schemes $S_{1,2,3}$.

3.2. Nonrealizability of the schemes $S_{1,2,3}$. In this subsection methods for proving the nonrealizability of a number of types of schemes $S_{1,2,3}$ by 6th degree curves are described (in all more than 40 types from the list of logically possible schemes). The simplest method is

3.2.1. *Reduction to a contradiction with the Bézout theorem by the small parameter method.* This is a classical method and it was applied essentially in the proof of Lemma 3.1.A, therefore, we only consider one example. Namely, consider scheme 3) in Figure 3. If there exists a 6th degree curve realizing it, then we remove all its double points by a small perturbation of degree 6 in order to obtain the nonsingular curve C_6 with $\mathbb{R}C_6$ shown in Figure 4 by the dotted line (the existence of the required perturbation is guaranteed by the Brusotti theorem). But $\mathbb{R}C_6$ has two nonempty ovals, α_1 and α_2. This contradicts Bézout's theorem: the straight line AB intersecting the inner ovals (with respect to different outer ones) have not less than eight common points with $\mathbb{R}C_6$.

Note that for this proof the position of $O(C_3)$ does not matter. Therefore, we have also proved the nonrealizability of scheme 4) in Figure 3.

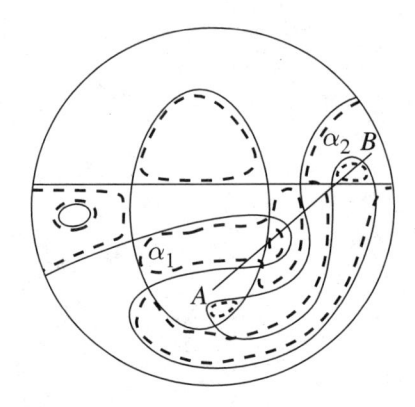

FIGURE 4

3.2.2. *Reduction to the classification of the schemes* $S_{3,3}$. Let some scheme $S_{1,2,3}$ be realized by a 6th degree curve having factors C_1, C_2, C_3. By a small perturbation of degree 3, we remove both double points of the curve $C_1 C_2$, so that it becomes a nonsingular curve $\widetilde{C}_3$. By the Brusotti theorem, this may be done in four ways. As the result, we obtain four curves of type $\widetilde{C}_3 C_3$ realizing some schemes $S_{3,3}$ (not necessarily of different types). If at least one of them is nonrealizable by virtue of the classification in [1, 2], we have a contradiction; this means that the initial scheme $S_{1,2,3}$ cannot be realized by a 6th degree curve. In this way we can prove the nonrealizability of 14 types of schemes $S_{1,2,3}$ which are not prohibited by the method of 3.2.1.

REMARK. There are other ways for proving the nonrealizability of the schemes $S_{m_1,\dots,m_k}$, namely: using a contradiction with the Bézout theorem by an immediate intersection with a straight line; an application of known congruences for curves with singularities (see for example [14, 15]); obtaining a contradiction by making use of quadratic transformations as in [1–3]. But here it turned out that it is sufficient to use the methods of 3.2.1 and 3.2.2 only.

3.3. Methods of realization of the schemes $S_{1,2,3}$.

3.3.1. *Different versions of the small parameter method.* The main idea of this subsection is to construct curves $C_{1,2,3}$ by using 6th degree curves decomposed into a larger number of factors.

3.3.1.A. *Construction of the curves $C_{1,2,3}$ from the curves $C_{1,1,2,2}$.* Let the factors of $C_{1,1,2,2}$ be the straight lines L_1, L_2 and the conics K_1, K_2. Choose one of the four pairs (i, j), where $i, j \in \{1, 2\}$, and remove both double points of $L_i K_j$ by a 3rd degree permutation so that as a result we get an M-curve C_3. There exist two collections of such removings. Thus, we obtain 8 curves $C_{1,2,3} = L_k K_m C_3$ ($k \neq i$, $m \neq j$) from every curve $C_{1,1,2,2}$, in all $109 \cdot 8$ curves $C_{1,2,3}$ (see row #4 of Table 1). So, we have realized 121 types of schemes $S_{1,2,3}$ (many types are repeated more than once). It remains to remark that obtaining the list of realizable types of schemes $S_{1,1,2,2}$ causes no difficulties.

To realize the schemes $S_{1,2,3}$, it is possible to use the curves $L_1 L_2 K_1 K_2$ for which the "maximality condition" (iv) from 1.1 is not fulfilled. As an example, let us consider the following construction.

Let $\mathbb{R}L_1$ intersect $\mathbb{R}K_1$ transversally and let $\mathbb{R}\widetilde{K}_2$ be inscribed in one of the halves into which the interior of K_1 is divided by $\mathbb{R}L_1$ (see Figure 5). Choose $\widetilde{L}_2$ so that

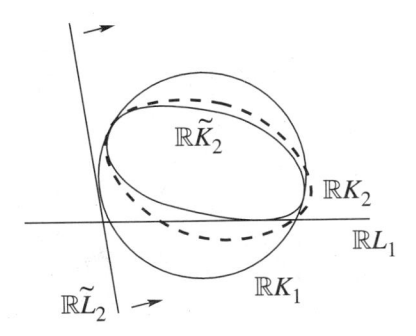

FIGURE 5

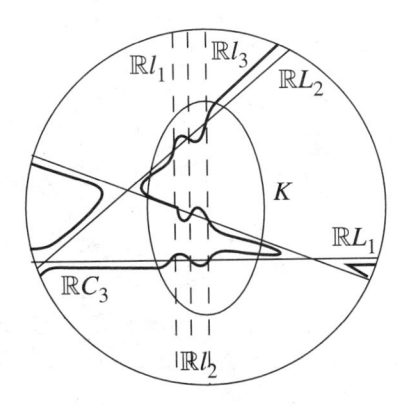

FIGURE 6

$\mathbb{R}\widetilde{L}_2$ touches K_1 as shown in the picture, and "inflate" $\widetilde{K}_2$ into a conic K_2 so that $\mathbb{R}K_2$ intersects $\mathbb{R}K_1$ in four points, $\mathbb{R}K_2$ intersects $\mathbb{R}L_1$, but $\mathbb{R}K_2 \cap \mathbb{R}\widetilde{L}_2 = \varnothing$. By a small perturbation of the tangent $\mathbb{R}\widetilde{L}_2$, we obtain a secant $\mathbb{R}L_2$ such that $\mathbb{R}K_2 \cap \mathbb{R}L_2 = \varnothing$ but $\mathbb{R}L_2$ intersects $\mathbb{R}L_1$ outside K_1 as before. It is clear that the curve $L_1 K_1 C_3$ realizes some scheme $S_{1,2,3}$, where C_3 is an indecomposable curve obtained by a small perturbation of coefficients of $K_2 L_2$ that does not change the isotopy type of $L_1 L_2 K_1 K_2$.

3.3.1.B. *Modifications of classical methods for constructing M-curves*. By "classical methods" here we mean the methods of Harnack and of Brusotti. Since they have been described repeatedly in detail (see for example [12, 16]) we shall consider only the example of constructing the curves $C_{1,2,3}$.

1) Let three straight lines L_1, L_2, L_3 and the conic K be arranged in $\mathbb{R}P_2$ as shown in Figure 6. We define C_3 by the formula

$$C_3 = L_1 L_2 L_3 + t l_1 l_2 l_3,$$

where l_i are auxiliary lines which are situated as shown in Figure 6 by dotted lines. If $|t|$ is sufficiently small and the sign of t is chosen in the proper way, then the curves $C_{1,2,3} = L_i K C_3$ ($i \in \{1, 2, 3\}$) realize some schemes $S_{1,2,3}$. If we choose l_i in a different way (so that some or all of the curves intersect L_1, L_2, L_3 outside K) and/or if double points of $L_1 L_2 L_3$ are situated in a different way with respect to K, we realize other types of schemes $S_{1,2,3}$ by the same formulas (some of them, but not all, were already realized by the method of 3.3.1.A).

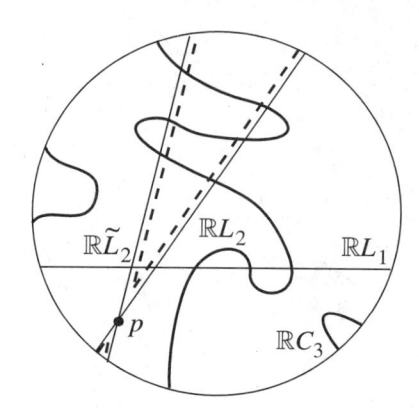

FIGURE 7

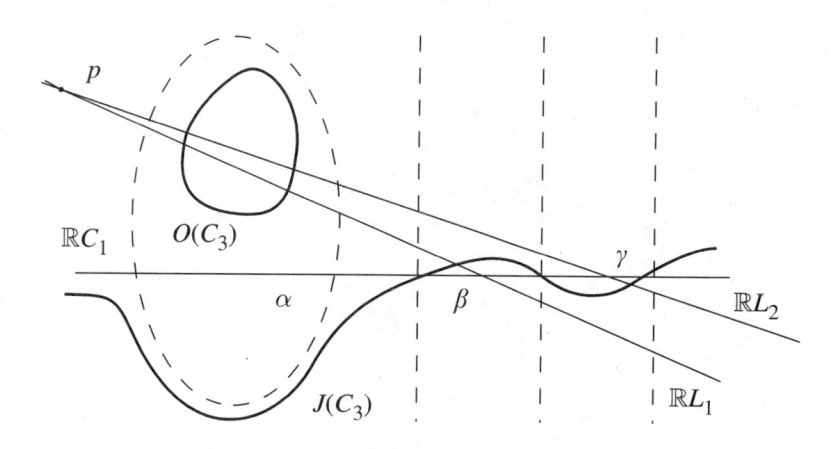

FIGURE 8

2) Define C_3 as in case 1) and C_2 by the formula

$$C_2 = L_2 \widetilde{L}_2 + \varepsilon D_2,$$

where $\mathbb{R}\widetilde{L}_2$ is the result of a sufficiently small rotation of $\mathbb{R}L_2$ about the point p and D_2 is a second degree curve with $\mathbb{R}D_2 = \varnothing$. By a suitable choice of sign of the small parameter ε, the curve $L_1 C_2 C_3$ realizes the scheme 8) in Figure 3 (see Figure 7, where $\mathbb{R}C_2$ is shown by a dotted line).

3) The mutual position of $\mathbb{R}C_1$ and $\mathbb{R}C_3$ shown in Figure 8 is obtained by the standard construction of Harnack. Choose L_1 as shown in Figure 8. Let $\mathbb{R}L_2$ be the result of rotation of $\mathbb{R}L_1$ about the point p. Define $C_{1,2,3}$ by the formulas

$$C_{1,2,3} = C_1 C_2 C_3, \qquad C_2 = L_1 L_2 + \varepsilon D_2,$$

where ε and D_2 are the same as in 2). By considering different combinations of dispositions of points $\mathbb{R}L_1 \cap \mathbb{R}C_1$ and $\mathbb{R}L_2 \cap \mathbb{R}C_1$ on the segments $\alpha - \gamma$ of $\mathbb{R}C_1$, we realize a number of schemes $S_{1,2,3}$, some of which were not realized by the methods described above.

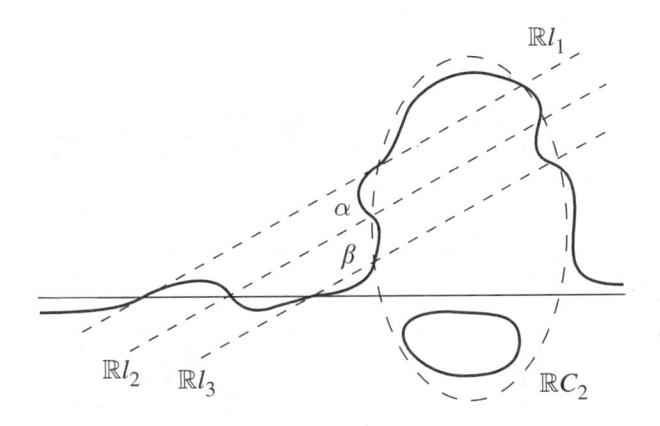

FIGURE 9

4) Define C_3 by the formula $L_1 C_2 + \varepsilon l_1 l_2 l_3$; the arrangement of curves in $\mathbb{R}P^2$ is shown in Figure 9. The curve $L_1 C_2 C_3$ realizes the scheme $S_{1,2,3}$, and its type is different from those realized above.

Further, let the line $\mathbb{R}\widetilde{L}_2$ be tangent to the arc α of a branch of $\mathbb{R}C_3$ and to the arc β of C_2. Such a common tangent exists due to the convexity of C_2 and to the transversality of the intersection of $\mathbb{R}C_2$ with $\mathbb{R}C_3$. By a small perturbation of the coefficients of $\widetilde{L}_2$ we get a line L_2 that intersects C_3 at two points on the arc α and C_2 at two points on the arc β. The curve $L_2 C_2 L_3$ gives one more type of scheme $S_{1,2,3}$ was not realized above.

In this way (by the displacement of a tangent to the position of a secant), it is easy to realize many schemes $S_{1,2,3}$ with card$(l \cap J(C_3)) = 3$ (i.e., with the fragment of the type of G in Figure 2).

3.3.2. *Application of quadratic transformations.* Here the transformation T of the plane $\mathbb{R}P^2(x)$ with coordinates $x = (x_0:x_1:x_2)$ into the plane $\mathbb{R}P^2(y)$ with coordinates $y = (y_0:y_1:y_2)$ defined by the formulas $y_0 = x_1 x_2$, $y_1 = x_0 x_2$, $y_2 = x_0 x_1$ is considered. All the necessary facts about this transformations may be found in [17]. Note only that T is a diffeomorphism outside the coordinate axes, and the image $TF(y)$ of the curve $F(x)$ is defined by the relation

$$\pi(y) TF(y) = F(y_1 y_2, y_0 y_2, y_0 y_1),$$

where the polynomials $\pi(y) = y_0^\alpha y_1^\beta y_2^\gamma$ and $TF(y)$ are relatively prime. The situation of $TF(y)$ in $\mathbb{R}P^2$ and its degree depend on the choice of the irregular points $(1:0:0)$, $(0:1:0)$, $(0:0:1)$ in $\mathbb{R}P^2(x)$.

By using quadratic transformations we were able to realize all types of schemes $S_{1,2,3}$ not prohibited in 3.2 and not realized in 3.3.1. For example, the scheme $S_{1,2,3}$ with $S_1 = \mathbb{R}T(C_2)$, $S_2 = \mathbb{R}T(C_1)$, $S_3 = \mathbb{R}T(C_3)$ (Figure 10, on the right) is obtained from the scheme $\widetilde{S}_{1,2,3}$ of type 8) of Figure 3 with $\widetilde{S}_1 = \mathbb{R}C_1$, $\widetilde{S}_2 = \mathbb{R}C_2$, $\widetilde{S}_3 = \mathbb{R}C_3$ (Figure 10, on the left), which was realized in case 2) of 3.3.1.B. In Figure 10, the arcs in the image and preimage corresponding to each other by T are denoted by the same numbers.

REMARK. Although quadratic transformations are a classical tool, Gudkov [5] was the first, to our knowledge, to apply this tool in the first part of the Hilbert 16th problem. In particular, the use of this method allowed Gudkov to construct

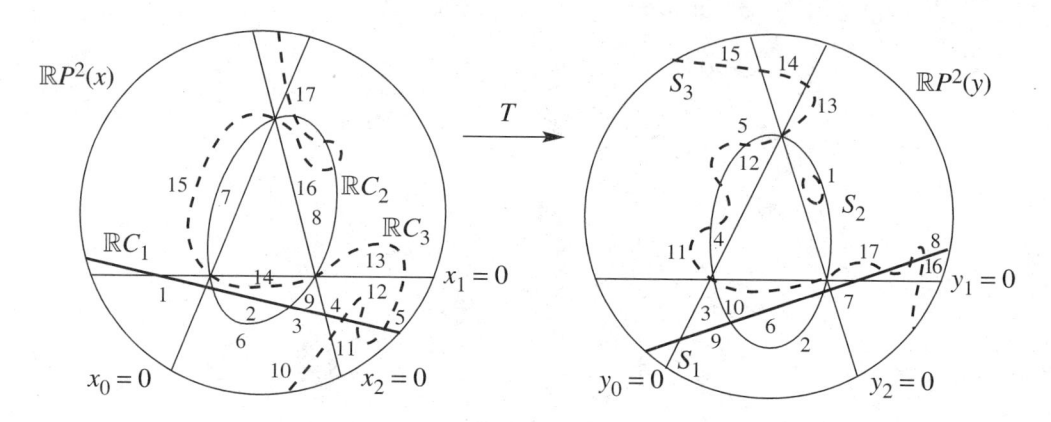

FIGURE 10

an M-curve of degree 6 exactly five ovals of which are enclosed by one of the six others. Note that Gudkov's initial construction was extraordinarily complicated. Quadratic transformations appeared to be necessary in [1–3] too. Later Viro introduced quadratic transformations of another kind into the Hilbert 16th problem; following Newton, he called them "hyperbolisms." Viro and some other authors effectively applied hyperbolisms to this problem (see [16, 4]).

3.4. The main result (classification of schemes $S_{1,2,3}$). Unfortunately, the formulation of the main result is a long enumeration, therefore, as in [1, 2], a coding of schemes is needed.

3.4.1. *Coding the schemes $S_{1,2,3}$.* The code is a substitution on the set $\{1, 2, \ldots, 9\}$ written as a product of cycles and provided with the following additional elements: the integer m, $0 \leqslant m \leqslant 5$, written before the substitution, and the symbols R, L and $*$, which may occur inside cycles. In addition, the symbols 7, 8, 9 may be underlined or overlined.

To restore a scheme from its code, one should draw two concentric circles: the outer one represents the line $l \subset \mathbb{R}P^2$ and the inner one represents the scheme S_2. Then it is necessary to draw the horizontal diameter (the scheme S_1) of the outer circle, to mark m points on the upper semicircle of S_2, $6 - m$ points on its lower semicircle, and enumerate these six points counter-clockwise, beginning with the leftmost point of the lower semicircle. Then mark three points 7, 8, 9 on the diameter S_1 from left to right, where the point lies on the left outside (resp., inside, on the right outside) the circle S_2 if its number in the substitution has a bar above (resp., has no bar, has a bar below).

The branch $J(S_3)$ ($O(S_3)$) connects the marked points in the order stated in the first cycle (resp., in the second cycle if it exists), and the arc joining the first point of the cycle with the second one is drawn inside the oval S_2 if this is possible. If there is no second cycle, $O(S_3)$ does not intersect $S_1 \cup S_2$. In this case between some symbols i_k and i_{k+1} of the substitution the symbol R (L) appears, which means that $O(S_3)$ must be situated in the domain adjoining to the arc $i_k i_{k+1}$ from the right (resp., from the left). Finally, the symbol $*$ between i and j means that the arc of $J(S_3)$ intersects l once on the way from i to j.

REMARK. In the case of code #105 from Table 2, the oval $O(S_3)$ must be situated so that Lemmas 3.1.A and 3.1.C are satisfied.

3.4.2. THE MAIN THEOREM. *Sixth degree curves realize the* 163 *pairwise different types of schemes* $S_{1,2,3}$ *whose codes are enumerated in Table 2, and only these types.*

TABLE 2

#	Code		#	Code		#	Code	
1	0(16$\underline{8}$952R34*$\overline{7}$)	A	36	0(2345*$\overline{7}$)(169$\underline{8}$)	A	71	0(34$\underline{9}$*)(256871)	A
2	0(16L52$\overline{8}$734*$\underline{9}$)	A	37	1(2345$\underline{9}$*)(186$\overline{7}$)	A	72	0(34$\underline{9}$*)(258761)	A
3	0(1652R34*$\overline{789}$)	A	38	2(2349$\underline{5}$*)(186$\overline{7}$)	A	73	0(34$\underline{9}$87*)(2561)	A
4	0(165234*$\overline{7}$)($\overline{89}$)	A	39	3(23$\underline{9}$45*)(186$\overline{7}$)	A	74	1(34$\underline{9}$*)(258671)	A
5	0(18$\underline{9}$652R34*$\overline{7}$)	A	40	0(345$\underline{6}$9*)(218$\overline{7}$)	A	75	2(34$\underline{9}$*)(285671)	A
6	0(16L5234*$\underline{9}$87)	B	41	0(345$\underline{6}$9*)(2871)	A	76	3(3$\underline{9}$4*)(285671)	A
7	1(176L8$\underline{5}$234$\underline{9}$*)	A	42	0(349$\underline{8}$567*)(12)	A	77	0(45$\underline{9}$*)(1236$\underline{8}$7)	A
8	2(176L58$\underline{2}$34$\underline{9}$*)	A	43	0(345$\underline{6}$987*)(12)	A	78	1(45$\underline{9}$*)(12386$\overline{7}$)	A
9	3(176L58$\underline{2}$394*)	A	44	0(34567*$\underline{9}$8)(12)	T	79	2(49$\underline{5}$*)(12386$\overline{7}$)	A
10	4(176L58$\underline{2}$934*)	A	45	1(3459$\underline{6}$*)(218$\overline{7}$)	A	80	3(45*$\underline{9}$)(12386$\overline{7}$)	A
11	5(17$\underline{6}$5234*)($\underline{89}$)	A	46	1(34596*)(2871)	A	81	2(23$\underline{9}$*)(148567)	A
12	5(19$\underline{6}\overline{8}$752R34*)	A	47	1(349$\underline{8}$576*)(12)	A	82	0(56$\underline{9}$*)(23418$\overline{7}$)	A
13	5(17652$\underline{8}$R$\underline{9}$34*)	A	48	1(34576*$\underline{9}$8)(12)	T	83	0(56$\underline{9}$87*)(2341)	A
14	5(17652R34*$\underline{9}$8)	A	49	1(3459$\underline{8}$76*)(12)	A	84	0(56$\underline{9}$*)(234871)	B
15	5(176L589$\underline{2}$34*)	A	50	2(3495$\underline{6}$*)(218$\overline{7}$)	A	85	1(5$\underline{9}$6*)(14327$\overline{8}$)	A
16	5(1987652R34*)	B	51	2(3495$\underline{6}$*)(2871)	A	86	1(59$\underline{6}\overline{8}$7*)(1432)	A
17	0(21$\overline{7}$863R45$\underline{9}$*)	A	52	2(349$\underline{5}$876*)(12)	A	87	1(59$\underline{6}$*)(234871)	A
18	2(23*65$\underline{7}$41)($\underline{89}$)	A	53	2(349$\underline{5}$687*)(12)	A	88	1(59876*)(1432)	A
19	1(21$\overline{7}$L683459*)	A	54	2(349$\underline{8}$756*)(12)	A	89	2(65$\underline{9}$*)(14327$\overline{8}$)	A
20	3(21$\overline{7}$L683$\underline{9}$45*)	A	55	3(39456*)(218$\overline{7}$)	A	90	2(65$\underline{9}$*)(234871)	A
21	2(23*65$\underline{9}$L4871)	A	56	3(39456*)(2871)	A	91	2(657$\underline{8}\underline{9}$*)(1432)	A
22	2(23*67859$\underline{4}$1R)	B	57	3(37456*$\underline{9}$8)(12)	A	92	2(6785$\underline{9}$*)(1432)	A
23	2(23*65$\underline{9}$L41$\overline{8}$7)	A	58	3(3945876*)(12)	A	93	0(16$\underline{8}$9*$\overline{7}$)(3452)	B
24	2(23*$\overline{7}$R865941)	A	59	3(3987456*)(12)	A	94	0(1896*$\overline{7}$)(3452)	B
25	2(23*65L78941)	B	60	2(1234*$\overline{7}$)(658$\underline{9}$)	A	95	0(16*$\underline{9}$)(34528$\overline{7}$)	A
26	2(23*659L8$\underline{7}$41)	B	61	2(1234*$\overline{7}$)(6895)	A	96	0($\underline{9}$*)(123456$\underline{8}$7)	A
27	0(1456$\underline{9}$*32)($\overline{78}$)	A	62	2(189234*$\overline{7}$)(56)	A	97	1($\underline{9}$*)(12345867)	A
28	0(14569*32$\overline{7}$L$\overline{8}$)	A	63	2(1234*$\overline{789}$)(56)	A	98	2($\underline{9}$*)(1234$\underline{8}$567)	A
29	0(149R8$\underline{5}$67*32)	T	64	0(1654*$\overline{789}$)(23)	T	99	3($\underline{9}$*)(12384567)	A
30	0(321784R569$\underline{~}$*)	T	65	0(1654*$\underline{9}$)(328$\overline{7}$)	A	100	0($\underline{9}$*)(23456871)	A
31	3(23*65471)($\underline{89}$)	A	66	0(189654*$\overline{7}$)(23)	A	101	0($\underline{9}$87*)(234561)	A
32	3(65L481239*$\overline{7}$)	A	67	0(16$\underline{8}$954*$\overline{7}$)(23)	A	102	0($\underline{9}$*)(23458761)	A
33	3(654918R723*)	A	68	1(186$\underline{9}$54*$\overline{7}$)(23)	A	103	2($\underline{9}$*)(23485671)	A
34	3(23*6785L491)	T	69	2(17658$\underline{4}\underline{9}$*)(23)	A	104	0(123456*$\overline{7}$)(89)	A
35	3(23L*6549871)	B	70	0(218$\underline{9}$65*$\overline{7}$)(34)	A	105	0(123456*$\overline{789}$)	B

TABLE 2 (CONTINUED)

#	Code		#	Code		#	Code	
106	0(18L923456*$\bar{7}$)	T	126	0(2L3456$\bar{1}$$\underline{98}$7*)	T	145	3(683$\underline{9}$*$\bar{7}$)(12*54*)	A
107	0(1L2$\bar{8}$73456*$\bar{9}$)	A	127	0(2L34569$\bar{8}$1$\bar{7}$*)	T	146	3(67893*)(12*54*)	A
108	0(1L2345$\overline{68}$7*$\bar{9}$)	T	128	1(2L3459$\underline{6}$81$\bar{7}$*)	A	147	2(14*$\bar{8}$)(23*65$\underline{9}$*$\bar{7}$)	A
109	1(1234596*)(78)	A	129	2(2L349566$\underline{1}$$\bar{7}$*)	A	148	2(14*$\bar{7}$)(23*6895*)	A
110	1(1L23459876*)	T	130	3(2L39456817*)	A	149	2(14$\underline{9}$*$\overline{78}$)(23*65*)	A
111	1(123459L6*$\overline{78}$)	T	131	2(2349L5*$\bar{7}$681)	A	150	2(1894*$\bar{7}$)(23*65*)	A
112	1(1L234$\underline{98}$576*)	A	132	1(2R1$\bar{8}$69543*$\bar{7}$)	A	151	0(12*$\underline{95}$L6$\underline{8}$*34*$\bar{7}$)	B
113	1(17L8234596*)	A	133	0(2R1$\bar{8}$96543*$\bar{7}$)	A	152	1(12*$\underline{95}$L86*34*$\bar{7}$)	B
114	1(123457L896*)	T	134	2(3L495$\overline{68}$12$\bar{7}$*)	A	153	2(21*$\underline{9}$43*65*)(78)	B
115	1(1L234576*$\underline{98}$)	T	135	0(3R21786549*)	A	154	2(21*$\underline{9}$4R3*6785*)	B
116	2(1234$\underline{9}$56*)(78)	A	136	0(3R2$\bar{7}$816549*)	A	155	3(12*56*394*)(78)	B
117	2(17L8234$\underline{9}$56*)	A	137	0(25$\underline{9}$*)(16$\underline{8}$*34*$\bar{7}$)	A	156	3(12*58L76*394*)	B
118	2(1L234$\underline{9}$5876*)	A	138	1(25$\underline{9}$*)(186*34*$\bar{7}$)	A	157	3(1L2*56*39874*)	B
119	2(1L234$\underline{98}$756*)	T	139	2(285*)(176*34$\underline{9}$*)	A	158	0(2L3*$\underline{8}$61$\bar{7}$*459*)	B
120	2(1L234756*$\underline{98}$)	T	140	3(285*)(176*394*)	A	159	2(2L3*671*$\underline{9}$485*)	B
121	2(123495L6*$\overline{78}$)	T	141	0(368*)(21$\bar{7}$*459*)	A	160	2(12*5L6*34*$\overline{789}$)	B
122	3(1239456*)(78)	A	142	1(386*)(21$\bar{7}$*459*)	A	161	3(2L3*6918$\bar{7}$*45*)	T
123	3(1L23945876*)	A	143	3(386*)(21$\bar{7}$*945*)	A	162	2(34*$\bar{7}$18L92*56*)	T
124	3(1L2394568$\bar{7}$*)	T	144	3(396*)(2871*45*)	A	163	2(1782*5L6*34$\underline{9}$*)	B
125	0(2L345$\underline{98}$61$\bar{7}$*)	B						

The scheme and some steps of the proof are described in the previous sections. The method of realization of each scheme is shown in Table 2 after the code: the letters A and B denote corresponding items of 3.3.1, and the letter T means the application of the quadratic transformations listed below. The arrow is directed from the preimage to the image, preimages have been realized before, and points chosen as irregular ones are marked by tildes.

$0(2L\tilde{3}\tilde{4}\tilde{5}\underline{98}6\tilde{1}\bar{7}*)$[#125] → $0(14\underline{9}R\underline{8}567*32)$[#29],

$2(\tilde{2}\tilde{3}*65\underline{9}L\tilde{4}18\bar{7})$[#23] → $0(3217\tilde{8}\tilde{4}R\tilde{5}\underline{6}\underline{9}*)$[#30] → $2(1L234\underline{98}756*)$[#119],

$5(\tilde{1}7\tilde{6}5\tilde{2}\underline{8}R\underline{9}34*)$[#13] → $3(23*6785L491)$[#34],

$0(\tilde{3}\tilde{4}\underline{9}*)(2\tilde{5}\underline{8}761)$[#72] → $0(34567*\underline{98})(12)$[#44],

$1(\tilde{3}\tilde{4}\underline{9}*)(2\tilde{5}\underline{8}671)$[#74] → $0(34576*\underline{98})(12)$[#48],

$0(\tilde{4}\tilde{5}\underline{9}*)(123\tilde{6}\underline{8}\bar{7})$[#77] → $0(1654*\overline{789})(23)$[#64],

$2(\tilde{1}2*5L6*\tilde{3}\tilde{4}*\overline{789})$[#160] → $0(18L9\tilde{2}\tilde{3}\tilde{4}5\tilde{6}*\bar{7})$[#106] → $2(123495L6*\overline{78})$[#121],

$0(165\tilde{2}R\tilde{3}\tilde{4}*\overline{789})$[#3] → $0(1L2345\overline{68}7*\bar{9})$[#108],

$3(65L48\tilde{1}\tilde{2}\tilde{3}\underline{9}*\bar{7})$[#32] → $1(1L23459876*)$[#110],

$3(2\tilde{3}L*65\tilde{4}987\tilde{1})$[#35] → $1(123459L6*\overline{78})$[#111],

$5(\tilde{1}987\tilde{6}\tilde{5}2R34*)$[#16] → $1(123457L896*)$[#114],

$$3(\tilde{1}L\tilde{2}*56*398\tilde{7}\tilde{4}*)[\#157] \to 1(1L234576*\underline{98})[\#115],$$
$$3(6\tilde{5}491\overline{8}R\overline{7}\tilde{2}\tilde{3}*)[\#33] \to 3(1L23945\overline{687}*)[\#124],$$
$$0(18965\tilde{2}R\tilde{3}\tilde{4}*\overline{7})[\#5] \to 2(1L234\underline{7}56*\underline{98})[\#120],$$
$$0(\tilde{1}\tilde{6}L\tilde{5}234*\overline{987})[\#6] \to 0(2L3456\overline{1987}*)[\#126] \quad \text{(see Figure 10)},$$
$$2(\tilde{2}\tilde{3}*\overline{7}R\overline{8}65\underline{9}4\tilde{1})[\#24] \to 0(2L3456981\overline{7}*)[\#127],$$
$$1(\tilde{1}7L8\tilde{2}34\tilde{5}96*)[\#113] \to 3(2L3*691\overline{87}*45*)[\#161],$$
$$2(\tilde{1}7L8\tilde{2}\tilde{3}4\underline{9}56*)[\#117] \to 2(34*\overline{7}18L92*56*)[\#162].$$

References

1. G. M. Polotovskiĭ, *Catalog of M-decomposed curves of 6th degree*, Dokl. Akad. Nauk SSSR **236** (1977), no. 3, 548–551; English transl., Soviet Math. Dokl. **18** (1977), no. 5, 1241–1245.
2. _____, *($M-1$)- and ($M-2$)-decomposed curves of 6th degree*, Methods of Qualitative Theory of Differential Equations, Gor'ky State Univ., Gor'ky, 1978, pp. 130–148. (Russian)
3. _____, *Topological classification of decomposed curves of 6th degree*, Ph. D. Thesis, Gor'ky State Univ., Gor'ky, 1979. (Russian)
4. _____, *On the classification of decomposed plane algebraic curves*, Lecture Notes in Math., vol. 1524, Springer-Verlag, Berlin, Heidelberg, and New York, 1992, pp. 52–74.
5. D. A. Gudkov, *Complete topological classification of the arrangement of ovals of a 6th degree curve in the projective plane*, Uch. Zap. Gor'kov. Gos. Univ. **87** (1969), Gor'ky, 118–153; English transl. in D. A. Gudkov and G. A. Utkin, *Nine Papers on Hilbert's 16th Problem*, Amer. Math. Soc. Transl. Ser. 2, vol. 112, Amer. Math. Soc., Providence, RI, 1978.
6. G. Gonzales-Sprinberg and V. Ruggiero, *Petites déformations de droites dans le plan projectif réél*, Ann. Univ. Ferrara Sez. VII, Sci. Mat. **XXIX** (1983), 179–210.
7. H. S. White, *The plane figure of seven real lines*, Bull. Amer. Math. Soc. **38** (1932), no. 2, 59–65.
8. L. D. Cammings, *Hexagonal systems of seven lines in a plane*, Bull. Amer. Math. Soc **38** (1932), no. 2, 105–110.
9. _____, *Heptagonal systems of eight lines in a plane*, Bull. Amer. Math. Soc. **38** (1932), 700–702.
10. _____, *On a method of comparison for straight-line nets*, Bull. Amer. Math. Soc. **39** (1933), 411–416.
11. B. Grünbaum, *Arrangements and spreads*, CBMS Regional Conference in Math., vol. 10, Amer. Math. Soc., Providence, RI, 1972.
12. D. A. Gudkov, *The topology of real projective algebraic varieties*, Uspekhi Mat. Nauk **29** (1974), no. 4, 3–79; English transl., Russian Math. Surveys **29** (1974), no. 4, 1–79.
13. V. A. Rokhlin, *Complex topological characteristics of real algebraic curves*, Uspekhi Mat. Nauk **33** (1978), no. 5, 77–89; English transl., Russian Math. Surveys **33** (1978), no. 5, 85–98.
14. O. Ya. Viro and V. M. Kharlamov, *Congruences for real algebraic curves with singularities*, Uspekhi Mat. Nauk **35** (1980), no. 4, 154–155; English transl. in Russian Math. Surveys **35** (1980).
15. V. M. Kharlamov and O. Ya. Viro, *Extensions of the Gudkov–Rokhlin congruence*, Lecture Notes in Math, vol. 1346, Springer-Verlag, Berlin, Heidelberg, and New York, 1988, pp. 357–406.
16. O. Ya. Viro, *Real algebraic plane curves: constructions with controlled topology*, Algebra i Analiz **1** (1989), no. 5, 1–73; English transl., Leningrad Math. J. **1** (1990), no. 5, 1059–1134.
17. R. J. Walker, *Algebraic curves*, Princeton Univ. Press, Princeton, NJ, 1950.

Translated by G. M. POLOTOVSKIĬ

23, Gagarin avenue, Nizhny Novgorod 603600, Russia

Spinors and Differentials of Real Algebraic Curves

S. M. Natanzon

ABSTRACT. In this paper we describe some properties of tensor fields of weight 1/2 (spinor field) and 1 (differentials) on real algebraic curves. We prove that there exists a one-to-one correspondence between the spinor fields on real curves P and some special type of representations $\pi_1(P) \to \mathrm{SL}(2,\mathbb{R})$, which determine quadratic forms $H_1(P, \mathbb{Z}_2) \to \mathbb{Z}_2$. Using these results, we investigate some properties of differentials on real curves.

The classical mathematics of the 19th century associates to any complex algebraic curve certain tensor fields and Abelian varieties. The structure of a real curve gives new topological invariants for these objects. The investigation of these invariants became especially important when it was found that they determine analytic properties of finite-gap solutions of differential equations [2]. In this paper we describe some properties of tensor fields of weight 1/2 (spinor field) and 1 (differentials) on real algebraic curves.

These questions are considered in [5, 8, 12, 13, 15], and part of the results of this paper follows from these papers. But the method of proof here is different. It is based on the one-to-one correspondence between the spinor fields on real curves P and a special type of representations $\pi_1(P) \to \mathrm{SL}(2, \mathbb{R})$, which determines quadratic forms $H_1(P, \mathbb{Z}_2) \to \mathbb{Z}_2$ (these results were announced in [6, 7]).

I thank V. Kharlamov for the discussion of the article [5] that stimulated the writing of this paper.

§1. Spinors on real algebraic curves

1. A line bundle $E\colon B \to P$ over a compact Riemann surface P is called a *theta characteristic* or a *spinor bundle* if its tensor square is a holomorphic cotangent bundle [1, 9]. We assume in the sequel that the group $\pi_1(P)$ is noncommutative, i.e., the genus of P is greater than 1.

The group $\mathrm{PSL}(2, \mathbb{R})$ acts on the upper half-plane $H = \{z \in \mathbb{C} \mid \mathrm{Im}\, z > 0\}$ by the automorphisms

$$\begin{pmatrix} a & b \\ c & d \end{pmatrix}(z) = \frac{az + b}{cz + d}$$

1991 *Mathematics Subject Classification*. Primary 14H99.

and the group $\mathrm{SL}(2, \mathbb{R})$ acts on $H \times \mathbb{C}$ by the automorphisms
$$\begin{pmatrix} a & b \\ c & d \end{pmatrix} (z, \theta) = \left(\frac{az+b}{cz+d}, \frac{\theta}{(cz+d)^{-1}} \right).$$

A subgroup $\Gamma \subset \mathrm{SL}(2, \mathbb{R})$ is called an SL-*Fuchsian group* if the natural projection
$$\varphi \colon \mathrm{SL}(2, \mathbb{R}) \to \mathrm{PSL}(2, \mathbb{R}) = \mathrm{Aut}(H)$$
is monomorphic on Γ and takes Γ to a Fuchsian group that acts freely on H (i.e., $\gamma p \ne p$ for all $\gamma \in \varphi(\Gamma)$ and $p \in H$).

Associate to any SL-Fuchsian group Γ the spinor bundle $E_\Gamma \colon B_\Gamma \to P_\Gamma$, where $P_\Gamma = H/\varphi(\Gamma)$, $B = (H \times \mathbb{C})/\Gamma$ and E_Γ is the map generated by the natural projection $\Phi \colon H \times \mathbb{C} \to H$.

Let $E \colon B \to P = H/\widetilde{\Gamma}$ be a spinor bundle, where $\widetilde{\Gamma}$ is a Fuchsian group that acts freely on H. Since any line bundle on H is trivial, we have $B = H \times \mathbb{C}/f(\widetilde{\Gamma})$, where $f(\widetilde{\gamma})(z, \theta) = (\widetilde{\gamma} z, \xi_{\widetilde{\gamma}}(z) \theta)$. Since E is a spinor bundle, it follows that
$$\xi_{\widetilde{\gamma}}^2(z) = \frac{\tilde{a}\tilde{d} - \tilde{b}\tilde{c}}{(\tilde{c}z + \tilde{d})^{-2}}, \quad \text{where } \widetilde{\gamma}(z) = \frac{\tilde{a}z + \tilde{b}}{\tilde{c}z + \tilde{d}}.$$

Therefore there exists an element $\begin{pmatrix} a & b \\ c & d \end{pmatrix} \in \mathrm{SL}(2, \mathbb{R})$ such that
$$f(\widetilde{\gamma})(z, \theta) = \left(\frac{az+b}{cz+d}, \frac{\theta}{(cz+d)^{-1}} \right).$$

Thus any spinor bundle is isomorphic to the spinor bundle $E_\Gamma \colon B_\Gamma \to P_\Gamma$ for some SL-Fuchsian group Γ.

For $\gamma = \begin{pmatrix} a & b \\ c & d \end{pmatrix}$, put $\omega_0(\gamma) = 0$ if $a + d < 0$ and $\omega_0(\gamma) = 1$ if $a + d > 0$. Since $|\mathrm{tr}\, \gamma| \geq 2$, this gives a function $\omega_\Gamma \colon \Gamma \to \mathbb{Z}_2 = \mathbb{Z}/2\mathbb{Z}$ for an SL-Fuchsian group Γ, which generates the function $\omega_\pi \colon \pi_1(P, p) \to \mathbb{Z}_2$ for $P = H/\varphi(\Gamma)$.

It can be proved (see [8]) that $\omega_\pi(a) = \omega_\pi(b)$ if a, b are any simple closed contours representing the same class in $H_1(P, \mathbb{Z}_2)$. Thus we get a function $\omega = \omega_E \colon H_1(P, \mathbb{Z}_2) \to \mathbb{Z}_2$, which is an Arf-function (this means [3] that $\omega(a + b) = \omega(a) + \omega(b) + (a, b)$, where $(a, b) \in \mathbb{Z}_2$ is the intersection index mod 2).

Thus we have described a construction that associates to any spinor bundle $E \colon B \to P$ a certain Arf-function $\omega \colon H_1(P, \mathbb{Z}_2) \to \mathbb{Z}_2$. It can be proved (see [8]) that this gives a one-to-one correspondence between the spinor bundles over P and the Arf-functions on $H_1(P, \mathbb{Z}_2)$.

Let $(a_i, b_i \mid i = 1, \ldots, n)$ be a symplectic basis of $H_1(P, \mathbb{Z}_2)$; that is, $(a_i, a_j) = (b_i, b_j) = 0$, $(a_i, b_j) = \delta_{ij}$. Then the Arf-invariant
$$\delta(E) = \delta(\omega) = \sum_{i=1}^{g} \omega(a_i) \omega(b_i) \in \mathbb{Z}_2$$
does not depend on the symplectic basic [3].

According to [10], for hyperelliptic curves the parity of $\delta(E)$ should be considered together with the parity $\delta'(E)$ of the dimension of the space of holomorphic sections of E. According to [1], $\delta'(E)$ is constant on each connected component of the moduli space of spinor bundles over curves of genus g. This connected component is the set of bundles with the same $\delta(E)$ [8]. Thus $\delta(E) = \delta'(E)$. There exists a direct proof

[9] of this classical result. In particular, if $\delta(E) = 1$, then there exists a holomorphic section of E.

2. Now consider a real algebraic curve, i.e., a pair (P, τ), where P is a compact Riemann surface and $\tau \colon P \to P$ is an antiholomorphic involution. The fixed points P^τ of τ are the *real points* of the curve. The connected components of the set of real points are called the *ovals*. We shall consider only real curves with real points.

By definition, a *real spinor bundle* is a spinor bundle $E \colon B \to P$ over a real algebraic curve (P, τ) with an antiholomorphic involution $\beta \colon B \to B$ which linearly moves $E^{-1}(p)$ to $E^{-1}(\tau p)$.

Let $\mathrm{SL}_\pm(2, \mathbb{R}) \subset \mathrm{GL}(2, \mathbb{R})$ be the group of 2×2 matrices with determinant ± 1. A subgroup $\Gamma \subset \mathrm{SL}_\pm(2, \mathbb{R})$ is called $\mathrm{SL}_\pm$-Fuchsian if $\Gamma_+ = \Gamma \cap \mathrm{SL}(2, \mathbb{R})$ is an SL-Fuchsian group and $\Gamma_+ \neq \Gamma$.

We suppose that $\begin{pmatrix} a & b \\ c & d \end{pmatrix} \in \Gamma \setminus \Gamma_+$ acts on $H \times \mathbb{C}$ by

$$\begin{pmatrix} a & b \\ c & d \end{pmatrix}(z, \theta) = \left(\frac{a\bar{z} + b}{c\bar{z} + d}, (c\bar{z} + d)\theta \right).$$

Then the set $\Gamma \setminus \Gamma_+$ defines involutions $\beta_\Gamma \colon B_{\Gamma_+} \to B_{\Gamma_+}$ and $\tau_\Gamma \colon P_{\Gamma_+} \to P_{\Gamma_+}$, which turn E_{Γ_+} into a real spinor bundle $(E_{\Gamma_+}, \beta_\Gamma)$ over $(P_{\Gamma_+}, \tau_\Gamma)$.

THEOREM 1.1. *Every real spinor bundle (E, β) over (P, τ) is isomorphic to a bundle $(E_{\Gamma_+}, \beta_\Gamma)$ for some $\mathrm{SL}_\pm$-Fuchsian group Γ.*

PROOF. The bundle $E \colon B \to P$ is isomorphic to $E_{\Gamma_+} \colon B_{\Gamma_+} \to P_{\Gamma_+}$ for some SL-Fuchsian group Γ_+. The involution τ has fixed points and is therefore generated by an involution $\alpha \colon H \to H$. Changing Γ_+ to $g\Gamma_+ g^{-1}$ ($g \in \mathrm{PSL}(2, \mathbb{R})$), we may assume that $\alpha z = -\bar{z}$.

Let Γ be the group generated by Γ_+ and $\begin{pmatrix} -1 & 0 \\ 0 & 1 \end{pmatrix}$. An isomorphism $E \to E_{\Gamma_+}$ moves β_Γ to β'. The automorphism $\beta\beta'$ preserves the base of E, and thus there exists some isomorphism $E \to E_{\Gamma_+}$ such that $\beta\beta' = 1$. □

Let (E, β) be a real spinor bundle, and a be an oval of the involution $\tau \colon P \to P$ which β generates on the base P of E. According to Theorem 1.1, there exists an $\mathrm{SL}_\pm$-Fuchsian group Γ such that (E, β) and $(E_{\Gamma_+}, \beta_\Gamma)$ are isomorphic. Changing Γ to $\Gamma' = h\Gamma h^{-1}$, it is possible to find a group Γ' such that the natural projection $H \to H/\Gamma'_+$ moves $I = \{x \in \mathbb{C} \mid \mathrm{Im}\, z = 0\}$ to a and $\beta_{\Gamma'}(z, \theta) = (-\bar{z}, \bar{\theta})$. The orientation of I corresponding to the increase of $\mathrm{Im}\, z$ gives an orientation on a. This orientation is called the *spinor orientation* of the oval.

LEMMA 1.2. *Let (E, β) be a real spinor bundle over a real algebraic curve (P, τ), let a_1, a_2 be ovals with spinor orientations and let $c \subset P$ be a simple oriented contour that intersects a_1 and a_2 so that $\tau c = -c$. Then c has the same intersection indices with a_1 and with a_2 if and only if $\omega_E(c) = 1$.*

PROOF. According to Theorem 1.1, (E, β) is isomorphic to $(E_{\Gamma_+}, \beta_\Gamma)$, where Γ is an $\mathrm{SL}_\pm$-Fuchsian group such that $\sigma \begin{pmatrix} \lambda & 0 \\ 0 & \lambda^{-1} \end{pmatrix} \in \Gamma_+$ ($\lambda > 0$, $\sigma = \pm 1$) corresponds to the contour c. In this case the oval a_i corresponds to

$$A_i = \sigma_i \begin{pmatrix} \alpha_i(\lambda_i + 1) & \alpha_i^2(\lambda_i - 1) \\ \lambda_i - 1 & \alpha_i(\lambda_i + 1) \end{pmatrix},$$

where $\lambda_i > 1$, $\alpha_1 > \alpha_2 > 0$ and $\sigma_i = \pm 1$. The intersection indices of a_1 and of a_2 with c are the same if and only if $\sigma_1 = \sigma_2$. But $C = C_1 C_2$, where $C_i = \sigma_i \begin{pmatrix} -\alpha_i & 0 \\ 0 & \alpha_i^{-1} \end{pmatrix}$. Thus $\sigma = \sigma_1 \sigma_2$. $\square$

3. A section of a spinor bundle is called a *spinor*. A section of a real spinor bundle (E, β) over a real curve (P, τ) is called a *real spinor* if $\xi(\tau p) = \beta \xi(p)$.

THEOREM 1.3. *Let ξ be a real meromorphic spinor of (E, β) over a curve (P, τ) and let a be an oval of (P, τ). Then $\omega_E(a) = 1$ if and only if the sum of the multiplicities of the zeros and poles of ξ on a is even.*

PROOF. According to Theorem 1.1, (E, β) is isomorphic to $(E_{\Gamma_+}, \beta_\Gamma)$, where Γ is a $SL_\pm$-Fuchsian group and $A = \sigma \begin{pmatrix} \lambda & 0 \\ 0 & \lambda^{-1} \end{pmatrix}$ corresponds to a. Then the spinor ξ generates the map $\eta \colon H \to \mathbb{C}$ such that $\eta(-\bar{z}) = \overline{\eta(z)}$ and in particular $\eta(I) \subset \mathbb{R}$. Moreover $\eta(\varphi(A)z) = \sigma\sqrt{\lambda}\eta(z)$. Thus the number of zeros and poles of ξ on a is equal to the number of zeros and poles of η on $[i, \lambda i] \subset I$, and it is even if and only if $\sigma = 1$, i.e., $\omega_E(a) = 1$. $\square$

§2. Real differentials

Now let (E, β) be a real cotangent bundle over a real algebraic curve (P, τ), i.e., $E \colon B \to P$ is a holomorphic cotangent bundle and $\beta \colon B \to B$ is the involution generated by τ. A section ω of E is said to be a *real differential* if $\beta\omega = \tau^*\omega$.

A local trivialization $f \colon B_0 \to U_0 \times \mathbb{C}$ of a real bundle (E, β) in a neighborhood of a fixed point $p \in P$ of the involution τ is called *real* if $f\beta f^{-1}(z, \theta) = (\bar{z}, \bar{\theta})$

In this case $Ef^{-1}(U_0 \cap \mathbb{R})$ is a part of some oval $a \subset P$. The usual orientation on $\mathbb{R}$ gives an orientation on a, which is called *adjusted* to f. A differential ξ on (P, τ) is real if for a local real trivialization $f(b) = (z, \xi_f(z)\theta)$ we have $\xi_f(\bar{z}) = \overline{\xi_f(z)}$.

A real differential is called *positive* (respectively, *negative*) on an oriented oval a if the function ξ_f has positive (resp., negative) values for a real trivialization f such that the orientation of a is adjusted to f. A real curve (P, τ) is called *orientable* if $P/\langle\tau\rangle$ is an orientable surface (with boundary).

THEOREM 2.1. *Let (P, τ) be a nonorientable real curve with oriented ovals $a_0, a_1, \ldots, a_k$ and $0 \leqslant n \leqslant k$. Then there exists a spinor ζ such that $\eta = \zeta^2$ is a holomorphic real differential that has zeros on a_i for $0 < i \leqslant k$, is nonnegative on $a_0, a_1, \ldots, a_n$ and nonpositive on $a_{n+1}, \ldots, a_k$.*

PROOF. According to [4], a uniformization group $\widetilde{\Gamma} \subset \mathrm{PSL}_\pm(2, \mathbb{R})$ (where $H/\widetilde{\Gamma} = P/\langle\tau\rangle$) and its system of generators $A_1, \ldots, A_g, B_1, \ldots, B_g, S$ may be chosen so that
1) $A_1, \ldots, A_k$ correspond to $a_1, \ldots, a_k$, and $A_0 = (\prod_{i=1}^g A_i)^{-1}$ corresponds to a_0;
2) the group $\widetilde{\Gamma}_+ = \widetilde{\Gamma} \cap \mathrm{Aut}(H)$ is generated by $A_1, \ldots, A_g, B_1, \ldots, B_g$, and the classes of $A_1, \ldots, A_g, B_1, \ldots, B_g$ form a symplectic basis on $H_1(P, \mathbb{Z}/2)$;
3) $Sz = -\bar{z}$, $SA_0 S = A_0$, $SA_i S = B_i A_i B_i^{-1}$, $SB_i S = B_i^{-1}$ for $0 < i \leqslant k$, and $A_i = (SB_i)^2$ for $i > k$.

Associate to A_i (where $A_i z = (\tilde{a}z + \tilde{b})/(\tilde{c}z + \tilde{d})$) the matrix

$$M(A_i) = \begin{pmatrix} a & b \\ c & d \end{pmatrix} = \lambda \begin{pmatrix} \tilde{a} & \tilde{b} \\ \tilde{c} & \tilde{d} \end{pmatrix} \in \mathrm{SL}(2, \mathbb{R}) \qquad (\lambda \cong \pm 1),$$

where $a + d < -2$ if $i \leqslant k$ and $a + d > 2$ if $i > k$. Associate to B_i (where $B_i z = (\tilde{a} z + \tilde{b})/(\tilde{c} z + \tilde{d})$) the matrix

$$M(B_i) = \begin{pmatrix} a & b \\ c & d \end{pmatrix} = \lambda \begin{pmatrix} \tilde{a} & \tilde{b} \\ \tilde{c} & \tilde{d} \end{pmatrix} \in \mathrm{SL}(2, \mathbb{R}) \qquad (\lambda = \pm 1),$$

where $a + d < -2$ if $i \leqslant n$ and $a + d > 2$ if $n < i < g$. Put $M(S) = \delta \begin{pmatrix} -1 & 0 \\ 0 & 1 \end{pmatrix}$, where $\delta = \pm 1$. For $v = g$ we choose $M(B_i)$ so that

$$\sum_{i=1}^{g} (1 + \operatorname{sign} \operatorname{tr} M(A_i))(1 + \operatorname{sign} \operatorname{tr} M(B_i)) \equiv 1 \pmod{2}.$$

Then

$$\prod_{i=1}^{g} M(A_i) \prod_{i=1}^{g} M(B_i)(M(A_i))^{-1}(M(B_i))^{-1} = 1,$$

$$M(S) M(A_i) M(S) = M(B_i) M(A_i) (M(B_i))^{-1},$$

$$M(S) M(B_i) M(S)^{-1} = (M(B_i))^{-1} \quad \text{if } i \leqslant k,$$
$$(M(S) M(B_i))^2 = M(A_i) \quad \text{if } i > k.$$

Thus the set $\{M(S), M(A_i), M(B_i)\}$ generates an $\mathrm{SL}_\pm$-Fuchsian group Γ whose image in $\mathrm{PSL}_\pm(2, \mathbb{R})$ is $\tilde{\Gamma}$.

Put $(E_{\Gamma_+}, \beta_\Gamma^{a_0}) = (E_{\Gamma_+}, \beta_\Gamma)$ if the orientation of a_0 coincides with the spinor orientation of a_0 corresponding to $(E_{\Gamma_+}, \beta_\Gamma)$.

In the opposite case, put $(E_{\Gamma_+}, \beta_\Gamma^{a_0}) = (E_{\Gamma_+}, -\beta_\Gamma)$. Since

$$\sum_{i=1}^{g} \omega_{\Gamma_+}(A_i) \omega_{\Gamma_+}(B_i) \equiv 1 \pmod{2},$$

the bundle $(E_{\Gamma_+}, \beta_\Gamma^{a_0})$ has a nonzero holomorphic section ξ_0. Thus we can choose a holomorphic section ξ of $(E_{\Gamma_+}, \beta_\Gamma^{a_0})$ so that $\beta_\Gamma^{a_0} \xi = \xi$. The spinor ξ gives a holomorphic function $\xi_H : H \to \mathbb{C}$ such that

$$\xi_H \left(\frac{az + b}{cz + d} \right) = \frac{\xi_H(z)}{(cz + d)^{-1}} \quad \text{for all} \quad \begin{pmatrix} a & b \\ c & d \end{pmatrix} \in \Gamma_+ \text{ and } \xi_H(Sz) = \overline{\xi_H(z)}.$$

Thus $\eta = \xi_H^2 \, dz$ descends to a real holomorphic differential on (P, τ), which is nonnegative on a_0.

According to Theorem 1.3, this differential has zeros on the ovals $a_1, \ldots, a_k$. According to Lemma 1.2 the spinor orientation of a_i coincides with orientation of a_i if and only if $i \leqslant n$. Thus η is nonnegative on $a_1, \ldots, a_n$ and nonpositive on $a_{n+1}, \ldots, a_k$. $\square$

According to Harnack's theorem, a real algebraic curve of genus g has no more than $g + 1$ ovals. If the number of ovals is equal to $g + 1$, the curve is said to be an M-curve. Such a curve is always orientable and $P \setminus P^\tau = P_1 \cup P_2$. The orientation of P_1 gives an orientation on P^τ. This is the standard orientation of the ovals.

LEMMA 2.2. *Let (P, τ) be an M-curve and let $a_0, \ldots, a_g$ be the ovals of (P, τ) with the standard orientation. Then for each $0 \leqslant n < g$ there exists a holomorphic spinor ξ such that $\eta = \xi^2$ is a real differential which is positive on a_0, nonnegative on $a_1, \ldots, a_n$, and nonpositive on $a_{n+1}, \ldots, a_g$.*

PROOF. According to [4], there exists a system of generators $A_1, \ldots, A_g, B_1, \ldots, B_g, S$ of a group $\widetilde{\Gamma} \subset \mathrm{PSL}_\pm(2, \mathbb{R})$ such that $H/\widetilde{\Gamma} = P/\langle \tau \rangle$ and:
1) the generators A_i correspond to the ovals a_i, and $A_0^{-1} = \prod_{i=1}^g A_i$ corresponds to a_0^{-1};
2) $Sz = -\bar{z}$, $SA_iS = BA_iB_i^{-1}$, $SB_iS = B_i^{-1}$ for $0 < i \leqslant g$.

Consider the mapping

$$Az = \frac{\tilde{a}z + \tilde{b}}{\tilde{c}z + \tilde{d}} \longmapsto M(A_i) = \begin{pmatrix} a & b \\ c & d \end{pmatrix} = \lambda \begin{pmatrix} \tilde{a} & \tilde{b} \\ \tilde{c} & \tilde{d} \end{pmatrix},$$

where $ad - bc = 1$ and $\delta(a+d) > 2$, while $\delta = -1$ for $1 \leqslant i < g$ and $\delta = 1$ for $i = 0, g$. Similarly

$$B_iz = \frac{\tilde{a}z + \tilde{b}}{\tilde{c}z + \tilde{d}} \longmapsto M(B_i) = \begin{pmatrix} a & b \\ c & d \end{pmatrix} = \lambda \begin{pmatrix} \tilde{a} & \tilde{b} \\ \tilde{c} & \tilde{d} \end{pmatrix},$$

where $ad - bc = 1$ and $\delta(a+d) > 2$, while $\delta = -1$ for $i \leqslant n$, $\delta = 1$ for $i > n$. Put $M(S) = \begin{pmatrix} -1 & 0 \\ 0 & 1 \end{pmatrix}$. Then $M(S), M(A_i), M(B_i)$ generate an $\mathrm{SL}_\pm$-Fuchsian group Γ. The proof is completed similarly to that of Theorem 2.1. □

Lemma 2.2 implies the following:

LEMMA 2.3. *Let (P, τ) be an M-curve with the ovals $a_0, a_1, \ldots, a_g$, which have the standard orientation. Then for any $0 \leqslant n < g$ there exists a real holomorphic differential η which is positive on $a_0, \ldots, a_n$ and negative on $a_{n+1}, \ldots, a_g$.* □

Let M_g be the moduli space of M-curves of genus g with a fixed ordering of the ovals and with the standard orientation. Let $E_g : B_g \to M_g$ be the bundle of real holomorphic differentials. Let $a_0, a_1, \ldots, a_g$ be the ovals of $(P, \tau) \in M_g$. Consider the set of differentials $\omega_1 = \omega_1^{(P,\tau)}, \ldots, \omega_g = \omega_g^{(P,\tau)}$ such that $\oint_{a_i} \omega_j = \delta_{ij}$. This is a basis of $E_g^{-1}(P, \tau)$. The transformation

$$\Phi : E_g^{-1}(P_1, \tau_1) \to E_g^{-1}(P_2, \tau_2),$$

where

$$\Phi((P_1, \tau_1), \omega_i^{(P_1,\tau_1)}) = ((P_2, \tau_2), \omega_i^{(P_2,\tau_2)}),$$

gives a connection on E.

THEOREM 2.4. *Let (P, τ) be an M-curve with ovals supplied with the standard orientation. Then for each real holomorphic differential on (P, τ) there exists an oval on which the differential is negative, and an oval on which the differential is positive.*

PROOF. Let $M'_g \subset M_g$ be the set of elements $(P, \tau) \in M_g$ such that there exists a real differential η on (P, τ) which is not negative on each oval. Let $M''_g \subset M'_g$ be the set of $(P, \tau) \in M_g$ such that there exists a real differential η on (P, τ) which is not negative on each oval and has only simple zeros on ovals.

Using the connection Φ, it is easy to prove that M_g'' is an open set. From Lemma 2.3 it follows that $M_g'' = M_g'$. But M_g' is a closed set and consequently M_g' is a connected component of M_g. According to [11, 4, 14], M_g is a connected set and consequently if $M_g - M_g' \neq \varnothing$, then $M_g' = \varnothing$.

Now let us prove that $M_g - M_g' \neq \varnothing$. Let P be the Riemann surface of the curve $y^2 = f(x)$, where $f(x) = (x - a_1) \cdots (x - a_{2g+2})$, $a_i \in \mathbb{R}$ and $a_i \neq a_j$. The transformation $(x, y) \mapsto (\overline{x}, \overline{y})$ gives an antiholomorphic involution $\tau \colon P \to P$. Every real differential on (P, τ) has the form $\eta h(x)(f(x))^{-1/2} dx$, where $h(z)$ is a polynomial of degree $\leq g - 1$ with real coefficients. Now it is easy to see that there exist ovals where η is negative. $\square$

Let $M_{g,k}$ be the moduli space of orientable real algebraic curves of genus g with $k < g + 1$ ovals.

THEOREM 2.5. *There exists a nonempty subset $M_0 \subset M_{g,k}$ such that any real holomorphic differential η on each $(P, \tau) \in M_0$ has either an oval where it is positive and an oval where it is negative, or more than $k/2$ ovals where it has a zero.*

PROOF. Consider the Riemann surface P of the curve
$$y^4 - 2y^2[(x - \beta_1) \cdots (x - \beta_m) - (x - \alpha_1) \cdots (x - \alpha_n)]$$
$$+ [(x - \beta_1) \cdots (x - \beta_m) + (x - \alpha_1) \cdots (x - \alpha_m)]^2 = 0,$$
where $\alpha_1 < \cdots < \alpha_n \leq \beta_1 < \cdots < \beta_m \in \mathbb{R}$, $n > 0$, $m > 2$, $n, m \equiv 0 \pmod{2}$, $\tau(x, y) = (\overline{x}, \overline{y})$. The surface P is obtained by resolving the singularities of the variety of pairs (x, y), where
$$y = \pm\sqrt{(x - \alpha_1) \cdots (x - \alpha_n)} \pm \sqrt{-(x - \beta_1) \cdots (x - \beta_n)}.$$

The surface admits the holomorphic involutions
$$\tau_\alpha \colon (x, \pm\sqrt{-(x - \alpha_1) \cdots (x - \alpha_n)} \pm \sqrt{(x - \beta_1) \cdots (x - \beta_m)})$$
$$\mapsto (x, \mp\sqrt{-(x - \alpha_1) \cdots (x - \alpha_n)} \pm \sqrt{(x - \beta_1) \cdots (x - \beta_m)}),$$
$$\tau_\beta \colon (x, \pm\sqrt{-(x - \alpha_1) \cdots (x - \alpha_n)} \mp \sqrt{(x - \beta_1) \cdots (x - \beta_m)})$$
$$\mapsto (x, \pm\sqrt{-(x - \alpha_1) \cdots (x - \alpha_n)} \pm \sqrt{(x - \beta_1) \cdots (x - \beta_m)}),$$
commuting with τ. The ovals of τ are projected on $\overline{a}_1 = [\alpha_1, \alpha_2], \ldots, \overline{a}_{n/2} = [a_{n-1}, a_n]$. For $1 \leq i < n/2$, two ovals are projected on $\overline{a}_i$: a_i and $a_{2i} = \tau_\beta a_i$. There are two ovals ($a_{n/2}$ and α_n) projected on the segment $\overline{a}_{n/2}$ if $\alpha_n < \beta_1$ and one oval ($a_{n/2}$) projected there if $a_n = \beta_1$.

Let us show that there does not exist a real holomorphic differential ω on (P, τ) that would be positive on $a_1, \ldots, a_{n/2}$ and nonnegative on the remaining ovals. Let us assume the contrary. Then $\widetilde{\omega} = \omega + \tau_\beta^* \omega$ is an invariant (under τ_β) and nonnegative (on all the ovals of τ) holomorphic real differential on (P, τ). It induces a real holomorphic differential on the M-curve (P', τ'), nonnegative on all the ovals. This contradicts Theorem 2.4. $\square$

References

1. M. F. Atiyah, *Riemann surfaces and spin structures*, Ann. Sci. École Norm. Sup. (4) **4** (1971), no. 1, 47–62.
2. B. A. Dubrovin, *Theta-functions and nonlinear equations*, Uspekhi Mat. Nauk **36** (1981), no. 2, 11–80; English transl., Russian Math. Surveys **36** (1981), no. 2, 11–92.
3. B. A. Dubrovin, S. P. Novikov, and A. T. Fomenko, *Modern geometry: methods and applications*, "Nauka", Moscow, 1983; English transl., Springer-Verlag, 1984.
4. S. M. Natanzon, *Moduli spaces of real curves*, Trudy Moskov. Mat. Obshch. **37** (1978), 219–253; English transl., Trans. Moscow Math. Soc. **1980**, no. 1, 233–272.
5. _____, *Prymians of real curves and their applications to the effectivization of Schrödinger operators*, Funktsional. Anal. i Prilozhen. **23** (1989), no. 1, 41–55; English transl., Functional Anal. Appl. **23** (1989), no. 1, 33–45.
6. _____, *Spinor bundles over real algebraic curves*, Uspekhi Mat. Nauk **44** (1989), no. 3, 165–166; English transl., Russian Math. Surveys **44** (1989), no. 3, 208–209.
7. _____, *Discrete subgroups of* $GL(2, C)$ *and spinor bundles over Riemann and Klein surfaces*, Funktsional. Anal. i Prilozhen. **25** (1991), no. 4, 76–78; English transl., Functional Anal. Appl. **25** (1991), no. 4, 293–294.
8. _____, *Moduli spaces of Riemann and Klein supersurfaces*, Development in Mathematics. The Moscow School (V. Arnold and M. Monastyrsky, eds.), Chapman & Hall, London, 1993, pp. 100–130.
9. D. Mumford, *Tata lectures on theta* I, II, Birkhäuser, Boston–Basel–Stuttgart, 1983, 1984.
10. E. Arbarello, M. Cornalba, P. A. Griffiths, and J. Harris, *Geometry of algebraic curves*, I, Springer-Verlag, New York–Berlin–Heidelberg–Tokyo, 1985.
11. C. J. Earle, *On moduli closed Riemann surfaces with symmetries*. Advances in the Theory of Riemann Surfaces, Ann. of Math. Stud. **66** (1971), 119–130.
12. J. D. Fay, *Theta functions in Riemann surfaces*, Lecture Notes in Math., vol. 352, Springer-Verlag, Berlin and New York, 1973.
13. B. H. Gross and J. Harris, *Real algebraic curves*, Ann. Sci. École Norm. Sup. (4) **14** (1981), 157–182.
14. M. Seppala, *Teichmuller spaces of Klein surfaces*, Ann. Acad. Sci. Fenn. Ser. A I Math. Dissertations **15** (1978), 1–37.
15. V. Vinnikov, *Selfadjoint determinantal representations of real plane curves*, Math. Ann. **296** (1993), 453–478.

Translated by THE AUTHOR

On the Topological Classification of Real Enriques Surfaces. I

Vyacheslav V. Nikulin

In memory of Dimitriĭ Andreevich Gudkov

ABSTRACT. This note contains our preliminary calculation of topological types of real Enriques surfaces. We prove the existence of 59 topological types of real Enriques surfaces (Theorem 6) and show that all other topological types belong to a list of 21 topological types (Theorem 7). In fact, our calculation contains much more information that can probably be used to construct or prohibit unknown topological types.

These notes contain our preliminary results on the topological classification of real Enriques surfaces.

The calculations were made during the author's stay at Bielefeld University, in May–June 1992. The author is grateful to this University and especially to Professor Heinz Helling for their hospitality. These notes were prepared during the author's stay at Kyoto University, November 1992–January 1993. The author is grateful to Kyoto University and especially to Professor Masaki Maruyama for their hospitalilty.

The author thanks Professor R. Sujatha for useful discussions.

The topological classification of real Enriques surfaces is a part of the general problem of the topological classification of real algebraic varieties. See the survey of D. A. Gudkov [**Gu**] about this problem.

We use the following notation for the topological type of a compact surface. We denote by T_g an orientable surface of genus g, and by U_g a nonorientable surface of genus g. Thus, $\chi(T_g) = 2 - 2g$ and $\chi(U_g) = 1 - g$. The 2-sheeted unramified orientizing covering of the surface U_g is the surface T_g.

Let Z be a smooth projective algebraic surface over $\mathbb{R}$. Then $Z(\mathbb{C})$ is a compact complex manifold of complex dimension 2, and $Z(\mathbb{R})$ is a compact surface with connected components of the type describe above. If q is the antiholomorphic involution on $X(\mathbb{C})$ defined by $G = \mathrm{Gal}(\mathbb{C}/\mathbb{R}) = \{1, q\}$, then $X(\mathbb{R}) = X(\mathbb{C})^q$ is the set of points fixed by q.

The classification of algebraic surfaces (for example, see [**A**]) tells us that algebraic surfaces over $\mathbb{C}$ are divided into the following types: (1) General type; (2) Elliptic

---

1991 *Mathematics Subject Classification.* Primary 14P25, 14J25.

©1996, American Mathematical Society

surfaces; (3) Abelian surfaces; (4) Hyperelliptic surfaces; (5) K3-surfaces; (6) Enriques surfaces; (7) Ruled surfaces; (8) Rational surfaces.

We say that a real surface $Z/\mathbb{R}$ has one of the above types if $Z \otimes \mathbb{C}$ has this type. Thus, we can speak about the topological classification of real surfaces of the above types. The first two classes (1) and (2) are too difficult for classification. But the classes (3)–(8) are not so large, and their topological classification is now known (for example, see [Si]), except for type (6) of real Enriques surfaces. In the book of Silhol [Si], only the topological type $U_{10} \amalg U_0$ of real Enriques surfaces was constructed, and we do not know any other publications on this subject. We apply [N3] and [N4] to get new results for this classification. In these papers a certain general method for working with real K3-surfaces with a condition on the Picard lattice were developed. We also apply results of [N-S] and [N5], where some general results on real Enriques surfaces were obtained.

First, we briefly describe our method of classification. At the end of this paper, we give the results of this classification. Unfortunately, the calculations are very long and delicate, and we can only hope that they are completely correct. We intend to publish details of these calculations elsewhere.

By definition (for example, see [C–D]), an Enriques surface is the quotient surface $Y_{\mathbb{C}} = X/\{\mathrm{id}, \tau\}$, where X is a complex $K3$-surface and τ an algebraic involution of X without fixed points. Let $\pi\colon X \to Y_{\mathbb{C}}$ be the quotient morphism. Since $X(\mathbb{C})$ is simply connected, $\pi\colon X(\mathbb{C}) \to Y(\mathbb{C})$ is the 2-sheeted universal covering with the holomorphic involution τ as the covering involution. Thus, $\pi\tau = \pi$.

Let θ be the antiholomorphic involution of the complex surface $Y(\mathbb{C})$ corresponding to the real surface Y. Below we use some results from [N-S].

Since $X(\mathbb{C})$ is simply connected, one can easily see that there are precisely two liftings σ and $\tau\sigma$ of θ to antiholomorphic automorphisms of $X(\mathbb{C})$. If $Y(\mathbb{R}) \ne \varnothing$, one can easily see that both these automorphisms are antiholomorphic involutions of $X(\mathbb{C})$. Further, we assume that if $Y(\mathbb{R}) = \varnothing$, then this is also the case; thus liftings σ and $\tau\sigma$ are involutions. For the topological classification of $Y(\mathbb{R})$, it suffices to consider only these two cases. (If $Y(\mathbb{R}) = \varnothing$, it might happen, in principle, that σ and $\tau\sigma$ have order 4, but we do not know any such examples and do not consider this case here.) In other words, we lift the group $G = \{\mathrm{id}, \theta\}$ of order 2 on $Y(\mathbb{C})$ to the group $\Gamma = \{\mathrm{id}, \tau, \sigma, \tau\sigma\}$ of order 4 on $X(\mathbb{C})$, with $\Gamma \cong (\mathbb{Z}/2)^2$.

We denote by X_σ and $X_{\tau\sigma}$ the real K3-surfaces defined by the antiholomorphic involutions σ and $\tau\sigma$ respectively. Thus, $X_\sigma(\mathbb{R}) = X(\mathbb{C})^\sigma$ and $X_{\tau\sigma}(\mathbb{R}) = X(\mathbb{C})^{\tau\sigma}$. Besides,
$$\tau(X_\sigma(\mathbb{R})) = X_\sigma(\mathbb{R}), \qquad \tau(X_{\tau\sigma}(\mathbb{R})) = X_{\tau\sigma}(\mathbb{R}).$$
We have $Y(\mathbb{R}) = \pi(X_\sigma(\mathbb{R})) \amalg \pi(X_{\tau\sigma}(\mathbb{R}))$, and we denote
$$Y(\mathbb{R})_\sigma = \pi(X_\sigma(\mathbb{R})) = X_\sigma(\mathbb{R})/\{1, \tau\},$$
$$Y(\mathbb{R})_{\tau\sigma} = \pi(X_{\tau\sigma}(\mathbb{R})) = X_{\tau\sigma}(\mathbb{R})/\{1, \tau\}.$$

Let ω_X be a nonzero holomorphic 2-form of X. The corresponding real parts ω_X^σ and $\omega_X^{\tau\sigma}$ of ω_X define the canonical volume forms on $X_\sigma(\mathbb{R})$ and $X_{\tau\sigma}(\mathbb{R})$. Since $\tau(\omega_X) = -\omega_X$ (this is well known), the canonical map

(1) $$\pi\colon X_\sigma(\mathbb{R}) \amalg X_{\tau\sigma}(\mathbb{R}) \to Y(\mathbb{R})_\sigma \amalg Y(\mathbb{R})_{\tau\sigma} = Y(\mathbb{R})$$

is the 2-sheeted unramified orientizing covering.

Thus, we can assign to a real Enriques surface Y the topological types of the five surfaces

(2) $\qquad (X_\sigma(\mathbb{R}), X_{\tau\sigma}(\mathbb{R}), Y(\mathbb{R})_\sigma, Y(\mathbb{R})_{\tau\sigma}, Y(\mathbb{R}) = Y(\mathbb{R})_\sigma \amalg Y(\mathbb{R})_{\tau\sigma}).$

We need more information about these five surfaces.

Here X_σ and $X_{\tau\sigma}$ are real K3-surfaces. Thus, we recall first the theory of real K3-surfaces.

Let X be a real K3-surface with antiholomorphic involution ϕ. We recall that the cohomology lattice (with the intersection pairing) $H^2(X(\mathbb{C}); \mathbb{Z})$ is isomorphic to a standard even unimodular lattice L of signature $(3, 19)$. Here we have the following result, which follows from the global Torelli theorem [PŠ-Š], the surjectivity of the Torelli map [Ku] for K3-surfaces, and the geometrical interpretation of the invariants.

THEOREM 1 (Kharlamov [Ha1], Nikulin [N3]). *The topological type $X(\mathbb{R}) = X(\mathbb{C})^\phi$ is defined by the action of ϕ on the cohomology lattice $H^2(X(\mathbb{C}); \mathbb{Z})$.*

An action of the abstract group $G = \{1, \phi\}$ of order two on the lattice L corresponds to a real K3-surface if and only if the lattice L^ϕ is hyperbolic (it has the signature $(1, t)$).

Below, using results of [Ha1] and [N3] (see also [N4]), we explain how the action of ϕ actually defines the topology of $X(\mathbb{R}) = X(\mathbb{C})^\phi$. Almost all the relations that we write out below are consequences of some general well-known relations, valid for arbitrary real algebraic surfaces and manifolds (see [Ha1] and [Ha2] on these general relations). But here, we restrict ourselves to the consideration of these relations for K3-surfaces only.

All actions of $G = \{1, \phi\}$ are classified by three invariants

(3) $\qquad (r(\phi), a(\phi), \delta(\phi))$

(see [N3, §3] and also [N4, §1]), defined as follows.

Let $L_+ = L^\phi$ and $L_- = L^{(-\phi)}$. Then we have an orthogonal decomposition up to a finite index $L_+ + L_- \subset L$. The invariant $r(\phi)$ is rk L_+.

We have

(4) $\qquad \begin{array}{ccccc} L_+/2L_+ & \subset & L/2L & \supset & L_-/2L_- \\ \cup & & \cup & & \cup \\ A(\phi)_+ & = & A(\phi) & = & A(\phi)_- \end{array}$

where $A(\phi) = L_+/2L_+ \cap L_-/2L_-$. Here $A(\phi) = (\mathbb{Z}/2\mathbb{Z})^{a(\phi)}$. This defines the invariant $a(\phi)$.

Let

(5) $\qquad q(\phi)(v_+ + 2L_+) = (1/2)v_+^2 \mod 4$

for $v_+ + 2L_+ \in A(\phi)_+$. This defines a quadratic form $q(\phi)$ on $A(\phi)$ with the nondegenerate symmetric bilinear form $b(\phi)$:

(6) $\qquad b(\phi)(v_+ + 2L_+, w_+ + 2L_+) = (1/2)v_+ \cdot w_+ \mod 2$

for $v_+ + 2L_+, w_+ + 2L_+ \in A(\phi)_+$. The invariant $\delta(\phi) = 0$ if $q(\phi)(A(\phi)) \subset 2\mathbb{Z}/4\mathbb{Z}$. Otherwise, $\delta(\phi) = 1$.

There exists another definition of the invariant $\delta(\phi)$. The element $s_\phi \in L/2L$ is called *characteristic* if

(7) $\qquad x \cdot \phi(x) \equiv s_\phi \cdot x \mod 2$

for any $x \in L$. Since the lattice L is unimodular, the characteristic element s_ϕ exists and is unique. One can see that $\delta(\phi) = 0$ iff the characteristic element s_ϕ is equal to 0.

Using these invariants, we can write

(8) $$X(\mathbb{C})^\phi = \begin{cases} \varnothing, & \text{if } (r(\phi), a(\phi), \delta(\phi)) = (10, 10, 0), \\ 2T_1, & \text{if } (r(\phi), a(\phi), \delta(\phi)) = (10, 8, 0), \\ T_{g(\phi)} \amalg k(\phi) T_0, & \text{where } g(\phi) = (22 - r(\phi) - a(\phi))/2, \\ & k(\phi) = (r(\phi) - a(\phi))/2, \text{ otherwise.} \end{cases}$$

We note that

(9) $$\chi(X(\mathbb{R})) = 2r(\phi) - 20,$$
(10) $$\dim H_*(X(\mathbb{R}); \mathbb{Z}/2) = 24 - 2a(\phi) \quad \text{if } X(\mathbb{R}) \neq \varnothing.$$

For the characteristic element, we have

(11) $$s_\phi \equiv X(\mathbb{R}) \mod 2$$

in $H_2(X(\mathbb{C}); \mathbb{Z})$ (here we identify cohomology and homology using the Poincaré duality). Thus,

(12) $$X(\mathbb{R}) \equiv 0 \mod 2 \iff \delta(\phi) = 0.$$

All the possibilities for the triplets (3) are known (see [**N3**]), and this gives the topological classification of real K3-surfaces (see [**Ha1**] and [**N3**]).

We want to apply a similar idea to study real Enriques surfaces, but for the action of the group

(13) $$\Gamma = \{\text{id}, \tau, \sigma, \tau\sigma\} \cong (\mathbb{Z}/2)^2$$

on L. Here the action of τ is standard. It is defined uniquely by the condition $L^\tau = S \cong E(2)$, where E is an even unimodular lattice of signature $(1, 9)$, and $E(2)$ means that we multiply the form of E by 2. The lattice $E = H^2(Y(\mathbb{C}); \mathbb{Z})/\text{Tor}$, and $S = \pi^* E$. Thus, the action is

(14) $$\sigma | S = \tau\sigma | S \cong \theta | E.$$

By the global Torelli theorem [**PŠ-Š**] and the surjectivity of the Torelli map [**Ku**], we have the following statement, similar to Theorem 1.

THEOREM 2. *An action of the group $\Gamma = \{\text{id}, \tau, \sigma, \tau\sigma\} \cong (\mathbb{Z}/2)^2$ on L corresponds to a real Enriques surface iff $L^\tau = S \cong E(2)$ and the lattices L^σ and $L^{\tau\sigma}$ are both hyperbolic.*

Unfortunately, we do not know that the action of Γ defines the topology of the five-tuple (2). We only have special results on the subject.

Thus, our problem of topological classification of real Enriques surfaces is divided into two:

(A) To classify actions of Γ on L satisfying the conditions of Theorem 2.

(B) To find the geometrical interpretation of invariants of the actions of Γ on L.

Problem (A). First, we must classify actions $\sigma | S = \tau\sigma | S$ that are equivalent to the action $\theta | E$. The lattice E is unimodular, and we have invariants similar to (3)

(15) $$(r(\theta), a(\theta), \delta(\theta))$$

for this action. Further, the lattice E^θ is negative definite. Using results of [N3] or [N4], we have the following possibilities for these invariants:

(16)
$$(r(\theta), a(\theta), \delta(\theta)) = (1,1,1),\ (2,2,1),\ (3,3,1),\ (4,4,1),\ (5,5,1),$$
$$(9,1,1),\ (8,2,1),\ (7,3,1),\ (6,4,1),\ (0,0,0),\ (8,2,0),$$
$$(5,3,1),\ (6,2,1),\ (7,1,1),\ (4,2,0),\ (8,0,0).$$

We fix one of these possibilities.

Second, we choose an involution σ from the set $\{\sigma, \tau\sigma\}$. In fact, one must choose between invariants $(r(\sigma), a(\sigma), \delta(\sigma))$ and $(r(\tau\sigma), a(\tau\sigma), \delta(\tau\sigma))$ of these involutions, and in almost all cases this choice can be done canonically. Formulas (24), (25), and (26) below are useful for that. For example, if $\delta(\theta) = 1$, using formula (26), we choose σ uniquely by the condition $\delta(\sigma) = 0$. If $\delta(\theta) = 0$, we choose σ by the condition $r(\sigma) \leq 6 + r(\theta)$ using formula (24). If $r(\sigma) \neq r(\tau\sigma)$, this choice is unique. If $\delta(\theta) = 0$ and $r(\sigma) = r(\tau\sigma) = 6 + r(\theta)$, we choose σ by the condition $a(\sigma) \leq 5 + a(\theta) + \gamma(\sigma) + \alpha(\sigma)$, using formula (25). If $a(\sigma) \neq a(\tau\sigma)$, this choice is unique. Cases when simultaneously $\delta(\theta) = 0$, $r(\sigma) = r(\tau\sigma)$, $a(\sigma) = a(\tau\sigma)$ are very rare and we then denote by σ one of the involutions from the set $\{\sigma, \tau\sigma\}$.

Third, we must study the possibility of extending an admissible involution $\sigma|S = \theta$ on S to an involution σ on the unimodular lattice L. The paper [N4] was devoted to this problem. In [N4], all invariants of the genus for these extensions for an arbitrary lattice S were found. Besides, all relations between these invariants that are necessary and sufficient for the existence of the extension were found in [N4] also. Let us describe these invariants.

We have the canonical subgroups $H(\sigma)_+$, $H(\sigma)_-$ (see (4)):

(17) $\quad (S_+/2S_+) \cap (S_-/2S_-) \subset H(\sigma)_\pm = S_\pm/2S_\pm \cap A(\sigma)_\pm \subset S_\pm/2S_\pm$.

Using the identifications $A(\sigma)_\pm = A(\sigma)$, we can regard these subgroups $H(\sigma)_\pm$ as subgroups in the space $A(\sigma)$ equipped with the bilinear form $b(\sigma)$. This defines the canonical pairing

(18) $\quad\quad\quad\quad\quad\quad p(\sigma)\colon H(\sigma)_+ \times H(\sigma)_- \to \mathbb{Z}/2\mathbb{Z}$.

The characteristic element s_σ may or may not belong to $S/2S$ or to $(S_+/2S_+) \cap (S_-/2S_-)$. This defines the invariants $\delta_{\sigma S}, \delta_{\sigma S_+ \cap S_-}$ similar to $\delta(\sigma)$, which are equal to 0 or 1. Thus, $\delta_{\sigma S} = 0$ iff $s_\sigma \in S/2S$, and $\delta_{\sigma S_+ \cap S_-} = 0$ iff $s_\sigma \in (S_+/2S_+) \cap (S_-/2S_-)$ (it would perhaps be more appropriate to use the more precise notation $\delta_{\sigma(S_+/2S_+) \cap (S_-/2S_-)}$ instead of $\delta_{\sigma S_+ \cap S_-}$, but that would be too cumbersome).

If $\delta_{\sigma S} = 0$, the element $s_\sigma \in S/2S$ gives an additional invariant of the extension.

It was proved in [N4] that the invariants

(19) $\quad a(\sigma),\ \delta(\sigma),\ H(\sigma)_+,\ H(\sigma)_-,\ p(\sigma),\ \delta_{\sigma S},\ s_\sigma \in S/2S\quad$ (if $\delta_{\sigma S} = 0$)

together with the real (over $\mathbb{R}$) invariants of σ on L (in our case, this is $r(\sigma)$) are a complete set of invariants for characterizing the genus of extensions of σ on even unimodular lattices L (in our case, L is defined by the signature $(3, 19)$). In addition, all the necessary and sufficient relations between these invariants for the existence of an extension were given. Thus, using results of [N4], we can describe all possible invariants (19) and $r(\sigma)$ for real Enriques surfaces.

Here we restrict ourselves to the most important invariants for topology, namely $(r(\sigma), a(\sigma), \delta(\sigma))$ and

(20) $$h(\sigma)_\pm = \dim H(\sigma)_\pm,$$
(21) $$\gamma(\sigma) = \dim H(\sigma)_- - \dim (H(\sigma)_+)^\perp \cap H(\sigma)_-$$

(we use the pairing $\rho(\sigma)$), and

(22) $$\delta_\sigma S, \ \delta_{\sigma S_+ \cap S_-}.$$

Using [N4] (see [N-S]), we can prove that

(23) $$a(\sigma) = h(\sigma)_+ + h(\sigma)_- + \alpha(\sigma),$$

where the invariant $\alpha(\sigma)$ is 0 or 1. The invariants $r(\sigma)$, $a(\sigma)$, $\alpha(\sigma)$ and $h(\sigma)_\pm$ are very closely related. In fact, (23) is a consequence of the inequalities

(23-1) $$h(\sigma)_+ + h(\sigma)_- \leqslant a(\sigma),$$

(23-2) $$a(\sigma) - r(\sigma) \leqslant 2h(\sigma)_+ - 2r(\theta),$$
$$a(\sigma) + r(\sigma) \leqslant 2h(\sigma)_- + 2r(\theta) + 2.$$

Besides, we have the congruence

(23-3) $$r(\sigma) \equiv a(\sigma) \mod 2$$

(by (8), for example). Thus, using (23) and (23-1), (23-2), (23-3), we have the following possibilities:

The case $(A(\sigma))$:

(23-4) $$\alpha(\sigma) = 1, \quad h(\sigma)_+ = r(\theta) - (r(\sigma) - a(\sigma))/2,$$
$$h(\sigma)_- = (r(\sigma) + a(\sigma))/2 - r(\theta) - 1.$$

The case $(B(\sigma))$:

(23-5) $$\alpha(\sigma) = 0, \quad h(\sigma)_+ = r(\theta) - (r(\sigma) - a(\sigma))/2,$$
$$h(\sigma)_- = (r(\sigma) + a(\sigma))/2 - r(\theta).$$

The case $(C(\sigma))$:

(23-6) $$\alpha(\sigma) = 0, \quad h(\sigma)_+ = r(\theta) - (r(\sigma) - a(\sigma))/2 + 1,$$
$$h(\sigma)_- = (r(\sigma) + a(\sigma))/2 - r(\theta) - 1.$$

Problem (B). First of all, we have the following formulas that relate the invariants above to the invariants $(r(\tau\sigma), a(\tau\sigma), \delta(\tau\sigma))$ (see [N-S]):

(24) $$r(\sigma) + r(\tau\sigma) = 12 + 2r(\theta),$$
(25) $$a(\sigma) + a(\tau\sigma) = 10 + 2a(\theta) + 2\gamma(\sigma) + 2\alpha(\sigma),$$
(26) $$\delta(\sigma) + \delta(\tau\sigma) \equiv \delta(\theta) \mod 2$$

(recall (12)). Thus, using the invariants (19), we can find the invariants $(r(\tau\sigma), a(\tau\sigma), \delta(\tau\sigma))$ of $\tau\sigma$.

For completeness, we also mention the relations:

(27) $$\alpha(\tau\sigma) = \alpha(\sigma), \quad \gamma(\tau\sigma) = \gamma(\sigma),$$

and

(28) $$A(\sigma) \Longrightarrow A(\tau\sigma), \quad B(\sigma) \Longrightarrow C(\tau\sigma), \quad C(\sigma) \Longrightarrow B(\tau\sigma).$$

By the formulas above, the invariants $(r(\theta), a(\theta), \delta(\theta))$, and $\alpha(\sigma), \delta(\sigma), \delta(\tau\sigma)$ and the topological type of $X_\sigma(\mathbb{R})$ and $X_{\tau\sigma}(\mathbb{R})$ define the invariants (3) for σ and $\tau\sigma$, and (20), (21). This is not quite clear for the invariants (20), and one must also use calculations of Theorem 5 below, where we consider the case $B(\sigma)$ or $C(\sigma)$, which takes place if the invariant $\alpha(\sigma) = 0$.

Thus, to formulate the results of our calculations, we can use these invariants instead of (20) and (21).

A very nontrivial formula is that for the number s_{nor} of nonorientable connected components of $Y(\mathbb{R})$ for a real Enriques surface Y. Here we use results of [**N-S**] and [**N5**].

THEOREM 3. *Let both $X_\sigma(\mathbb{R})$ and $X_{\tau\sigma}(\mathbb{R})$ be nonempty. Then*

(29) $$s_{\text{nor}} = 1 + \alpha(\sigma)(2\delta_{\sigma S_+ \cap S_-} - 1) + \gamma(\sigma).$$

When $X_\sigma(\mathbb{R}) = \varnothing$, we only have an inequality.

THEOREM 4 (see [**N-S**]). *Let $X_\sigma(\mathbb{R}) = \varnothing$ (then it follows that $\delta_{\sigma S_+ \cap S_-} = 0$) and $X_{\tau\sigma}(\mathbb{R}) \neq \varnothing$. Then*

$$s_{\text{nor}} \leqslant 2 - \alpha(\sigma) + \gamma(\sigma).$$

To use these theorems, we add the invariant $\delta_{\sigma S_+ \cap S_-}$ to the invariants described above. Evidently, $\delta_{\sigma S} = \delta_{\sigma S_+ \cap S_-} = 0$ if $\delta(\sigma) = 0$, and $\delta(\sigma) = \delta_{\sigma S_+ \cap S_-} = 1$ if $\delta_{\sigma S} = 1$. In many cases, Theorems 3 and 4 allow us to find the topological type of $Y(\mathbb{R})$ and $Y(\mathbb{R})_\sigma$, $Y(\mathbb{R})_{\tau\sigma}$ using the orientizing map (1).

Below, we give results of calculations using the method described above. The notation "A or B" means that we are proving that one of possibilities A, B, (A and B) holds, but we do not know which.

THEOREM 5. *We have the following and only the following possibilities for the invariants*

$$(r(\theta), a(\theta), \delta(\theta)); \quad \alpha(\sigma); \quad X_\sigma(\mathbb{R}), \delta(\sigma), \delta_{\sigma S}, \delta_{\sigma S_+ \cap S_-};$$
$$X_{\tau\sigma}(\mathbb{R}), \delta(\tau\sigma), \delta_{\tau\sigma S}, \delta_{\tau\sigma S_+ \cap S_-}; \quad Y(\mathbb{R})_\sigma, Y(\mathbb{R})_{\tau\sigma}, Y(\mathbb{R})$$

of real Enriques surfaces Y:

The case $(r(\theta), a(\theta), \delta(\theta)) = (1, 1, 1)$:
  $\alpha(\sigma) = 1$; $X_\sigma(\mathbb{R}) = \varnothing$, $\delta(\sigma) = 0$, $X_{\tau\sigma}(\mathbb{R}) = T_7$, $\delta(\tau\sigma) = 1$,
  $Y(\mathbb{R})_\sigma = \varnothing$, $Y(\mathbb{R})_{\tau\sigma} = U_7$, $Y(\mathbb{R}) = U_7$;
  $\alpha(\sigma) = 0$ $(B(\sigma))$; $X_\sigma(\mathbb{R}) = \varnothing$, $\delta(\sigma) = 0$, $X_{\tau\sigma}(\mathbb{R}) = T_8 \amalg T_0$, $\delta(\tau\sigma) = 1$,
  $Y(\mathbb{R})_\sigma = \varnothing$, $Y(\mathbb{R})_{\tau\sigma} = U_8 \amalg U_0$, $Y(\mathbb{R}) = U_8 \amalg U_0$;
  $\alpha(\sigma) = 0$ $(C(\sigma))$; $X_\sigma(\mathbb{R}) = 2T_1$, $\delta(\sigma) = 0$, $X_{\tau\sigma}(\mathbb{R}) = T_7$, $\delta(\tau\sigma) = 1$,
  $Y(\mathbb{R})_\sigma = T_1$, $Y(\mathbb{R})_{\tau\sigma} = U_7$, $Y(\mathbb{R}) = U_7 \amalg T_1$;
  $\alpha(\sigma) = 0$ $(B(\sigma))$; $X_\sigma(\mathbb{R}) = T_9$, $\delta(\sigma) = 0$, $X_{\tau\sigma}(\mathbb{R}) = 2T_0$, $\delta(\tau\sigma) = 1$,
  $Y(\mathbb{R})_\sigma = U_9$, $Y(\mathbb{R})_{\tau\sigma} = T_0$, $Y(\mathbb{R}) = U_9 \amalg T_0$.

The case $(r(\theta), a(\theta), \delta(\theta)) = (2, 2, 1)$:
  $\alpha(\sigma) = 1$; $X_\sigma(\mathbb{R}) = \varnothing$, $\delta(\sigma) = 0$, $X_{\tau\sigma}(\mathbb{R}) = T_5$, $\delta(\tau\sigma) = 1$,
  $Y(\mathbb{R})_\sigma = \varnothing$, $Y(\mathbb{R})_{\tau\sigma} = U_5$, $Y(\mathbb{R}) = U_5$;
  $\alpha(\sigma) = 0$ $(B(\sigma))$; $X_\sigma(\mathbb{R}) = \varnothing$, $\delta(\sigma) = 0$, $X_{\tau\sigma}(\mathbb{R}) = T_6 \amalg T_0$, $\delta(\tau\sigma) = 1$,

$Y(\mathbb{R})_\sigma = \varnothing,\ Y(\mathbb{R})_{\tau\sigma} = U_6 \amalg U_0,\ Y(\mathbb{R}) = U_6 \amalg U_0;$
$\alpha(\sigma) = 0\ (C(\sigma));\ X_\sigma(\mathbb{R}) = 2T_1,\ \delta(\sigma) = 0,\ X_{\tau\sigma}(\mathbb{R}) = T_5,\ \delta(\tau\sigma) = 1,$
$Y(\mathbb{R})_\sigma = T_1,\ Y(\mathbb{R})_{\tau\sigma} = U_5,\ Y(\mathbb{R}) = U_5 \amalg T_1.$

The case $(r(\theta), a(\theta), \delta(\theta)) = (3, 3, 1)$:
$\alpha(\sigma) = 1;\ X_\sigma(\mathbb{R}) = \varnothing,\ \delta(\sigma) = 0,\ X_{\tau\sigma}(\mathbb{R}) = T_3,\ \delta(\tau\sigma) = 1,$
$Y(\mathbb{R})_\sigma = \varnothing,\ Y(\mathbb{R})_{\tau\sigma} = U_3,\ Y(\mathbb{R}) = U_3;$
$\alpha(\sigma) = 0\ (B(\sigma));\ X_\sigma(\mathbb{R}) = \varnothing,\ \delta(\sigma) = 0,\ X_{\tau\sigma}(\mathbb{R}) = T_4 \amalg T_0,\ \delta(\tau\sigma) = 1,$
$Y(\mathbb{R})_\sigma = \varnothing,\ Y(\mathbb{R})_{\tau\sigma} = U_4 \amalg U_0,\ Y(\mathbb{R}) = U_4 \amalg U_0;$
$\alpha(\sigma) = 0\ (C(\sigma));\ X_\sigma(\mathbb{R}) = 2T_1,\ \delta(\sigma) = 0,\ X_{\tau\sigma}(\mathbb{R}) = T_3,\ \delta(\tau\sigma) = 1,$
$Y(\mathbb{R})_\sigma = T_1,\ Y(\mathbb{R})_{\tau\sigma} = U_3,\ Y(\mathbb{R}) = U_3 \amalg T_1.$

The case $(r(\theta), a(\theta), \delta(\theta)) = (4, 4, 1)$:
$\alpha(\sigma) = 1;\ X_\sigma(\mathbb{R}) = \varnothing,\ \delta(\sigma) = 0,\ X_{\tau\sigma}(\mathbb{R}) = T_1,\ \delta(\tau\sigma) = 1,$
$Y(\mathbb{R})_\sigma = \varnothing,\ Y(\mathbb{R})_{\tau\sigma} = U_1,\ Y(\mathbb{R}) = U_1;$
$\alpha(\sigma) = 0\ (B(\sigma));\ X_\sigma(\mathbb{R}) = \varnothing,\ \delta(\sigma) = 0,\ X_{\tau\sigma}(\mathbb{R}) = T_2 \amalg T_0,\ \delta(\tau\sigma) = 1,$
$Y(\mathbb{R})_\sigma = \varnothing,\ Y(\mathbb{R})_{\tau\sigma} = U_2 \amalg U_0,\ Y(\mathbb{R}) = U_2 \amalg U_0;$
$\alpha(\sigma) = 0\ (C(\sigma));\ X_\sigma(\mathbb{R}) = 2T_1,\ \delta(\sigma) = 0,\ X_{\tau\sigma}(\mathbb{R}) = T_1,\ \delta(\tau\sigma) = 1,$
$Y(\mathbb{R})_\sigma = T_1,\ Y(\mathbb{R})_{\tau\sigma} = U_1,\ Y(\mathbb{R}) = U_1 \amalg T_1.$

The case $(r(\theta), a(\theta), \delta(\theta)) = (5, 5, 1)$:
$\alpha(\sigma) = 0\ (B(\sigma));\ X_\sigma(\mathbb{R}) = \varnothing,\ \delta(\sigma) = 0,\ X_{\tau\sigma}(\mathbb{R}) = 2T_0,\ \delta(\tau\sigma) = 1,$
$Y(\mathbb{R})_\sigma = \varnothing,\ Y(\mathbb{R})_{\tau\sigma} = Y(\mathbb{R}) = 2U_0\ \text{or}\ T_0.$

The case $(r(\theta), a(\theta), \delta(\theta)) = (9, 1, 1)$:
$\alpha(\sigma) = 0\ (B(\sigma));\ X_\sigma(\mathbb{R}) = \varnothing,\ \delta(\sigma) = 0,\ X_{\tau\sigma}(\mathbb{R}) = 10T_0,\ \delta(\tau\sigma) = 1,$
$Y(\mathbb{R})_\sigma = \varnothing,\ Y(\mathbb{R})_{\tau\sigma} = Y(\mathbb{R}) = 5T_0\ \text{or}\ 2U_0 \amalg 4T_0;$
$\alpha(\sigma) = 1;\ X_\sigma(\mathbb{R}) = 8T_0,\ \delta(\sigma) = 0,\ X_{\tau\sigma}(\mathbb{R}) = 2T_0,\ \delta(\tau\sigma) = 1,$
$Y(\mathbb{R})_\sigma = 4T_0,\ Y(\mathbb{R})_{\tau\sigma} = T_0,\ Y(\mathbb{R}) = 5T_0;$
$\alpha(\sigma) = 0\ (B(\sigma));\ X_\sigma(\mathbb{R}) = T_1 \amalg 8T_0,\ \delta(\sigma) = 0,\ X_{\tau\sigma}(\mathbb{R}) = 2T_0,\ \delta(\tau\sigma) = 1,$
$Y(\mathbb{R})_\sigma = U_1 \amalg 4T_0,\ Y(\mathbb{R})_{\tau\sigma} = T_0,\ Y(\mathbb{R}) = U_1 \amalg 5T_0;$
$\alpha(\sigma) = 0\ (C(\sigma));\ X_\sigma(\mathbb{R}) = 8T_0,\ \delta(\sigma) = 0,\ X_{\tau\sigma}(\mathbb{R}) = T_1 \amalg 2T_0,\ \delta(\tau\sigma) = 1,$
$Y(\mathbb{R})_\sigma = 4T_0,\ Y(\mathbb{R})_{\tau\sigma} = U_1 \amalg T_0,\ Y(\mathbb{R}) = U_1 \amalg 5T_0.$

The case $(r(\theta), a(\theta), \delta(\theta)) = (8, 2, 1)$:
$\alpha(\sigma) = 1;\ X_\sigma(\mathbb{R}) = 4T_0,\ \delta(\sigma) = 0,\ X_{\tau\sigma}(\mathbb{R}) = 4T_0,\ \delta(\tau\sigma) = 1,$
$Y(\mathbb{R})_\sigma = 2T_0,\ Y(\mathbb{R})_{\tau\sigma} = 2T_0,\ Y(\mathbb{R}) = 4T_0;$
$\alpha(\sigma) = 0\ (B(\sigma));\ X_\sigma(\mathbb{R}) = T_1 \amalg 4T_0,\ \delta(\sigma) = 0,\ X_{\tau\sigma}(\mathbb{R}) = 4T_0,\ \delta(\tau\sigma) = 1,$
$Y(\mathbb{R})_\sigma = U_1 \amalg 2T_0,\ Y(\mathbb{R})_{\tau\sigma} = 2T_0,\ Y(\mathbb{R}) = U_1 \amalg 4T_0;$
$\alpha(\sigma) = 0\ (C(\sigma));\ X_\sigma(\mathbb{R}) = 4T_0,\ \delta(\sigma) = 0,\ X_{\tau\sigma}(\mathbb{R}) = T_1 \amalg 4T_0,\ \delta(\tau\sigma) = 1,$
$Y(\mathbb{R})_\sigma = 2T_0,\ Y(\mathbb{R})_{\tau\sigma} = U_1 \amalg 2T_0,\ Y(\mathbb{R}) = U_1 \amalg 4T_0.$

The case $(r(\theta), a(\theta), \delta(\theta)) = (7, 3, 1)$:
$\alpha(\sigma) = 0\ (B(\sigma));\ X_\sigma(\mathbb{R}) = \varnothing,\ \delta(\sigma) = 0,\ X_{\tau\sigma}(\mathbb{R}) = 6T_0,\ \delta(\tau\sigma) = 1,$
$Y(\mathbb{R})_\sigma = \varnothing,\ Y(\mathbb{R})_{\tau\sigma} = Y(\mathbb{R}) = 3T_0\ \text{or}\ 2U_0 \amalg 2T_0.$

The case $(r(\theta), a(\theta), \delta(\theta)) = (6, 4, 1)$:
$\alpha(\sigma) = 0\ (C(\sigma));\ X_\sigma(\mathbb{R}) = 4T_0,\ \delta(\sigma) = 0,\ X_{\tau\sigma}(\mathbb{R}) = T_1,\ \delta(\tau\sigma) = 1,$
$Y(\mathbb{R})_\sigma = 2T_0,\ Y(\mathbb{R})_{\tau\sigma} = U_1,\ Y(\mathbb{R}) = U_1 \amalg 2T_0.$

The case $(r(\theta), a(\theta), \delta(\theta)) = (0, 0, 0)$:
$\alpha(\sigma) = 1;\ X_\sigma(\mathbb{R}) = T_9,\ \delta(\sigma) = 0,\ X_{\tau\sigma}(\mathbb{R}) = \varnothing,\ \delta(\tau\sigma) = 0,$
$Y(\mathbb{R})_\sigma = U_9,\ Y(\mathbb{R})_{\tau\sigma} = \varnothing,\ Y(\mathbb{R}) = U_9;$
$\alpha(\sigma) = 0\ (C(\sigma));\ X_\sigma(\mathbb{R}) = T_{10} \amalg T_0,\ \delta(\sigma) = 0,\ X_{\tau\sigma}(\mathbb{R}) = \varnothing,\ \delta(\tau\sigma) = 0,$
$Y(\mathbb{R})_\sigma = U_{10} \amalg U_0,\ Y(\mathbb{R})_{\tau\sigma} = \varnothing,\ Y(\mathbb{R}) = U_{10} \amalg U_0;$

$\alpha(\sigma) = 0 \ (B(\sigma)); X_\sigma(\mathbb{R}) = T_9, \delta(\sigma) = 0, X_{\tau\sigma}(\mathbb{R}) = 2T_1, \delta(\tau\sigma) = 0,$
$Y(\mathbb{R})_\sigma = U_9, Y(\mathbb{R})_{\tau\sigma} = T_1, Y(\mathbb{R}) = U_9 \amalg T_1;$
$\alpha(\sigma) = 1; X_\sigma(\mathbb{R}) = T_{10-2t}, \delta_{\sigma S} = 1, X_{\tau\sigma}(\mathbb{R}) = T_{2t}, \delta_{\tau\sigma S} = 1,$
$Y(\mathbb{R})_\sigma = U_{10-2t}, Y(\mathbb{R})_{\tau\sigma} = U_{2t}, Y(\mathbb{R}) = U_{10-2t} \amalg U_{2t},$ where $t = 0, 1, 2;$
$\alpha(\sigma) = 1; X_\sigma(\mathbb{R}) = T_{9-2t}, \delta_{\sigma S} = 0, \delta_{\sigma S_+ \cap S_-} = 1,$
$X_{\tau\sigma}(\mathbb{R}) = T_{1+2t}, \delta_{\tau\sigma S} = 0, \delta_{\tau\sigma S_+ \cap S_-} = 1,$
$Y(\mathbb{R})_\sigma = U_{9-2t}, Y(\mathbb{R})_{\tau\sigma} = U_{1+2t}, Y(\mathbb{R}) = U_{9-2t} \amalg U_{1+2t},$ where $t = 0, 1, 2.$

The case $(r(\theta), a(\theta), \delta(\theta)) = (8, 2, 0)$:
$\alpha(\sigma) = 0 \ (B(\sigma)); X_\sigma(\mathbb{R}) = \varnothing, \delta(\sigma) = 0, X_{\tau\sigma}(\mathbb{R}) = 8T_0, \delta(\tau\sigma) = 0,$
$Y(\mathbb{R})_\sigma = \varnothing, Y(\mathbb{R})_{\tau\sigma} = Y(\mathbb{R}) = 4T_0$ or $2U_0 \amalg 3T_0;$
$\alpha(\sigma) = 1; X_\sigma(\mathbb{R}) = (1+2t)T_0, \delta_{\sigma S} = 1, X_{\tau\sigma}(\mathbb{R}) = (7-2t)T_0, \delta_{\tau\sigma S} = 1,$
$Y(\mathbb{R})_\sigma = U_0 \amalg tT_0, Y(\mathbb{R})_{\tau\sigma} = U_0 \amalg (3-t)T_0, Y(\mathbb{R}) = 2U_0 \amalg 3T_0,$ where $t = 0, 1;$
$\alpha(\sigma) = 1; X_\sigma(\mathbb{R}) = 2T_0, \delta_{\sigma S} = 0, \delta_{\sigma S_+ \cap S_-} = 1,$
$X_{\tau\sigma}(\mathbb{R}) = 6T_0, \delta_{\tau\sigma S} = 0, \delta_{\tau\sigma S_+ \cap S_-} = 1,$
$(Y(\mathbb{R})_\sigma, Y(\mathbb{R})_{\tau\sigma}) = (2U_0, 3T_0)$ or $(T_0, 2U_0 \amalg 2T_0), Y(\mathbb{R}) = 2U_0 \amalg 3T_0;$
$\alpha(\sigma) = 1; X_\sigma(\mathbb{R}) = 4T_0, \delta_{\sigma S} = 0, \delta_{\sigma S_+ \cap S_-} = 1,$
$X_{\tau\sigma}(\mathbb{R}) = 4T_0, \delta_{\tau\sigma S} = 0, \delta_{\tau\sigma S_+ \cap S_-} = 1,$
$Y(\mathbb{R})_\sigma = 2U_0 \amalg T_0, Y(\mathbb{R})_{\tau\sigma} = 2T_0, Y(\mathbb{R}) = 2U_0 \amalg 3T_0;$
$\alpha(\sigma) = 1; X_\sigma(\mathbb{R}) = 4T_0, \delta(\sigma) = 1, \delta_{\sigma S_+ \cap S_-} = 0,$
$X_{\tau\sigma}(\mathbb{R}) = 4T_0, \delta(\tau\sigma) = 1, \delta_{\tau\sigma S_+ \cap S_-} = 0,$
$Y(\mathbb{R})_\sigma = 2T_0, Y(\mathbb{R})_{\tau\sigma} = 2T_0, Y(\mathbb{R}) = 4T_0;$
$\alpha(\sigma) = 0 \ (B(\sigma)); X_\sigma(\mathbb{R}) = T_1 \amalg (4-4t)T_0, \delta(\sigma) = 1, \delta_{\sigma S_+ \cap S_-} = 0,$
$X_{\tau\sigma}(\mathbb{R}) = (4+4t)T_0, \delta(\tau\sigma) = 1, \delta_{\tau\sigma S_+ \cap S_-} = 0,$
$Y(\mathbb{R})_\sigma = U_1 \amalg (2-2t)T_0, Y(\mathbb{R})_{\tau\sigma} = (2+2t)T_0, Y(\mathbb{R}) = U_1 \amalg 4T_0,$ where $t = 0, 1.$

The case $(r(\theta), a(\theta), \delta(\theta)) = (5, 3, 1)$:
$\alpha(\sigma) = 1; X_\sigma(\mathbb{R}) = \varnothing, \delta(\sigma) = 0, X_{\tau\sigma}(\mathbb{R}) = 2T_0, \delta(\tau\sigma) = 1,$
$Y(\mathbb{R})_\sigma = \varnothing, Y(\mathbb{R})_{\tau\sigma} = Y(\mathbb{R}) = 2U_0$ or $T_0;$
$\alpha(\sigma) = 0 \ (B(\sigma)); X_\sigma(\mathbb{R}) = \varnothing, \delta(\sigma) = 0, X_{\tau\sigma}(\mathbb{R}) = 2T_0, \delta(\tau\sigma) = 1,$
$Y(\mathbb{R})_\sigma = \varnothing, Y(\mathbb{R})_{\tau\sigma} = Y(\mathbb{R}) = 2U_0$ or $T_0;$
$\alpha(\sigma) = 0 \ (B(\sigma)); X_\sigma(\mathbb{R}) = \varnothing, \delta(\sigma) = 0, X_{\tau\sigma}(\mathbb{R}) = T_1 \amalg 2T_0, \delta(\tau\sigma) = 1,$
$Y(\mathbb{R})_\sigma = \varnothing, Y(\mathbb{R})_{\tau\sigma} = Y(\mathbb{R}) = U_1 \amalg 2U_0$ or $U_1 \amalg T_0;$
$\alpha(\sigma) = 1; X_\sigma(\mathbb{R}) = \varnothing, \delta(\sigma) = 0, X_{\tau\sigma}(\mathbb{R}) = T_1 \amalg 2T_0, \delta(\tau\sigma) = 1,$
$Y(\mathbb{R})_\sigma = \varnothing, Y(\mathbb{R})_{\tau\sigma} = Y(\mathbb{R}) = U_1 \amalg T_0;$
$\alpha(\sigma) = 0 \ (B(\sigma)); X_\sigma(\mathbb{R}) = \varnothing, \delta(\sigma) = 0, X_{\tau\sigma}(\mathbb{R}) = T_2 \amalg 3T_0, \delta(\tau\sigma) = 1,$
$Y(\mathbb{R})_\sigma = \varnothing, Y(\mathbb{R})_{\tau\sigma} = Y(\mathbb{R}) = U_2 \amalg U_0 \amalg T_0;$
$\alpha(\sigma) = 0 \ (B(\sigma)); X_\sigma(\mathbb{R}) = 2T_1, \delta(\sigma) = 0, X_{\tau\sigma}(\mathbb{R}) = 2T_0, \delta(\tau\sigma) = 1,$
$(Y(\mathbb{R})_\sigma, Y(\mathbb{R})_{\tau\sigma}, Y(\mathbb{R})) = (2U_1, T_0, 2U_1 \amalg T_0)$ or $(T_1, 2U_0, 2U_0 \amalg T_1);$
$\alpha(\sigma) = 1; X_\sigma(\mathbb{R}) = 2T_1, \delta(\sigma) = 0, X_{\tau\sigma}(\mathbb{R}) = 2T_0, \delta(\tau\sigma) = 1,$
$Y(\mathbb{R})_\sigma = T_1, Y(\mathbb{R})_{\tau\sigma} = T_0, Y(\mathbb{R}) = T_1 \amalg T_0;$
$\alpha(\sigma) = 0 \ (C(\sigma)); X_\sigma(\mathbb{R}) = 2T_1, \delta(\sigma) = 0, X_{\tau\sigma}(\mathbb{R}) = T_1 \amalg 2T_0, \delta(\tau\sigma) = 1,$
$Y(\mathbb{R})_\sigma = T_1, Y(\mathbb{R})_{\tau\sigma} = U_1 \amalg T_0, Y(\mathbb{R}) = U_1 \amalg T_1 \amalg T_0;$
$\alpha(\sigma) = 0 \ (B(\sigma)); X_\sigma(\mathbb{R}) = T_3 \amalg 2T_0, \delta(\sigma) = 0, X_{\tau\sigma}(\mathbb{R}) = 2T_0, \delta(\tau\sigma) = 1,$
$Y(\mathbb{R})_\sigma = U_3 \amalg T_0, Y(\mathbb{R})_{\tau\sigma} = T_0, Y(\mathbb{R}) = U_3 \amalg 2T_0;$

The case $(r(\theta), a(\theta), \delta(\theta)) = (6, 2, 1)$:
$\alpha(\sigma) = 1; X_\sigma(\mathbb{R}) = \varnothing, \delta(\sigma) = 0, X_{\tau\sigma}(\mathbb{R}) = 4T_0, \delta(\tau\sigma) = 1,$
$Y(\mathbb{R})_\sigma = \varnothing, Y(\mathbb{R})_{\tau\sigma} = Y(\mathbb{R}) = 2U_0 \amalg T_0$ or $2T_0;$
$\alpha(\sigma) = 0 \ (B(\sigma)); X_\sigma(\mathbb{R}) = \varnothing, \delta(\sigma) = 0, X_{\tau\sigma}(\mathbb{R}) = 4T_0, \delta(\tau\sigma) = 1,$
$Y(\mathbb{R})_\sigma = \varnothing, Y(\mathbb{R})_{\tau\sigma} = Y(\mathbb{R}) = 2U_0 \amalg T_0$ or $2T_0;$

$\alpha(\sigma) = 0$ $(B(\sigma))$; $X_\sigma(\mathbb{R}) = \varnothing, \delta(\sigma) = 0$, $X_{\tau\sigma}(\mathbb{R}) = T_1 \amalg 4T_0, \delta(\tau\sigma) = 1$,
$Y(\mathbb{R})_\sigma = \varnothing$, $Y(\mathbb{R})_{\tau\sigma} = Y(\mathbb{R}) = U_1 \amalg 2U_0 \amalg T_0$ or $U_1 \amalg 2T_0$;
$\alpha(\sigma) = 1$; $X_\sigma(\mathbb{R}) = \varnothing, \delta(\sigma) = 0$, $X_{\tau\sigma}(\mathbb{R}) = T_1 \amalg 4T_0, \delta(\tau\sigma) = 1$,
$Y(\mathbb{R})_\sigma = \varnothing$, $Y(\mathbb{R})_{\tau\sigma} = U_1 \amalg 2T_0$, $Y(\mathbb{R}) = U_1 \amalg 2T_0$;
$\alpha(\sigma) = 0$ $(B(\sigma))$; $X_\sigma(\mathbb{R}) = \varnothing, \delta(\sigma) = 0$, $X_{\tau\sigma}(\mathbb{R}) = T_2 \amalg 5T_0, \delta(\tau\sigma) = 1$,
$Y(\mathbb{R})_\sigma = \varnothing$, $Y(\mathbb{R})_{\tau\sigma} = U_2 \amalg U_0 \amalg 2T_0$, $Y(\mathbb{R}) = U_2 \amalg U_0 \amalg 2T_0$;
$\alpha(\sigma) = 0$ $(B(\sigma))$; $X_\sigma(\mathbb{R}) = 2T_1, \delta(\sigma) = 0$, $X_{\tau\sigma}(\mathbb{R}) = 4T_0, \delta(\tau\sigma) = 1$,
$(Y(\mathbb{R})_\sigma, Y(\mathbb{R})_{\tau\sigma}, Y(\mathbb{R})) = (2U_1, 2T_0, 2U_1 \amalg 2T_0)$ or $(T_1, 2U_0 \amalg T_0, 2U_0 \amalg T_1 \amalg T_0)$;
$\alpha(\sigma) = 1$; $X_\sigma(\mathbb{R}) = 2T_1, \delta(\sigma) = 0$, $X_{\tau\sigma}(\mathbb{R}) = 4T_0, \delta(\tau\sigma) = 1$,
$Y(\mathbb{R})_\sigma = T_1$, $Y(\mathbb{R})_{\tau\sigma} = 2T_0$, $Y(\mathbb{R}) = T_1 \amalg 2T_0$;
$\alpha(\sigma) = 0$ $(C(\sigma))$; $X_\sigma(\mathbb{R}) = 2T_1, \delta(\sigma) = 0$, $X_{\tau\sigma}(\mathbb{R}) = T_1 \amalg 4T_0, \delta(\tau\sigma) = 1$,
$Y(\mathbb{R})_\sigma = T_1$, $Y(\mathbb{R})_{\tau\sigma} = U_1 \amalg 2T_0$, $Y(\mathbb{R}) = U_1 \amalg T_1 \amalg 2T_0$;
$\alpha(\sigma) = 0$ $(B(\sigma))$; $X_\sigma(\mathbb{R}) = T_3 \amalg 2T_0, \delta(\sigma) = 0$, $X_{\tau\sigma}(\mathbb{R}) = 4T_0, \delta(\tau\sigma) = 1$,
$Y(\mathbb{R})_\sigma = U_3 \amalg T_0$, $Y(\mathbb{R})_{\tau\sigma} = 2T_0$, $Y(\mathbb{R}) = U_3 \amalg 3T_0$.

The case $(r(\theta), a(\theta), \delta(\theta)) = (7, 1, 1)$:

$\alpha(\sigma) = 1$; $X_\sigma(\mathbb{R}) = \varnothing, \delta(\sigma) = 0$, $X_{\tau\sigma}(\mathbb{R}) = 6T_0, \delta(\tau\sigma) = 1$,
$Y(\mathbb{R})_\sigma = \varnothing$, $Y(\mathbb{R})_{\tau\sigma} = Y(\mathbb{R}) = 2U_0 \amalg 2T_0$ or $3T_0$;
$\alpha(\sigma) = 0$ $(B(\sigma))$; $X_\sigma(\mathbb{R}) = \varnothing, \delta(\sigma) = 0$, $X_{\tau\sigma}(\mathbb{R}) = 6T_0, \delta(\tau\sigma) = 1$,
$Y(\mathbb{R})_\sigma = \varnothing$, $Y(\mathbb{R})_{\tau\sigma} = Y(\mathbb{R}) = 4U_0 \amalg T_0$ or $2U_0 \amalg 2T_0$ or $3T_0$;
$\alpha(\sigma) = 0$ $(B(\sigma))$; $X_\sigma(\mathbb{R}) = \varnothing, \delta(\sigma) = 0$, $X_{\tau\sigma}(\mathbb{R}) = T_1 \amalg 6T_0, \delta(\tau\sigma) = 1$,
$Y(\mathbb{R})_\sigma = \varnothing$, $Y(\mathbb{R})_{\tau\sigma} = Y(\mathbb{R}) = U_1 \amalg 2U_0 \amalg 2T_0$ or $U_1 \amalg 3T_0$;
$\alpha(\sigma) = 1$; $X_\sigma(\mathbb{R}) = \varnothing, \delta(\sigma) = 0$, $X_{\tau\sigma}(\mathbb{R}) = T_1 \amalg 6T_0, \delta(\tau\sigma) = 1$,
$Y(\mathbb{R})_\sigma = \varnothing$, $Y(\mathbb{R})_{\tau\sigma} = U_1 \amalg 3T_0$, $Y(\mathbb{R}) = U_1 \amalg 3T_0$;
$\alpha(\sigma) = 0$ $(B(\sigma))$; $X_\sigma(\mathbb{R}) = \varnothing, \delta(\sigma) = 0$, $X_{\tau\sigma}(\mathbb{R}) = T_2 \amalg 7T_0, \delta(\tau\sigma) = 1$,
$Y(\mathbb{R})_\sigma = \varnothing$, $Y(\mathbb{R})_{\tau\sigma} = U_2 \amalg U_0 \amalg 3T_0$, $Y(\mathbb{R}) = U_2 \amalg U_0 \amalg 3T_0$;
$\alpha(\sigma) = 0$ $(B(\sigma))$; $X_\sigma(\mathbb{R}) = 2T_1, \delta(\sigma) = 0$, $X_{\tau\sigma}(\mathbb{R}) = 6T_0, \delta(\tau\sigma) = 1$,
$(Y(\mathbb{R})_\sigma, Y(\mathbb{R})_{\tau\sigma}, Y(\mathbb{R})) = (2U_1, 3T_0, 2U_1 \amalg 3T_0)$ or $(T_1, 2U_0 \amalg 2T_0, 2U_0 \amalg T_1 \amalg 2T_0)$;
$\alpha(\sigma) = 1$; $X_\sigma(\mathbb{R}) = 2T_1, \delta(\sigma) = 0$, $X_{\tau\sigma}(\mathbb{R}) = 6T_0, \delta(\tau\sigma) = 1$,
$Y(\mathbb{R})_\sigma = T_1$, $Y(\mathbb{R})_{\tau\sigma} = 3T_0$, $Y(\mathbb{R}) = T_1 \amalg 3T_0$;
$\alpha(\sigma) = 0$ $(C(\sigma))$; $X_\sigma(\mathbb{R}) = 2T_1, \delta(\sigma) = 0$, $X_{\tau\sigma}(\mathbb{R}) = T_1 \amalg 6T_0, \delta(\tau\sigma) = 1$,
$Y(\mathbb{R})_\sigma = T_1$, $Y(\mathbb{R})_{\tau\sigma} = U_1 \amalg 3T_0$, $Y(\mathbb{R}) = U_1 \amalg T_1 \amalg 3T_0$;
$\alpha(\sigma) = 0$ $(B(\sigma))$; $X_\sigma(\mathbb{R}) = T_3 \amalg 2T_0, \delta(\sigma) = 0$, $X_{\tau\sigma}(\mathbb{R}) = 6T_0, \delta(\tau\sigma) = 1$,
$Y(\mathbb{R})_\sigma = U_3 \amalg T_0$, $Y(\mathbb{R})_{\tau\sigma} = 3T_0$, $Y(\mathbb{R}) = U_3 \amalg 4T_0$;
$\alpha(\sigma) = 0$ $(C(\sigma))$; $X_\sigma(\mathbb{R}) = 8T_0, \delta(\sigma) = 0$, $X_{\tau\sigma}(\mathbb{R}) = T_3, \delta(\tau\sigma) = 1$,
$Y(\mathbb{R})_\sigma = 4T_0$, $Y(\mathbb{R})_{\tau\sigma} = U_3$, $Y(\mathbb{R}) = U_3 \amalg 4T_0$.

The case $(r(\theta), a(\theta), \delta(\theta)) = (4, 2, 0)$:

$\alpha(\sigma) = 1$; $X_\sigma(\mathbb{R}) = 2T_1, \delta(\sigma) = 0$, $X_{\tau\sigma}(\mathbb{R}) = \varnothing, \delta(\tau\sigma) = 0$,
$Y(\mathbb{R})_{\tau\sigma} = \varnothing$, $Y(\mathbb{R})_\sigma = Y(\mathbb{R}) = 2U_1$ or T_1;
$\alpha(\sigma) = 0$ $(C(\sigma))$; $X_\sigma(\mathbb{R}) = 2T_1, \delta(\sigma) = 0$, $X_{\tau\sigma}(\mathbb{R}) = \varnothing, \delta(\tau\sigma) = 0$,
$Y(\mathbb{R})_{\tau\sigma} = \varnothing$, $Y(\mathbb{R})_\sigma = Y(\mathbb{R}) = 2U_1$ or T_1;
$\alpha(\sigma) = 0$ $(C(\sigma))$; $X_\sigma(\mathbb{R}) = T_3 \amalg 2T_0, \delta(\sigma) = 0$, $X_{\tau\sigma}(\mathbb{R}) = \varnothing, \delta(\tau\sigma) = 0$,
$Y(\mathbb{R})_{\tau\sigma} = \varnothing$, $Y(\mathbb{R})_\sigma = Y(\mathbb{R}) = U_3 \amalg 2U_0$ or $U_3 \amalg T_0$;
$\alpha(\sigma) = 1$; $X_\sigma(\mathbb{R}) = \varnothing, \delta(\sigma) = 0$, $X_{\tau\sigma}(\mathbb{R}) = \varnothing, \delta(\tau\sigma) = 0$,
$Y(\mathbb{R})_\sigma = \varnothing$, $Y(\mathbb{R})_{\tau\sigma} = \varnothing$, $Y(\mathbb{R}) = \varnothing$;
$\alpha(\sigma) = 1$; $X_\sigma(\mathbb{R}) = T_3 \amalg 2T_0, \delta(\sigma) = 0$, $X_{\tau\sigma}(\mathbb{R}) = \varnothing, \delta(\tau\sigma) = 0$,
$Y(\mathbb{R})_\sigma = U_3 \amalg T_0$, $Y(\mathbb{R})_{\tau\sigma} = \varnothing$, $Y(\mathbb{R}) = U_3 \amalg T_0$;
$\alpha(\sigma) = 0$ $(C(\sigma))$; $X_\sigma(\mathbb{R}) = T_4 \amalg 3T_0, \delta(\sigma) = 0$, $X_{\tau\sigma}(\mathbb{R}) = \varnothing, \delta(\tau\sigma) = 0$,
$Y(\mathbb{R})_\sigma = U_4 \amalg U_0 \amalg T_0$, $Y(\mathbb{R})_{\tau\sigma} = \varnothing$, $Y(\mathbb{R}) = U_4 \amalg U_0 \amalg T_0$;

$\alpha(\sigma) = 0$ $(B(\sigma))$; $X_\sigma(\mathbb{R}) = 2T_1, \delta(\sigma) = 0, X_{\tau\sigma}(\mathbb{R}) = 2T_1, \delta(\tau\sigma) = 0,$
$(Y(\mathbb{R})_\sigma, Y(\mathbb{R})_{\tau\sigma}) = (2U_1, T_1)$ or $(T_1, 2U_1), Y(\mathbb{R}) = 2U_1 \amalg T_1;$
$\alpha(\sigma) = 1; X_\sigma(\mathbb{R}) = 2T_1, \delta(\sigma) = 0, X_{\tau\sigma}(\mathbb{R}) = 2T_1, \delta(\tau\sigma) = 0,$
$Y(\mathbb{R})_\sigma = T_1, Y(\mathbb{R})_{\tau\sigma} = T_1, Y(\mathbb{R}) = 2T_1;$
$\alpha(\sigma) = 0$ $(B(\sigma))$; $X_\sigma(\mathbb{R}) = T_3 \amalg 2T_0, \delta(\sigma) = 0, X_{\tau\sigma}(\mathbb{R}) = 2T_1, \delta(\tau\sigma) = 0,$
$Y(\mathbb{R})_\sigma = U_3 \amalg T_0, Y(\mathbb{R})_{\tau\sigma} = T_1, Y(\mathbb{R}) = U_3 \amalg T_1 \amalg T_0;$
$\alpha(\sigma) = 1; X_\sigma(\mathbb{R}) = T_2, \delta_{\sigma S} = 1, X_{\tau\sigma}(\mathbb{R}) = T_2 \amalg 2T_0, \delta_{\tau\sigma S} = 1,$
$Y(\mathbb{R})_\sigma = U_2, Y(\mathbb{R})_{\tau\sigma} = U_2 \amalg T_0, Y(\mathbb{R}) = 2U_2 \amalg T_0;$
$\alpha(\sigma) = 1; X_\sigma(\mathbb{R}) = T_4 \amalg 2T_0, \delta_{\sigma S} = 1, X_{\tau\sigma}(\mathbb{R}) = T_0, \delta_{\tau\sigma S} = 1,$
$Y(\mathbb{R})_\sigma = U_4 \amalg T_0, Y(\mathbb{R})_{\tau\sigma} = U_0, Y(\mathbb{R}) = U_4 \amalg U_0 \amalg T_0;$
$\alpha(\sigma) = 1; X_\sigma(\mathbb{R}) = T_3 \amalg T_0, \delta_{\sigma S} = 1, X_{\tau\sigma}(\mathbb{R}) = T_0, \delta_{\tau\sigma S} = 1,$
$Y(\mathbb{R})_\sigma = U_3 \amalg U_0, Y(\mathbb{R})_{\tau\sigma} = U_0, Y(\mathbb{R}) = U_3 \amalg 2U_0;$
$\alpha(\sigma) = 1; X_\sigma(\mathbb{R}) = T_2, \delta_{\sigma S} = 1, X_{\tau\sigma}(\mathbb{R}) = T_1 \amalg T_0, \delta_{\tau\sigma S} = 1,$
$Y(\mathbb{R})_\sigma = U_2, Y(\mathbb{R})_{\tau\sigma} = U_1 \amalg U_0, Y(\mathbb{R}) = U_2 \amalg U_1 \amalg U_0;$
$\alpha(\sigma) = 1; X_\sigma(\mathbb{R}) = T_2, \delta_{\sigma S} = 1, X_{\tau\sigma}(\mathbb{R}) = T_2 \amalg 2T_0, \delta_{\tau\sigma S} = 1,$
$Y(\mathbb{R})_\sigma = U_2, Y(\mathbb{R})_{\tau\sigma} = U_2 \amalg 2U_0, Y(\mathbb{R}) = 2U_2 \amalg 2U_0;$
$\alpha(\sigma) = 1; X_\sigma(\mathbb{R}) = T_3 \amalg 2T_0, \delta_{\sigma S} = 0, \delta_{\sigma S_+ \cap S_-} = 1,$
$X_{\tau\sigma}(\mathbb{R}) = T_1, \delta_{\tau\sigma S} = 0, \delta_{\tau\sigma S_+ \cap S_-} = 1,$
$Y(\mathbb{R})_\sigma = U_3 \amalg T_0, Y(\mathbb{R})_{\tau\sigma} = U_1, Y(\mathbb{R}) = U_3 \amalg U_1 \amalg T_0;$
$\alpha(\sigma) = 1; X_\sigma(\mathbb{R}) = T_3, \delta_{\sigma S} = 0, \delta_{\sigma S_+ \cap S_-} = 1,$
$X_{\tau\sigma}(\mathbb{R}) = T_1 \amalg 2T_0, \delta_{\tau\sigma S} = 0, \delta_{\tau\sigma S_+ \cap S_-} = 1,$
$Y(\mathbb{R})_\sigma = U_3, Y(\mathbb{R})_{\tau\sigma} = U_1 \amalg T_0, Y(\mathbb{R}) = U_3 \amalg U_1 \amalg T_0;$
$\alpha(\sigma) = 1; X_\sigma(\mathbb{R}) = T_4 \amalg T_0, \delta_{\sigma S} = 0, \delta_{\sigma S_+ \cap S_-} = 1,$
$X_{\tau\sigma}(\mathbb{R}) = 2T_0, \delta_{\tau\sigma S} = 0, \delta_{\tau\sigma S_+ \cap S_-} = 1,$
$Y(\mathbb{R})_\sigma = U_4 \amalg U_0, Y(\mathbb{R})_{\tau\sigma} = T_0, Y(\mathbb{R}) = U_4 \amalg U_0 \amalg T_0;$
$\alpha(\sigma) = 1; X_\sigma(\mathbb{R}) = T_2 \amalg T_0, \delta_{\sigma S} = 0, \delta_{\sigma S_+ \cap S_-} = 1,$
$X_{\tau\sigma}(\mathbb{R}) = T_1, \delta_{\tau\sigma S} = 0, \delta_{\tau\sigma S_+ \cap S_-} = 1,$
$Y(\mathbb{R})_\sigma = U_2 \amalg U_0, Y(\mathbb{R})_{\tau\sigma} = U_1, Y(\mathbb{R}) = U_2 \amalg U_1 \amalg U_0;$
$\alpha(\sigma) = 1; X_\sigma(\mathbb{R}) = T_3, \delta_{\sigma S} = 0, \delta_{\sigma S_+ \cap S_-} = 1,$
$X_{\tau\sigma}(\mathbb{R}) = 2T_0, \delta_{\tau\sigma S} = 0, \delta_{\tau\sigma S_+ \cap S_-} = 1,$
$Y(\mathbb{R})_\sigma = U_3, Y(\mathbb{R})_{\tau\sigma} = 2U_0, Y(\mathbb{R}) = U_3 \amalg 2U_0;$
$\alpha(\sigma) = 0$ $(B(\sigma))$; $X_\sigma(\mathbb{R}) = T_3, \delta(\sigma) = 1, \delta_{\sigma S_+ \cap S_-} = 0,$
$X_{\tau\sigma}(\mathbb{R}) = 4T_0, \delta(\tau\sigma) = 1, \delta_{\tau\sigma S_+ \cap S_-} = 0,$
$Y(\mathbb{R})_\sigma = U_3, Y(\mathbb{R})_{\tau\sigma} = 2T_0, Y(\mathbb{R}) = U_3 \amalg 2T_0.$

The case $(r(\theta), a(\theta), \delta(\theta)) = (8, 0, 0)$:
$\alpha(\sigma) = 1; X_\sigma(\mathbb{R}) = 8T_0, \delta(\sigma) = 0, X_{\tau\sigma}(\mathbb{R}) = \varnothing, \delta(\tau\sigma) = 0,$
$Y(\mathbb{R})_{\tau\sigma} = \varnothing, Y(\mathbb{R})_\sigma = Y(\mathbb{R}) = 2U_0 \amalg 3T_0$ or $4T_0;$
$\alpha(\sigma) = 0$ $(C(\sigma))$; $X_\sigma(\mathbb{R}) = 8T_0, \delta(\sigma) = 0, X_{\tau\sigma}(\mathbb{R}) = \varnothing, \delta(\tau\sigma) = 0,$
$Y(\mathbb{R})_{\tau\sigma} = \varnothing, Y_\sigma(\mathbb{R}) = Y(\mathbb{R}) = 4U_0 \amalg 2T_0$ or $2U_0 \amalg 3T_0$ or $4T_0;$
$\alpha(\sigma) = 0$ $(C(\sigma))$; $X_\sigma(\mathbb{R}) = T_1 \amalg 8T_0, \delta(\sigma) = 0, X_{\tau\sigma}(\mathbb{R}) = \varnothing, \delta(\tau\sigma) = 0,$
$Y(\mathbb{R})_{\tau\sigma} = \varnothing, Y_\sigma(\mathbb{R}) = Y(\mathbb{R}) = U_1 \amalg 2U_0 \amalg 3T_0$ or $U_1 \amalg 4T_0;$
$\alpha(\sigma) = 1; X_\sigma(\mathbb{R}) = T_1 \amalg 8T_0, \delta(\sigma) = 0, X_{\tau\sigma}(\mathbb{R}) = \varnothing, \delta(\tau\sigma) = 0,$
$Y(\mathbb{R})_\sigma = U_1 \amalg 4T_0, Y(\mathbb{R})_{\tau\sigma} = \varnothing, Y(\mathbb{R}) = U_1 \amalg 4T_0;$
$\alpha(\sigma) = 0$ $(C(\sigma))$; $X_\sigma(\mathbb{R}) = T_2 \amalg 9T_0, \delta(\sigma) = 0, X_{\tau\sigma}(\mathbb{R}) = \varnothing, \delta(\tau\sigma) = 0,$
$Y(\mathbb{R})_\sigma = U_2 \amalg U_0 \amalg 4T_0, Y(\mathbb{R})_{\tau\sigma} = \varnothing, Y(\mathbb{R}) = U_2 \amalg U_0 \amalg 4T_0;$
$\alpha(\sigma) = 0$ $(C(\sigma))$; $X_\sigma(\mathbb{R}) = 8T_0, \delta(\sigma) = 0, X_{\tau\sigma}(\mathbb{R}) = 2T_1, \delta(\tau\sigma) = 0,$
$(Y(\mathbb{R})_\sigma, Y(\mathbb{R})_{\tau\sigma}, Y(\mathbb{R})) = (4T_0, 2U_1, 2U_1 \amalg 4T_0)$ or $(2U_0 \amalg 3T_0, T_1, 2U_0 \amalg T_1 \amalg 3T_0);$
$\alpha(\sigma) = 1; X_\sigma(\mathbb{R}) = 8T_0, \delta(\sigma) = 0, X_{\tau\sigma}(\mathbb{R}) = 2T_1, \delta(\tau\sigma) = 0,$

$Y(\mathbb{R})_\sigma = 4T_0,\ Y(\mathbb{R})_{\tau\sigma} = T_1,\ Y(\mathbb{R}) = T_1 \amalg 4T_0;$
$\alpha(\sigma) = 0\ (B(\sigma));\ X_\sigma(\mathbb{R}) = T_1 \amalg 8T_0,\ \delta(\sigma) = 0,\ X_{\tau\sigma}(\mathbb{R}) = 2T_1,\ \delta(\tau\sigma) = 0,$
$Y(\mathbb{R})_\sigma = U_1 \amalg 4T_0,\ Y(\mathbb{R})_{\tau\sigma} = T_1,\ Y(\mathbb{R}) = U_1 \amalg T_1 \amalg 4T_0;$
$\alpha(\sigma) = 0\ (C(\sigma));\ X_\sigma(\mathbb{R}) = 8T_0,\ \delta(\sigma) = 0,\ X_{\tau\sigma}(\mathbb{R}) = T_3 \amalg 2T_0,\ \delta(\tau\sigma) = 0,$
$Y(\mathbb{R})_\sigma = 4T_0,\ Y(\mathbb{R})_{\tau\sigma} = U_3 \amalg T_0,\ Y(\mathbb{R}) = U_3 \amalg 5T_0;$
$\alpha(\sigma) = 1;\ X_\sigma(\mathbb{R}) = 5T_0, \delta_\sigma S = 1,\ X_{\tau\sigma}(\mathbb{R}) = 3T_0, \delta_{\tau\sigma}S = 1,$
$(Y(\mathbb{R})_\sigma, Y(\mathbb{R})_{\tau\sigma}) = (U_0 \amalg 2T_0, 3U_0)\ or\ (3U_0 \amalg T_0, U_0 \amalg T_0),\ Y(\mathbb{R}) = 4U_0 \amalg 2T_0;$
$\alpha(\sigma) = 1;\ X_\sigma(\mathbb{R}) = 5T_0, \delta_\sigma S = 1,\ X_{\tau\sigma}(\mathbb{R}) = T_2 \amalg 4T_0, \delta_{\tau\sigma}S = 1,$
$Y(\mathbb{R})_\sigma = U_0 \amalg 2T_0,\ Y(\mathbb{R})_{\tau\sigma} = U_2 \amalg 2T_0,\ Y(\mathbb{R}) = U_2 \amalg U_0 \amalg 4T_0;$
$\alpha(\sigma) = 1;\ X_\sigma(\mathbb{R}) = T_2 \amalg 6T_0, \delta_\sigma S = 1,\ X_{\tau\sigma}(\mathbb{R}) = 3T_0, \delta_{\tau\sigma}S = 1,$
$Y(\mathbb{R})_\sigma = U_2 \amalg 3T_0,\ Y(\mathbb{R})_{\tau\sigma} = U_0 \amalg T_0,\ Y(\mathbb{R}) = U_2 \amalg U_0 \amalg 4T_0;$
$\alpha(\sigma) = 1;\ X_\sigma(\mathbb{R}) = 7T_0, \delta_\sigma S = 1,\ X_{\tau\sigma}(\mathbb{R}) = T_2 \amalg 2T_0, \delta_{\tau\sigma}S = 1,$
$Y(\mathbb{R})_\sigma = U_0 \amalg 3T_0,\ Y(\mathbb{R})_{\tau\sigma} = U_2 \amalg T_0,\ Y(\mathbb{R}) = U_2 \amalg U_0 \amalg 4T_0;$
$\alpha(\sigma) = 1;\ X_\sigma(\mathbb{R}) = T_1 \amalg 5T_0, \delta_\sigma S = 1,\ X_{\tau\sigma}(\mathbb{R}) = 3T_0, \delta_{\tau\sigma}S = 1,$
$Y(\mathbb{R})_\sigma = U_1 \amalg U_0 \amalg 2T_0,\ Y(\mathbb{R})_{\tau\sigma} = U_0 \amalg T_0,\ Y(\mathbb{R}) = U_1 \amalg 2U_0 \amalg 3T_0;$
$\alpha(\sigma) = 1;\ X_\sigma(\mathbb{R}) = 5T_0, \delta_\sigma S = 1,\ X_{\tau\sigma}(\mathbb{R}) = T_1 \amalg 3T_0, \delta_{\tau\sigma}S = 1,$
$Y(\mathbb{R})_\sigma = U_0 \amalg 2T_0,\ Y(\mathbb{R})_{\tau\sigma} = U_1 \amalg U_0 \amalg T_0,\ Y(\mathbb{R}) = U_1 \amalg 2U_0 \amalg 3T_0;$
$\alpha(\sigma) = 1;\ X_\sigma(\mathbb{R}) = T_1 \amalg 7T_0, \delta_\sigma S = 1,\ X_{\tau\sigma}(\mathbb{R}) = T_0, \delta_{\tau\sigma}S = 1,$
$Y(\mathbb{R})_\sigma = U_1 \amalg U_0 \amalg 3T_0,\ Y(\mathbb{R})_{\tau\sigma} = U_0,\ Y(\mathbb{R}) = U_1 \amalg 2U_0 \amalg 3T_0;$
$\alpha(\sigma) = 1;\ X_\sigma(\mathbb{R}) = 7T_0, \delta_\sigma S = 1,\ X_{\tau\sigma}(\mathbb{R}) = T_1 \amalg T_0, \delta_{\tau\sigma}S = 1,$
$Y(\mathbb{R})_\sigma = U_0 \amalg 3T_0,\ Y(\mathbb{R})_{\tau\sigma} = U_1 \amalg U_0,\ Y(\mathbb{R}) = U_1 \amalg 2U_0 \amalg 3T_0;$
$\alpha(\sigma) = 1;\ X_\sigma(\mathbb{R}) = 7T_0, \delta_\sigma S = 1,\ X_{\tau\sigma}(\mathbb{R}) = T_0, \delta_{\tau\sigma}S = 1,$
$Y(\mathbb{R})_\sigma = 3U_0 \amalg 2T_0,\ Y(\mathbb{R})_{\tau\sigma} = U_0,\ Y(\mathbb{R}) = 4U_0 \amalg 2T_0;$
$\alpha(\sigma) = 1;\ X_\sigma(\mathbb{R}) = T_1 \amalg 4T_0, \delta_\sigma S = 0, \delta_{\sigma S_+ \cap S_-} = 1,$
$X_{\tau\sigma}(\mathbb{R}) = 4T_0, \delta_{\tau\sigma}S = 0, \delta_{\tau\sigma S_+ \cap S_-} = 1,$
$(Y(\mathbb{R})_\sigma, Y(\mathbb{R})_{\tau\sigma}) = (U_1 \amalg 2U_0 \amalg T_0, 2T_0)\ or\ (U_1 \amalg 2T_0, 2U_0 \amalg T_0),$
$Y(\mathbb{R}) = U_1 \amalg 2U_0 \amalg 3T_0;$
$\alpha(\sigma) = 1;\ X_\sigma(\mathbb{R}) = T_1 \amalg 6T_0, \delta_\sigma S = 0, \delta_{\sigma S_+ \cap S_-} = 1,$
$X_{\tau\sigma}(\mathbb{R}) = 2T_0, \delta_{\tau\sigma}S = 0, \delta_{\tau\sigma S_+ \cap S_-} = 1,$
$(Y(\mathbb{R})_\sigma, Y(\mathbb{R})_{\tau\sigma}) = (U_1 \amalg 2U_0 \amalg 2T_0, T_0)\ or\ (U_1 \amalg 3T_0, 2U_0),$
$Y(\mathbb{R}) = U_1 \amalg 2U_0 \amalg 3T_0;$
$\alpha(\sigma) = 1;\ X_\sigma(\mathbb{R}) = 6T_0, \delta_\sigma S = 0, \delta_{\sigma S_+ \cap S_-} = 1,$
$X_{\tau\sigma}(\mathbb{R}) = T_1 \amalg 2T_0, \delta_{\tau\sigma}S = 0, \delta_{\tau\sigma S_+ \cap S_-} = 1,$
$(Y(\mathbb{R})_\sigma, Y(\mathbb{R})_{\tau\sigma}) = (2U_0 \amalg 2T_0, U_1 \amalg T_0)\ or\ (3T_0, U_1 \amalg 2U_0),$
$Y(\mathbb{R}) = U_1 \amalg 2U_0 \amalg 3T_0;$
$\alpha(\sigma) = 1;\ X_\sigma(\mathbb{R}) = 4T_0, \delta_\sigma S = 0, \delta_{\sigma S_+ \cap S_-} = 1,$
$X_{\tau\sigma}(\mathbb{R}) = 4T_0, \delta_{\tau\sigma}S = 0, \delta_{\tau\sigma S_+ \cap S_-} = 1,$
$(Y(\mathbb{R})_\sigma, Y(\mathbb{R})_{\tau\sigma}) = (2U_0 \amalg T_0, 2U_0 \amalg T_0)\ or\ (4U_0, 2T_0),$
$Y(\mathbb{R}) = 4U_0 \amalg 2T_0;$
$\alpha(\sigma) = 1;\ X_\sigma(\mathbb{R}) = 6T_0, \delta_\sigma S = 0, \delta_{\sigma S_+ \cap S_-} = 1,$
$X_{\tau\sigma}(\mathbb{R}) = 2T_0, \delta_{\tau\sigma}S = 0, \delta_{\tau\sigma S_+ \cap S_-} = 1,$
$(Y(\mathbb{R})_\sigma, Y(\mathbb{R})_{\tau\sigma}) = (4U_0 \amalg T_0, T_0)\ or\ (2U_0 \amalg 2T_0, 2U_0),$
$Y(\mathbb{R}) = 4U_0 \amalg 2T_0;$
$\alpha(\sigma) = 1;\ X_\sigma(\mathbb{R}) = T_1 \amalg 4T_0, \delta_\sigma S = 0, \delta_{\sigma S_+ \cap S_-} = 1,$
$X_{\tau\sigma}(\mathbb{R}) = T_1 \amalg 4T_0, \delta_{\tau\sigma}S = 0, \delta_{\tau\sigma S_+ \cap S_-} = 1,$
$Y(\mathbb{R})_\sigma = U_1 \amalg 2T_0,\ Y(\mathbb{R})_{\tau\sigma} = U_1 \amalg 2T_0,\ Y(\mathbb{R}) = 2U_1 \amalg 4T_0;$
$\alpha(\sigma) = 1;\ X_\sigma(\mathbb{R}) = T_1 \amalg 6T_0, \delta_\sigma S = 0, \delta_{\sigma S_+ \cap S_-} = 1,$
$X_{\tau\sigma}(\mathbb{R}) = T_1 \amalg 2T_0, \delta_{\tau\sigma}S = 0, \delta_{\tau\sigma S_+ \cap S_-} = 1,$

$Y(\mathbb{R})_\sigma = U_1 \amalg 3T_0$, $Y(\mathbb{R})_{\tau\sigma} = U_1 \amalg T_0$, $Y(\mathbb{R}) = 2U_1 \amalg 4T_0$;
$\alpha(\sigma) = 1$; $X_\sigma(\mathbb{R}) = T_1 \amalg 8T_0$, $\delta_\sigma S = 0$, $\delta_{\sigma S_+ \cap S_-} = 1$,
$X_{\tau\sigma}(\mathbb{R}) = T_1$, $\delta_{\tau\sigma} S = 0$, $\delta_{\tau\sigma S_+ \cap S_-} = 1$,
$Y(\mathbb{R})_\sigma = U_1 \amalg 4T_0$, $Y(\mathbb{R})_{\tau\sigma} = U_1$, $Y(\mathbb{R}) = 2U_1 \amalg 4T_0$;
$\alpha(\sigma) = 1$; $X_\sigma(\mathbb{R}) = T_2 \amalg 5T_0$, $\delta_\sigma S = 0$, $\delta_{\sigma S_+ \cap S_-} = 1$,
$X_{\tau\sigma}(\mathbb{R}) = 4T_0$, $\delta_{\tau\sigma} S = 0$, $\delta_{\tau\sigma S_+ \cap S_-} = 1$,
$Y(\mathbb{R})_\sigma = U_2 \amalg U_0 \amalg 2T_0$, $Y(\mathbb{R})_{\tau\sigma} = 2T_0$, $Y(\mathbb{R}) = U_2 \amalg U_0 \amalg 4T_0$;
$\alpha(\sigma) = 1$; $X_\sigma(\mathbb{R}) = T_2 \amalg 7T_0$, $\delta_\sigma S = 0$, $\delta_{\sigma S_+ \cap S_-} = 1$,
$X_{\tau\sigma}(\mathbb{R}) = 2T_0$, $\delta_{\tau\sigma} S = 0$, $\delta_{\tau\sigma S_+ \cap S_-} = 1$,
$Y(\mathbb{R})_\sigma = U_2 \amalg U_0 \amalg 3T_0$, $Y(\mathbb{R})_{\tau\sigma} = T_0$, $Y(\mathbb{R}) = U_2 \amalg U_0 \amalg 4T_0$;
$\alpha(\sigma) = 1$; $X_\sigma(\mathbb{R}) = 6T_0$, $\delta_\sigma S = 0$, $\delta_{\sigma S_+ \cap S_-} = 1$,
$X_{\tau\sigma}(\mathbb{R}) = T_2 \amalg 3T_0$, $\delta_{\tau\sigma} S = 0$, $\delta_{\tau\sigma S_+ \cap S_-} = 1$,
$Y(\mathbb{R})_\sigma = 3T_0$, $Y(\mathbb{R})_{\tau\sigma} = U_2 \amalg U_0 \amalg T_0$, $Y(\mathbb{R}) = U_2 \amalg U_0 \amalg 4T_0$;
$\alpha(\sigma) = 1$; $X_\sigma(\mathbb{R}) = 8T_0$, $\delta_\sigma S = 0$, $\delta_{\sigma S_+ \cap S_-} = 1$,
$X_{\tau\sigma}(\mathbb{R}) = T_2 \amalg T_0$, $\delta_{\tau\sigma} S = 0$, $\delta_{\tau\sigma S_+ \cap S_-} = 1$,
$Y(\mathbb{R})_\sigma = 4T_0$, $Y(\mathbb{R})_{\tau\sigma} = U_2 \amalg U_0$, $Y(\mathbb{R}) = U_2 \amalg U_0 \amalg 4T_0$;
$\alpha(\sigma) = 1$; $X_\sigma(\mathbb{R}) = 8T_0$, $\delta_\sigma S = 0$, $\delta_{\sigma S_+ \cap S_-} = 1$,
$X_{\tau\sigma}(\mathbb{R}) = T_1$, $\delta_{\tau\sigma} S = 0$, $\delta_{\tau\sigma S_+ \cap S_-} = 1$,
$Y(\mathbb{R})_\sigma = 2U_0 \amalg 3T_0$, $Y(\mathbb{R})_{\tau\sigma} = U_1$, $Y(\mathbb{R}) = U_1 \amalg 2U_0 \amalg 3T_0$.

As a result, we get the following

THEOREM 6. *There exist real Enriques surfaces with real part $Y(\mathbb{R})$ having one of the 59 topological types indicated below*:

$s_{\text{nor}} = 0$: $\varnothing$; $T_1 \amalg T_0$, $2T_1$; $T_1 \amalg 2T_0$; $4T_0$, $T_1 \amalg 3T_0$; $5T_0$, $T_1 \amalg 4T_0$;

$s_{\text{nor}} = 1$: U_{1+2k}, $k = 0, 1, 2, 3, 4$;
$U_1 \amalg T_0$, $U_1 \amalg T_1$, $U_3 \amalg T_0$, $U_3 \amalg T_1$,
$U_5 \amalg T_1$, $U_7 \amalg T_1$, $U_9 \amalg T_0$, $U_9 \amalg T_1$;
$U_1 \amalg 2T_0$, $U_1 \amalg T_1 \amalg T_0$, $U_3 \amalg 2T_0$, $U_3 \amalg T_1 \amalg T_0$;
$U_1 \amalg 3T_0$, $U_1 \amalg T_1 \amalg 2T_0$, $U_3 \amalg 3T_0$;
$U_1 \amalg 4T_0$, $U_1 \amalg T_1 \amalg 3T_0$, $U_3 \amalg 4T_0$;
$U_1 \amalg 5T_0$, $U_1 \amalg T_1 \amalg 4T_0$;
$U_3 \amalg 5T_0$;

$s_{\text{nor}} = 2$: $U_{2k} \amalg U_0$, $k = 1, 2, 3, 4, 5$,
$U_{10-k} \amalg U_k$, $k = 1, 2, 3, 4, 5$;
$2U_1 \amalg T_1$, $U_2 \amalg U_0 \amalg T_0$, $2U_2 \amalg T_0$, $U_3 \amalg U_1 \amalg T_0$, $U_4 \amalg U_0 \amalg T_0$;
$U_2 \amalg U_0 \amalg 2T_0$;
$2U_0 \amalg 3T_0$, $U_2 \amalg U_0 \amalg 3T_0$;
$2U_1 \amalg 4T_0$, $U_2 \amalg U_0 \amalg 4T_0$;

$s_{\text{nor}} = 3$: $U_2 \amalg U_1 \amalg U_0$, $U_3 \amalg 2U_0$;
$U_1 \amalg 2U_0 \amalg 3T_0$;

$s_{\text{nor}} = 4$: $2U_2 \amalg 2U_0$; $4U_0 \amalg 2T_0$.

THEOREM 7. *If Y is a real Enriques surface and $Y(\mathbb{R})$ is not of one of types listed in Theorem 6, then $Y(\mathbb{R})$ is of one of the following* 21 *topological types*:

$s_{\text{nor}} = 0$: T_0, T_1; $2T_0$; $3T_0$;

$s_{\text{nor}} = 2$: $2U_0$, $2U_1$;

$\qquad\qquad 2U_0 \amalg T_0$, $2U_0 \amalg T_1$, $2U_1 \amalg T_0$;

$\qquad\qquad 2U_0 \amalg 2T_0$, $2U_0 \amalg T_1 \amalg T_0$, $2U_1 \amalg 2T_0$;

$\qquad\qquad 2U_0 \amalg T_1 \amalg 2T_0$, $2U_1 \amalg 3T_0$;

$\qquad\qquad 2U_0 \amalg 4T_0$, $2U_0 \amalg T_1 \amalg 3T_0$;

$s_{\text{nor}} = 3$: $U_1 \amalg 2U_0$; $U_1 \amalg 2U_0 \amalg T_0$; $U_1 \amalg 2U_0 \amalg 2T_0$;

$s_{\text{nor}} = 4$: $4U_0$; $4U_0 \amalg T_0$.

We think that the calculations above (Theorem 5) of additional invariants will be useful to forbid or construct some topological types listed in Theorem 7 for real Enriques surfaces.

It would be very interesting to calculate arithmetic invariants of real Enriques surfaces, for example the Brauer group and the Witt group, and compare these invariants with similar invariants for real rational surfaces. See [**CT–P, Kr, Ma, Mi, N5, N-S, Si, Su**] on this subject. We hope that our results here will be useful for the calculation of these invariants.

We are grateful to the referee for very useful remarks.

References

[A]   I. R. Shafarevich (ed.), *Algebraic surfaces*, Trudy Mat. Inst. Steklov **75** (1965); English transl. in Proc. Steklov Inst. Math. **75** (1967), Amer. Math. Soc., Providence, RI.

[Br] G. E. Bredon, *Introduction to compact transformation groups*, Academic Press, New York and London, 1972.

[CT-P] J.-L. Colliot-Thélène and R. Parimala, *Real components of algebraic varieties and étale cohomology*, Invent. Math. **101** (1990), 81–99.

[C-D] F. R. Cossec and I. Dolgachev, *Enriques surfaces*. I, Progress in Mathematics, vol. 76, Birkhäuser, Basel, 1989.

[Gu] D. A. Gudkov, *The topology of real projective algebraic varieties*, Uspekhi Mat. Nauk **29** (1974), no. 4, 3–79; English transl., Russian Math. Surveys **29** (1974), no. 4, 1–79.

[Ha1] V. M. Kharlamov, *Topological types of nonsingular surfaces of degree* 4 *in* $\mathbb{R}P^3$, Funktsional. Anal. i Prilozhen. **10** (1976), no. 4, 55–68; English transl., Functional Anal. Appl. **10** (1976), no. 4, 295–305.

[Ha2] _____, *Real algebraic surfaces*, Proc. Internat. Congress Math. (Helsinki, 1978), vol. 1, Acad. Sci. Fennica, Helsinki, 1980, pp. 421–428; English transl., Amer. Math. Soc. Transl. (2) **117** (1981), 15–21.

[Ho] E. Horikawa, *On the periods of Enriques surfaces*, I, II, Math. Ann. **234** (1978), 73–108; **235** (1978), 217–246.

[Kr] V. A. Krasnov, *Harnack–Thom inequalities for mappings of real algebraic varieties*, Izv. Akad. Nauk SSSR Ser. Mat. **47** (1983), 268–297; English transl., Math. USSR-Izv. **22** (1984), 247–275.

[Ku] V. S. Kulikov, *Degenerations of K3-surfaces and Enriques surfaces*, Izv. Akad. Nauk SSSR Ser. Mat. **41** (1977), 1008–1042; English transl., Math. USSR-Izv. **11** (1978), 957–989.

[Ma] Yu. I. Manin, *Le groupe de Brauer–Grothendieck en géometrie diophantienne*, Actes du Congrès Intern. Math. Nice, 1970, vol. 1, Gauthier-Villars, Paris, 1971, pp. 401–411.

[Mi] J. Milne, *Étale cohomology*, Princeton Univ. Press, Princeton, NJ, 1980.

[N1] V. V. Nikulin, *Finite groups of automorphisms of Kählerian surfaces of type* K3, Trudy Moskov. Mat. Obshch. **38** (1979), 75–138; English transl. in Trans. Moscow Math. Soc. **38** (1980), 71–135.

[N2] _____, *On the quotient groups of the automorphism groups of hyperbolic forms by the subgroups generated by* 2*-reflections. Algebraic-geometric applications*, Contemporary Problems in Math., vol. 18, VINITI, Moscow, 1981, pp. 3–114; English transl. in J. Soviet Math. **22** (1983), 1401–1476.

[N3] _____, *Integral symmetric bilinear forms and some of their geometric applications*, Izv. Akad. Nauk SSSR Ser. Mat. **43** (1979), 111–177; English transl. in Math. USSR-Izv. **14** (1980), no. 1, 103–167.

[N4] _____, *Involutions of integral quadratic forms and their application to real algebraic geometry*, Izv. Akad. Nauk SSSR Ser. Mat. **47** (1983), 109–188; English transl. in Math. USSR-Izv. **22** (1984), 99–172.

[N5] _____, *On the Brauer group of real algebraic surfaces*, Algebraic Geometry and its Applications (Yaroslavl', 1992), Viewig, Braunschweig, 1994, pp. 113–136.

[N-S] V. V. Nikulin and R. Sujatha, *On Brauer groups of real Enriques surfaces*, J. Reine Angew. Math. **444** (1993), 115–154.

[PŠ-Š] I. I. Piateckiĭ-Šapiro and I. R. Šafarevič, *A Torelli theorem for algebraic surfaces of type* K3, Izv. Akad. Nauk SSSR Ser. Mat. **35** (1971), 530–572; English transl. in Math. USSR-Izv. **5** (1971), 547–588.

[Si] R. Silhol, *Real algebraic surfaces*, Lecture Notes in Math., vol. 1392, Springer-Verlag, Berlin, Heidelberg, and New York, 1989.

[Sp] E. H. Spanier, *Algebraic topology*, McGraw-Hill, New York, 1966.

[Su] R. Sujatha, *Witt groups of real projective surfaces*, Math. Ann. **288** (1990), 89–101.

Translated by THE AUTHOR

STEKLOV MATHEMATICAL INSTITUTE, UL. VAVILOVA 42, MOSCOW 117966, GSP-1, RUSSIA
E-mail address: slava@nikulin.mian.su

Critical Points of Real Polynomials, Subdivisions of Newton Polyhedra and Topology of Real Algebraic Hypersurfaces

Evgeniĭ Shustin

Dedicated to the memory of my teacher
Dmitriĭ Andreevich Gudkov

ABSTRACT. By means of Viro's method of gluing real polynomials, we construct polynomials with prescribed collections of critical points. Using this technique, we solve the problem on the numbers of maxima, minima, and saddles of a generic real polynomial in two variables, and establish new restrictions on the topology of some real algebraic hypersurfaces.

Introduction

Let F be a real polynomial of degree d in n variables $x_1, \ldots, x_n$. The solutions of the system

$$(0.1) \qquad F_{x_1}(x_1, \ldots, x_n) = \cdots = F_{x_n}(x_1, \ldots, x_n) = 0$$

are called *critical points* of F. A critical point $\overline{x}$ is called *nondegenerate* if the Hessian

$$(0.2) \qquad M_F(\overline{x}) = (F_{x_i x_j}(\overline{x}))_{1 \leqslant i, j \leqslant n}$$

is nondegenerate. Assume that all critical points of F are nondegenerate. Denote by p the number of pairs of conjugate imaginary critical points of F, and by c_i the number of critical points of index $i = 0, \ldots, n$ (the *index* is the number of negative eigenvalues of the Hessian). In this paper we deal with the classification problem of the vectors $\overline{c}(F) = (c_0, \ldots, c_n, p)$ for polynomials F of a given degree d without singularities at infinity (further on such vectors are called *index distributions*).

For $n = 1$, the problem is trivial. For $n \geqslant 2$ it is open (see [**Ar, Du**]). Our first result is a new construction of real polynomials with a given index distribution (§1). Then we apply it to the classification of index distributions of real polynomials in two variables (§3). The answer is: the following upper bound on the number of all critical points and the Morse formula are the only restrictions to index distributions, except some special cases. This answer is similar to the one for the distributions of

---

1991 *Mathematics Subject Classification.* Primary 26C99, 52B20, 57R70; Secondary 14N10, 14P25.

©1996, American Mathematical Society

nodes of different types for real plane curves [S4]. We note here that such a result could not be obtained by the methods used before [ChV, Du], which consisted in the following: take the complement to a level line of a real polynomial in the real plane, and then estimate the numbers of minima and maxima by the numbers of components of the complement with positive or negative values of the original polynomial. Indeed, the Comessatti–Petrovskiĭ inequality [Gu, Wi] provides a new nontrivial restriction to the numbers of the components mentioned, and does not allow us to get a complete answer.

The second application of our construction is the specification of new upper bounds on Betti numbers of real algebraic hypersurfaces of T-type (§4).

§1. Gluing theorem for critical points

1.1. Preliminaries.
Here we introduce some notions and notation used below.

Below, the word *polyhedron* stands for a convex polyhedron in $\mathbb{R}^n$ with integral vertices having only nonnegative coordinates. It is called *nondegenerate* if its dimension is n. For a nondegenerate polyhedron $\Delta \subset \mathbb{R}^n$, define
$$V(\Delta) = n!\,\text{vol}(\Delta),$$
where vol is the usual Euclidean volume in $\mathbb{R}^n$.

A *subdivision* of a polyhedron Δ is a set of polyhedra $\Delta_1, \ldots, \Delta_r$ of the same dimension such that $\bigcup_i \Delta_i = \Delta$ and any intersection $\Delta_i \cap \Delta_j$ is either empty or is a common face of Δ_i and Δ_j.

Let Δ be a nondegenerate polyhedron, and let $\mathcal{P}(\Delta)$ be the space of complex polynomials with Newton polyhedron Δ. We say that $F \in \mathcal{P}(\Delta)$ is *nondegenerate* (or, briefly, NDP) if all its critical points in $(\mathbb{C}^*)^n$ are nondegenerate. For a real NDP F, define the vector $\overline{c}(F) = (c_0, \ldots, c_n, p)$ as in the Introduction, counting critical points in $(\mathbb{C}^*)^n$. Put
$$|\overline{c}(F)| = c_0 + \cdots + c_n + 2p.$$
Denote by $C(\Delta)$ the maximum of $|\overline{c}(F)|$ taken over all NDP from $\mathcal{P}(\Delta)$.

PROPOSITION 1.1.1. *The NDP's with $C(\Delta)$ critical points in $(\mathbb{C}^*)^n$ form a Zariski open subset of $\mathcal{P}(\Delta)$.*

PROOF. It suffices to show that any degenerate critical point can be deformed into a number of nondegenerate critical points by a variation of a given polynomial in $\mathcal{P}(\Delta)$. Let $\overline{x}_0 \in (\mathbb{C}^*)^n$ be a degenerate critical point of a polynomial $F_0 \in \mathcal{P}(\Delta)$. Let us embed $\mathcal{P}(\Delta)$ into the space Λ of polynomials in n variables of sufficiently high degree. Consider the germ $\mathcal{M}$ at F_0 of the variety in Λ that consists of polynomials F with one degenerate critical point $\overline{x}$ in a neighborhood of $\overline{x}_0$ and $\text{rk}\,M_F(\overline{x}) = \text{rk}\,M_{F_0}(\overline{x}_0) = \rho$. Denote by W_F, $F \in \mathcal{M}$, the linear eigenspace in $\mathbb{R}^n$ corresponding to the eigenvalue 0 of $M_F(\overline{x})$. It is clear that W_F, $F \in \mathcal{M}$, forms a germ of a variety in the Grassmannian $\text{Gr}_n(n-\rho)$. From this it is easy to derive that the tangent space $T_{F_0}\mathcal{M}$ is contained in the space
$$\{F \in \mathcal{P}(\Delta) \mid \text{grad}\,F(\overline{x}_0) \perp W_{F_0}\}.$$
The nondegeneracy of Δ implies the existence of an integer point $(i_1, \ldots, i_n) \in \Delta$ such that
$$(i_1/x_{01}, \ldots, i_n/x_{0n}) \notin V_{F_0}^\perp, \qquad (x_{01}, \ldots, x_{0n}) = \overline{x}_0.$$
This immediately implies that the deformation $F_t = F_0 + t x_1^{i_1} \cdots x_n^{i_n}$, $t \in \mathbb{C}$, increases the rank of the Hessian at critical points of F_t, and we are done. □

Below we shall use the following properties of the number $C(\Delta)$: for any polyhedron Δ, we have

(1.1.2) $$C(\Delta) \leqslant V(\Delta),$$

which follows from the Koushnirenko theorem [**Ko**], and for the standard simplex τ_d^n with vertices $(0, \ldots, 0), (d, 0, \ldots, 0), (0, d, 0, \ldots, 0), \ldots, (0, \ldots, 0, d)$ we have

(1.1.3) $$C(\tau_d^n) = (d-1)^n.$$

1.2. Gluing theorem. Let

$$F^{(k)}(x_1, \ldots, x_n) = \sum_{(j_1, \ldots, j_n) \in \Delta_k} A_{j_1 \ldots j_n} x_1^{j_1} \cdots x_n^{j_n}, \quad k = 1, \ldots, r,$$

be a set of real polynomials with nondegenerate Newton polyhedra $\Delta_1, \ldots, \Delta_r$. Assume that $\Delta_1, \ldots, \Delta_r$ form a subdivision of a polyhedron Δ, and that this subdivision is *regular*, i.e., there is a continuous, piecewise linear, convex down function $v \colon \Delta \to \mathbb{R}$ which is linear on each Δ_k, $k = 1, \ldots, r$, and is not linear on each union $\Delta_k \cup \Delta_s$, $k \neq s$. Introduce the family of polynomials

(1.2.1) $$F_t(x_1, \ldots, x_n) = \sum_{(j_1, \ldots, j_n) \in \Delta_k} A_{j_1 \ldots j_n} x_1^{j_1} \cdots x_n^{j_n} t^{v(j_1, \ldots, j_n)}, \quad t > 0.$$

THEOREM 1.2.2. *If $F^{(1)}, \ldots, F^{(r)}$ are NDP's and*

(1.2.3) $$\sum_{k=1}^r |\bar{c}(F^{(k)})| = C(\Delta),$$

then, for a sufficiently small $t > 0$, F_t is an NDP with

$$\bar{c}(F_t) = \sum_{k=1}^r \bar{c}(F^{(k)}).$$

PROOF. Denote by $\lambda_k \colon \Delta \to \mathbb{R}$ the linear function equal to v on Δ_k, and put $v_k = v - \lambda_k$. Note that the substitution of v for v_k in (1.2.1) is the composition of the linear coordinate change

$$T_k(x_1, \ldots, x_n) = (x_1 t^{\alpha_{1k}}, \ldots, x_n t^{\alpha_{nk}}),$$

with the multiplication of F_t by a positive number. Obviously, this operation does not influence $\bar{c}(F_t)$, but only moves critical points in $(\mathbb{C}^*)^n$. Note also that, according to the definition of v,

(1.2.4) $$|\alpha_{1k} - \alpha_{1j}| + \cdots + |\alpha_{nk} - \alpha_{nj}| > 0, \quad k \neq j.$$

Let $Q \subset (\mathbb{C}^*)^n$ be a compact set whose interior contains all the critical points of $F^{(1)}, \ldots, F^{(r)}$ from $(\mathbb{C}^*)^n$. Since (1.2.4) holds, there is $t_1 > 0$ such that, for any $t \in (0, t_1)$, the compact sets $T_1^{-1}(Q), \ldots, T_r^{-1}(Q)$ are disjoint. Since v is convex down, for any $k = 1, \ldots, r$ we have

$$F_t(T_k(x_1, \ldots, x_n)) = t^{\alpha_k}(F^{(k)}(x_1, \ldots, x_n) + G_t^{(k)}(x_1, \ldots, x_n)),$$

where each coefficient of $G_t^{(k)}$ contains t to a positive power. Then there is a $t_2 > 0$ such that, for any $t \in (0, t_2)$ and $k = 1, \ldots, r$, the polynomial $F_t \circ T_k$ has exactly

$|\bar{c}(F^{(k)})|$ nondegenerate critical points in Q with the index distribution $\bar{c}(F^{(k)})$. Thus, relation (1.2.3) completes the proof. □

§2. Polynomials with Newton simplices

2.1. Preliminaries. Let $F \in \mathbb{R}[x_1, \ldots, x_n]$ be a polynomial with a nondegenerate Newton polyhedron Δ. For any vector $\bar{\varepsilon} = (\varepsilon_1, \ldots, \varepsilon_n) \in \{+1; -1\}^n$ denote by $\Delta(\bar{\varepsilon})$ the corresponding symmetric image of Δ in the orthant

$$\mathbb{R}(\bar{\varepsilon}) = \{\bar{x} \mid x_i \varepsilon_i \geq 0, \ i = 1, \ldots, n\}.$$

The *chart* $\mathrm{Ct}(F)$ of F is the closure of the image of $\{F = 0\} \cap (\mathbb{R}^*)^n$ under some homeomorphism

$$(\mathbb{R}^*)^n \to \bigcup \mathrm{Int}(\Delta(\bar{\varepsilon})).$$

For details we refer to [S1, St, V2, V3].

Let us formulate two statements needed below:

THEOREM 2.1.1 ([V2, V3]). *Keeping the notation and assumptions of* 1.2, *let* $F^{(1)}, \ldots, F^{(r)}$ *be NDP's with* $|\bar{c}(F^{(i)})| = C(\Delta_i)$, $i = 1, \ldots, r$, *and suppose these polynomials define nonsingular hypersurfaces in* $(\mathbb{R}^*)^n$. *Then, for a sufficiently small* $t > 0$,

$$\mathrm{Ct}(F_t) = \bigcup_{k=1}^r \mathrm{Ct}(F^{(k)}).$$

Now assume that Δ is an n-dimensional simplex and $F \in \mathcal{P}(\Delta)$ is a real $(n+1)$-nomial (this means that the nonzero coefficients in F correspond to the vertices of Δ only). Denote by $M(\bar{\varepsilon})$ the set of midpoints of the edges of $\Delta(\bar{\varepsilon})$. To any point $v \in M(\bar{\varepsilon})$ on an edge with endpoints $(i_1, \ldots, i_n)$, $(j_1, \ldots, j_n)$ we assign

$$\delta(v) = \mathrm{sgn}(A_{\varepsilon_1 i_1, \ldots, \varepsilon_n i_n} \cdot A_{\varepsilon_1 j_1, \ldots, \varepsilon_n j_n}) \varepsilon_1^{i_1 + j_1} \cdots \varepsilon_n^{i_n + j_n},$$

where $A_{\varepsilon_1 i_1, \ldots, \varepsilon_n i_n}$, $A_{\varepsilon_1 j_1, \ldots, \varepsilon_n j_n}$ are the corresponding coefficients in F. Put $M_-(\bar{\varepsilon}) = \{v \in M(\bar{\varepsilon}) \mid \delta(v) = -1\}$.

PROPOSITION 2.1.2 (see [St]). *If* $M_-(\bar{\varepsilon}) = \varnothing$, *then* $\mathrm{Ct}(F) \cap \Delta(\bar{\varepsilon}) = \varnothing$. *If* $M_-(\bar{\varepsilon}) \neq \varnothing$, *then* $\mathrm{Ct}(F) = \Delta(\bar{\varepsilon}) \cap H$, *where* H *is the unique hyperplane passing through* $M_-(\bar{\varepsilon})$.

PROOF. Let us apply the diffeomorphism

$$y_k = x_1^{\alpha_{1k}} \cdots x_n^{\alpha_{nk}}, \qquad k = 1, \ldots, n, \ \{\alpha_{ij}\} \subset \mathbb{Q},$$

of the open positive orthant. It transforms F into $F' = y_1^{j_1} \cdots y_n^{j_n} L(y_1, \ldots, y_n)$, where L is a linear polynomial with Newton polygon τ_1^n. This implies the desired statement for $\mathrm{Ct}(F)$ in the positive orthant. Using analogous operations in other orthants, we complete the proof.

2.2. Critical points of polynomials with Newton simplices. In this subsection, Δ will denote a nondegenerate simplex in $\mathbb{R}^n$. We shall call Δ *elementary* if $V(\Delta) = 1$. Let $K(\Delta)$ be the minimal cone with vertex at the origin, containing Δ. If $\partial K(\Delta)$ contains at least one facet of Δ (a face of codimension 1), then put $i_0(\Delta) = 1$, otherwise $i_0(\Delta) = 0$. In the last case $K(\Delta) \setminus \Delta$ consists of two components: the bounded one $K_+(\Delta)$ (possibly empty), and the unbounded one $K_-(\Delta)$. Denote by $i_+(\Delta)$ the number of facets of Δ contained in $\partial K_+(\Delta)$.

PROPOSITION 2.2.1. *Let F be a real $(n+1)$-nomial with Newton simplex Δ. If $i_0(\Delta) = 1$, then $|\bar{c}(F)| = 0$. If $i_0(\Delta) = 0$, then $|\bar{c}(F)| = V(\Delta)$. In the last case, if $V(\Delta)$ is odd, then F has exactly one critical point in $(\mathbb{R}^*)^n$, and if $V(\Delta)$ is even, then F has an even number of critical points in $(\mathbb{R}^*)^n$, at most one in each orthant.*

PROOF. Let $v_k = (\alpha_{1k}, \ldots, \alpha_{nk})$, $k = 0, \ldots, n$, be the vertices of Δ. By $\bar{x}^{v_k}$ we denote the monomial $x_1^{\alpha_{1k}} \cdots x_n^{\alpha_{nk}}$. Let

$$F = \sum_{k=0}^{n} A_k \bar{x}^{v_k}.$$

In $(\mathbb{C}^*)^n$ system (0.1) is equivalent to $x_1 F_{x_1} = \cdots = x_n F_{x_n} = 0$, which is actually a linear system in $n+1$ unknowns $\bar{x}^{v_k}$, $k = 0, \ldots, n$:

$$\text{(2.2.2)} \qquad \sum_{k=0}^{n} A_k \alpha_{jk} \bar{x}^{v_k} = 0, \qquad j = 1, \ldots, n.$$

Assume that $i_0(\Delta) \neq 0$, and, for example, that the hypersurface containing $v_1, \ldots, v_n$ passes through the origin. Let $\bar{a} = (a_1, \ldots, a_n)$ be its normal vector; then

$$(\bar{a}, v_1) = \cdots = (\bar{a}, v_n) = 0, \qquad (\bar{a}, v_0) \neq 0.$$

Hence the linear combination of equations (2.2.2) with coefficients $a_1, \ldots, a_n$ gives $\bar{x}^{v_0} = 0$, i.e., $|\bar{c}(F)| = 0$.

Assume that $i_0(\Delta) = 0$. Then (2.2.2) can be easily transformed into the system

$$\bar{x}^{v_k} = B_k \bar{x}^{v_0}, \qquad k = 1, \ldots, n, \; B_1 \neq 0, \ldots, B_n \neq 0,$$

or

$$\bar{x}^{w_k} = B_k, \qquad k = 1, \ldots, n,$$

where $w_k = v_k - v_0$, $k = 1, \ldots, n$. Elementary transformations of the last system lead to the system

$$\text{(2.2.3)} \qquad \begin{aligned} x_1^{d_1} x_2^{p_{12}} x_3^{p_{13}} \cdots x_n^{p_{1n}} &= C_1, \\ x_2^{d_2} x_3^{p_{23}} \cdots x_n^{p_{2n}} &= C_2, \\ &\cdots\cdots\cdots \\ x_n^{d_n} &= C_n, \end{aligned}$$

where $C_1 \cdots C_n \neq 0$ and $d_1 \cdots d_n = V(\Delta)$. This immediately implies the rest of Proposition 2.2.1.

PROPOSITION 2.2.4. *Let F be a real $(n+1)$-nomial with Newton simplex Δ, and let $i_0(\Delta) = 0$. Then the index of any critical point of F in $(\mathbb{R}^*)^n$ is either equal to $i_+(\Delta)$ or to $n - i_+(\Delta)$.*

PROOF. First assume that Δ is elementary. Then F has a unique real critical point $\bar{x}$. Note that the action of $(\mathbb{R}^*)^{n+1}$ on the set of real polynomials in $\mathcal{P}(\Delta)$ defined by

$$(\lambda_0, \lambda_1, \ldots, \lambda_n)(F) = \lambda_0 F(\lambda_1 x_1, \ldots, \lambda_n x_n)$$

is transitive. Hence it suffices to consider only one arbitrary real polynomial $F \in \mathcal{P}(\Delta)$. More precisely, we must study the bifurcation of $\text{Ct}(F + \alpha)$, when α runs through $\mathbb{R}$. Namely, passing through a critical value that corresponds to a point of index q means that a quadric of homotopy type S^{q-1} is transformed into a quadric of homotopy type S^{n-q-1}.

Let us fix the coefficients of F. Let $i_+(\Delta) = q$. Denote by σ_j^+, $j = 1, \ldots, q$, the facets of Δ contained in $\partial K_+(\Delta)$, and denote by Σ_j^+ the simplex with a facet σ_j^+ and a vertex at the origin, $j = 1, \ldots, q$. Analogously, let σ_j^-, $j = 1, \ldots, n+1-q$, be the facets of Δ contained in $\partial K_-(\Delta)$, and let Σ_j^-, $j = 1, \ldots, n+1-q$, be the corresponding simplices. We distinguish between three situations:
  (A) Let $q = n$. Then there is a vertex v_0 of Δ not belonging to $\partial K_-(\Delta)$. Set the corresponding coefficient A_{v_0} equal to -1, and other coefficients equal to $+1$.
  (B) Let $q = 1$. Then there is a vertex v_0 of Δ not belonging to $\partial K_+(\Delta)$. Set $A_{v_0} = 1$, and the other coefficients equal to -1.
  (C) Let $1 < q < n$. Then there is a unique face $\sigma \subset \Delta$ of dimension $n - q$ intersecting $\mathrm{Int}\,(\Delta \cup K_+(\Delta))$. Set $A_v = -1$ if the vertex v belongs to σ, and $A_v = 1$ otherwise.

According to Theorem 2.1.1, for $\alpha > 0$ small enough, $\mathrm{Ct}\,(F + \alpha)$ is the gluing of charts of $(n+1)$-nomials with Newton simplices $\Sigma_1^+, \ldots, \Sigma_q^+, \Delta$, and, for $\alpha > 0$ large enough, $\mathrm{Ct}\,(F + \alpha)$ is homeomorphic to the gluing of charts of $(n+1)$-nomials with Newton simplices $\Sigma_1^-, \ldots, \Sigma_{n+1-q}^-$. According to Proposition 2.1.2, these subdivisions and the given coefficient signs determine the chart of $F + \alpha$ for small and large $\alpha > 0$, which allows us to describe the evolution of the hypersurface $F + \alpha = 0$ completely, because of the uniqueness of the critical point of F. Now denote by Ct_+ the part of the chart Ct in the positive orthant $\mathbb{R}_+^n$.

In case (A), according to Theorem 2.1.1 and Proposition 2.1.2, for small $\alpha > 0$, $\mathrm{Ct}_+(F + \alpha)$ is homeomorphic to S^{n-1}, and, for large $\alpha > 0$, $\mathrm{Ct}_+(F + \alpha) = \varnothing$. This means F has a minimum in $\mathbb{R}_+^n$.

In cases (B) and (C), we have two transversal sections of the polyhedron $Q = K_+(\Delta) \cup \Delta$: the section Q_- cut off by the $(n-q)$-plane passing through $n-q+1$ vertices with negative coefficients, and the section Q_+ cut off by the q-plane passing through q vertices with positive coefficients and through the origin. According to Theorem 2.1.1 and Proposition 2.1.2, for small $\alpha > 0$, the chart $\mathrm{Ct}_+(F + \alpha)$ is a strict deformation retract of $Q \setminus Q_-$, hence is of homotopy type S^{q-1}, and for large $\alpha > 0$, the chart $\mathrm{Ct}_+(F + \alpha)$ is a strict deformation retract of $Q \setminus Q_+$, hence is of homotopy type S^{n-q-1}. If $n \neq 2q$, this means F has a critical point of index q in $\mathbb{R}_+^n$. If $n = 2q$, then both charts mentioned are homeomorphic but not isotopic in Q with fixed boundary ∂Q, which implies F has a critical point of index q.

Assume that Δ is not elementary. In any orthant, F has at most one real critical point (Proposition 2.2.1). Then the above-presented geometric description of the evolution of $\mathrm{Ct}\,(F + \alpha)$ in any orthant coincides with the one for an elementary simplex Δ', having the same number i_+, and we are done. $\square$

Let $n = 2$ and Δ be an elementary triangle with $i_0(\Delta) = 0$. If $i_+(\Delta) = 1$, we shall call Δ *odd*. If $i_+(\Delta) = 2$, we shall call Δ *even*. An even triangle Δ is said to be of *type* 1 if one of its vertices has both even coordinates, and of *type* 2 otherwise.

PROPOSITION 2.2.5. *Let $n = 2$ and Δ be an elementary triangle with $i_0(\Delta) = 0$ and vertices v_1, v_2, v_3. Let $F \in \mathcal{P}(\Delta)$ be a real trinomial with coefficients $A_{v_1}, A_{v_2}, A_{v_3}$ and a critical point $z \in (\mathbb{R}^*)^2$.*
  (i) *If Δ is odd, then z is a saddle.*

(ii) *If Δ is even of type 2, then z is a maximum or a minimum according as $A_{v_1}A_{v_2}A_{v_3}$ is positive or negative.*
(iii) *If Δ is even of type 1, the coordinates of v_1 are even and $v_1 \in \partial K(\Delta)$, then z is a maximum or a minimum according as A_{v_1} is negative or positive.*
(iv) *If Δ is even of type 1, the coordinates of v_1 are even and $v_1 \notin \partial K(\Delta)$, then z is a maximum or a minimum according as A_{v_1} is positive or negative.*

PROOF. Statement (i) follows from Proposition 2.2.4. To prove the other statements, consider the coordinates of v_1, v_2, v_3 modulo 2. Because of the minimality of Δ, we have two possibilities: either $(0;0), (1;0), (0;1),$ or $(1;0), (0;1), (1;1)$. It is easy to see that in the first case the signs of A_{v_2}, A_{v_3} can be independently changed by means of a suitable transformation

$$(2.2.6) \qquad (x,y) \mapsto (\varepsilon_1 x, \varepsilon_2 y), \qquad \varepsilon_1 = \pm 1, \ \varepsilon_2 = \pm 1.$$

Similarly, in the second case any two signs of $A_{v_1}, A_{v_2}, A_{v_3}$ can be changed by means of (2.2.6). Now the construction from the proof of Proposition 2.2.4 completes the present proof. □

§3. Real polynomials in two variables

3.1. Formulation of the result. Let $F(x,y)$ be a real NDP with Newton triangle τ_d^2, $d \geqslant 2$, and with $(d-1)^2$ critical points in $(\mathbb{C}^*)^2$.

PROPOSITION 3.1.1. *The vector $\overline{c}(F) = (c_0, c_1, c_2, p)$ satisfies the equalities*

$$(3.1.2) \qquad c_0 + c_1 + c_2 + 2p = (d-1)^2,$$
$$(3.1.3) \qquad c_0 - c_1 + c_2 = 1 - m,$$

where m is the number of real intersection points of the projective closure of $\{F(x,y) = 0\}$ by the line at infinity.

PROOF. The first equation follows from (1.1.3), the second one is, in fact, Morse's formula (see also [**Du**]). Indeed, let $m > 0$. Assume that all real critical values of F belong to the interval $(a;b)$. Then, according to the Proposition from [**ChB**] and Lemma 1 from [**S2**], the set $\{F(x,y) < a\}$ consists of m components homeomorphic to $\mathbb{R}^2$, and the set $\{F(x,y) < b\}$ is homeomorphic to $\mathbb{R}^2$ (Figure 1).

Let t run over $(a;b)$. Note that the passage of t through a maximal value or through a minimal value adds 1 to the Euler characteristic $\chi\{F < t\}$, and the passage of t through a saddle value subtracts 1 from $\chi\{F < t\}$. This yields (3.1.3). The case $m = 0$ can be considered similarly.

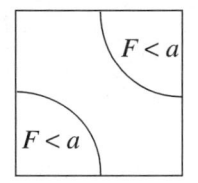

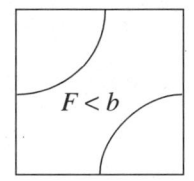

FIGURE 1

REMARK 3.1.4. It is clear that

(3.1.5) $$d \geqslant m \geqslant 0, \qquad m \equiv d \pmod 2.$$

THEOREM 3.1.6. *For any positive integer d and any nonnegative integers m, c_0, c_1, c_2, p satisfying (3.1.2), (3.1.3), (3.1.5) there is a real NDP $F(x, y)$ with Newton triangle τ_d^2 and $\overline{c}(F) = (c_0, c_1, c_2, p)$, except for the following cases:*

(3.1.7) $$d \equiv 1 \pmod 2, \qquad \min\{c_0, c_2\} < (d - m)/2,$$

(3.1.8) $$\begin{gathered} d \equiv 0 \pmod 2, \quad m = 0, \\ ((d-3)^2 + 1)/2 \leqslant p \leqslant ((d-1)^2 - 1)/2, \quad \max\{c_0, c_2\} < d - 1, \end{gathered}$$

(3.1.9) $$\begin{gathered} d = 4, \quad m = 0, \\ (c_0, c_1, c_2, p) = (3, 4, 2, 0), \text{ or } (2, 4, 3, 0), \text{ or } (2, 3, 2, 1). \end{gathered}$$

REMARK 3.1.10. The inequality $\min\{c_0, c_2\} \geqslant (d - m)/2$ is necessary for $d = 3$. Indeed, for $m = 1$, we have $c_0 + c_2 = 2$ from (3.1.2), (3.1.3), and there are no cubic polynomials with two maxima: assuming that z_1, z_2 are two maximum points of a cubic polynomial F, and restricting F to the straight line $(z_1 z_2)$, we obtain a cubic polynomial in one variable with two maxima, which is impossible. Also there are no quartic polynomials satisfying (3.1.9). Indeed, $m = 0$ implies that such a polynomial has the same sign, say plus, in a neighborhood of infinity. Then its restriction to any straight line has at most one maximum.

COROLLARY 3.1.11. *For any $d \geqslant 2$ there exists a real polynomial of degree d with $(d-1)^2$ real critical points and without minima.*

Indeed, if $m = d$, then, according to Theorem 3.1.6, there exists a polynomial with $c_1 = d(d-1)/2$ saddles and $c_2 = (d-1)(d-2)/2$ maxima. Moreover, such a polynomial can be defined by (1.2.1), where $A_{ij} = 1, 0 \leqslant i, j \leqslant d$, and ν corresponds to the regular triangulations of τ_d^2 described at the first and third steps of the proof of Theorem 3.1.6 below.

The proof of Theorem 3.1.6 consists of an explicit construction of polynomials with a given index distribution, based on Theorem 1.2.2. The proof is divided into several steps. The first is the principal one: we describe the main construction in a particular case. Actually, it implies the desired statement under the additional assumption

(3.1.12) $$\min\{c_0, c_2\} \geqslant O(d).$$

In the next steps we extend the construction to other cases and weaken the restriction (3.1.12) to (3.1.7)–(3.1.9).

3.2. First step. Assume that $d \geqslant 5$ is odd, $p = 0$ and $m = d$.

First we describe the regular subdivision of τ_d^2 that we use. Let $T \subset \tau_d^2$ be the triangle with vertices $(1; 1)$, $(1; d-1)$, $(d-1; 1)$. Introduce convex continuous piecewise linear functions

$$(i; j) \in T \mapsto \nu_k(i, j) \in \mathbb{R}, \qquad k = 1, 2, 3,$$

such that
 (i) the linearity domains of ν_1 are the strips

$$2l - 1 \leqslant i \leqslant 2l + 1, \qquad l = 1, \ldots, (d-1)/2,$$

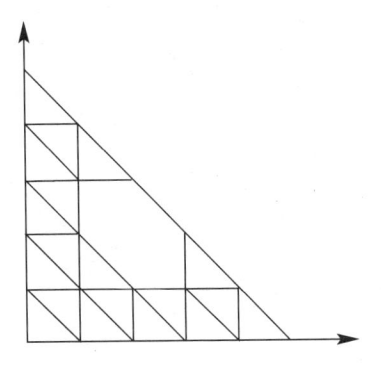

FIGURE 2

(ii) the linearity domains of v_2 are the strips
$$2l - 1 \leqslant j \leqslant 2l + 1, \qquad l = 1, \ldots, (d-1)/2,$$

(iii) the linearity domains of v_3 are the strips
$$2l - 1 \leqslant i + j \leqslant 2l + 1, \qquad l = 1, \ldots, (d-1)/2.$$

Put $v = v_1 + v_2 + v_3$. This function defines a regular subdivision of T into elementary triangles and into the hexagons H_{kl}, $(2k; 2l) \in T$, with vertices

$$v_1 = (2k - 1; 2l), \quad v_2 = (2k - 1; 2l + 1), \quad v_3 = (2k; 2l + 1),$$
$$v_4 = (2k + 1; 2l), \quad v_5 = (2k + 1; 2l - 1), \quad v_6 = (2k; 2l - 1).$$

Note that all integral points in $\text{Int}(\tau_d^2)$ with both even coordinates are the centers of the hexagons H_{kl}. Also the restriction of v to the edges

$$[(1;1), (1;d-1)], \quad [(1;1), (d-1;1)]$$

of T is linear exactly on the unit segments

$$[(1;k), (1;k+1)], \quad [(k;1), (k+1;1)], \qquad k = 1, \ldots, d-2.$$

This means that we can extend the regular subdivision of T to τ_d^2 by adding the following elementary triangles (see Figure 2):

$$\{(0;i), (1;i), (0;i+1)\}, \qquad i = 0, \ldots, d-1,$$
$$\{(0;i), (1;i), (1;i-1)\}, \qquad i = 1, \ldots, d-1,$$
$$\{(i;0), (i;1), (i+1;0)\}, \qquad i = 1, \ldots, d-1,$$
$$\{(i;0), (i;1), (i-1;0)\}, \qquad i = 2, \ldots, d-1.$$

LEMMA 3.2.1. *For any hexagon H_{kl} there exists a real NDP in $\mathcal{P}(H_{kl})$ with prescribed nonzero coefficients A_{v_1}, A_{v_6}, A_{v_5}, and 6 critical points in $(\mathbb{R}^*)^2$.*

PROOF. According to (1.2.1), $C(H_{kl}) \leq V(H_{kl}) = 6$. We shall construct a real NDP $\Phi \in \mathcal{P}(H_{kl})$ with 6 critical points in $(\mathbb{R}^*)^2$ by gluing 6 trinomials with elementary Newton triangles. Namely, define the following regular triangulation of H_{kl}. Let σ_{kl} denote the triangle with vertices v_1, v_3, v_5 if $k \leq l$ or with vertices v_2, v_4, v_6 if $k > l$. It is easy to see that the complement of σ_{kl} in H_{kl} consists of three elementary triangles. Finally, we divide σ_{kl} into three triangles connecting the point $(2k; 2l)$ with the vertices of σ_{kl}. Here, $i_0(\delta) = 0$ for any triangle δ, hence, according to Theorem 1.2.2 and Proposition 2.2.1, gluing 6 trinomials we obtain a real NDP in $\mathcal{P}(H_{kl})$. Finally, by means of a suitable transformation

$$\Phi(x, y) \mapsto \Phi'(x, y) = a\Phi(bx, cy), \qquad a, b, c > 0,$$

we can set arbitrary nonzero coefficients A_{v_1}, A_{v_6}, A_{v_5}. □

Now let us take a collection of compatible real NDP's with Newton polygons from the given subdivision. Namely, first we order the hexagons H_{kl} by

$$(k, l) < (k', l') \iff \text{either } k < k', \text{ or } k = k', l < l'.$$

Then, using Lemma 3.2.1, we take polynomials compatible with the Newton polygons H_{kl}, and, finally, we take suitable trinomials with Newton triangles.

Note that our subdivision contains exactly $2d - 1$ triangles δ with $i_0(\delta) > 0$. Hence, gluing the chosen polynomials, we obtain a real NDP with $(d-1)^2$ critical points. Further, Proposition 2.2.1 and the construction from Lemma 3.2.1 show that all these points are real. In addition, all integral points on the edge $[(d; 0), (0; d)]$ of τ_d^2 are vertices of the subdivision, hence $m = d$. Therefore, according to Proposition 3.1.1, we have

(3.2.2) $\qquad c_1 = d(d-1)/2, \qquad c_0 + c_2 = (d-1)(d-2)/2.$

Now we shall regard the constructed polynomial to be glued from trinomials with elementary Newton triangles; this means that first we glue polynomials with Newton hexagons from trinomials as described in Lemma 3.2.1, and then we add the remaining trinomials. This viewpoint shows that we can construct a polynomial with Newton triangle τ_d^2 and prescribed signs of coefficients, and according to Proposition 2.2.5, the sign distribution determines $\bar{c} = (c_0, c_1, c_2)$ uniquely. Therefore, the following lemma completes the proof.

LEMMA 3.2.3. *For any nonnegative integers c_0, c_1, c_2 satisfying (3.2.2), there is a sign distribution providing the index distribution $\bar{c} = (c_0, c_1, c_2)$.*

PROOF. The equalities (3.2.2) mean that our subdivision contains $d(d-1)/2$ odd triangles and $(d-1)(d-2)/2$ even ones. Each odd triangle gives one saddle independently of the signs of the coefficients. So we must deal with even triangles only. Let us prescribe the critical point index (either 0, or 2) for any even triangle. The problem is to find a suitable sign distribution of the coefficients (below we shall speak of *integral point signs*) providing the given index distribution.

According to our construction, vertices of even triangles have only positive coordinates, and for any integral point in τ_d^2 with both coordinates even and positive, there is exactly one even triangle having this point as a vertex. Hence, according to Proposition 2.2.5, we get the prescribed indices of critical points for even triangles of type 1 by assigning suitable signs to the points $(2k; 2l)$ independently of all other signs.

Consider even triangles of type 2. Introduce the set

$$S = \{(i; j) \in \text{Int}(\tau_d^2) \cap \mathbb{Z}^2 \mid i > 0, \ j > 0, \ (i-1)(j-1) \equiv 0 \pmod{2}\}.$$

For any even triangle δ of type 2, we denote by $v(\delta) \in S$ its vertex that does not belong to $\partial K(\delta)$. This yields, evidently, a one-to-one correspondence between S and the set of even triangles of type 2, and we write $\delta = \delta(v)$ for $v = v(\delta)$. In any even triangle δ of type 2, let us mark each edge $[v', v'']$ such that $v', v'' \in S$ and $v' = v(\delta)$, and let us orient this edge from v' to v''. Consider the oriented graph Γ whose vertex set is S and whose arcs are all the marked edges. Let us show that Γ has no oriented cycles. Indeed, assume that $L \subset \Gamma$ is the shortest cycle, oriented clockwise, and $v = (i; j) \in L$ has the minimal slope j/i. Then v belongs to two arcs $[v, v']$, $[v'', v]$, and we see that $[v'', v]$ must intersect the triangle $\delta(v) \supset [v, v']$. Similarly we can prohibit cycles oriented counterclockwise. Thus, the orientation of Γ turns S into a poset. Define the order $N(v)$ of $v \in S$ to be the maximal path length from one of the minimal points to v, and put $S_k = \{v \in S \mid N(v) = k\}$.

Now we shall construct the sign distribution by decreasing induction on the order of points. Let us fix the signs of all integer points outside S. Assume that the signs of points of S with order $\geqslant k$ are fixed too. According to Proposition 2.2.5, we provide the prescribed critical point indices for triangles $\delta(v)$, $v \in S_{k-1}$, by setting suitable signs at $v \in S_{k-1}$. Finally, note that any $v \in S_{k-1}$ does not belong to the triangles $\delta(w)$, $N(w) \geqslant k$, hence the previous operation does not influence these triangles, and we are done. □

3.3. Second step. Assume that $d \geqslant 5$ is odd, $p = 0$, and $m < d$.

We shall modify the previous construction as follows. In the triangulation of τ_d^2 presented above, we substitute the triangles

$$\{(0; d), (0; d-1), (1; d-1)\}, \quad \{(0; d-1), (1; d-1), (1; d-2)\},$$
$$\{(d; 0), (d-1; 0), (d-1; 1)\}, \quad \{(d-1; 0), (d-1; 1), (d-2; 1)\},$$

for the triangles

$$\{(0; d), (0; d-1), (1; d-2)\}, \quad \{(0; d), (1; d-1), (1; d-2)\},$$
$$\{(d; 0), (d-1; 0), (d-2; 1)\}, \quad \{(d; 0), (d-1; 1), (d-2; 1)\},$$

(see Figure 3). Now consider the triangles with an edge on the closed interval $[(0; d), (d; 0)]$. From these elementary triangles we make up $(d-1)/2$ triangles of area $V = 2$ (shown by hatching in Figure 3):

$$\delta_i = \{(2i-2; d-2i+2), (2i; d-2i), (2i-1; d-2i)\}, \quad 1 \leqslant i \leqslant (d+1)/4,$$
$$\delta_i = \{(2i-1; d-2i+1), (2i+1; d-2i-1), (2i-1; d-2i)\},$$
$$(d+1)/4 < i \leqslant (d-1)/2.$$

Now put

$$\tau = \tau_d^2 \setminus \bigcup_{i=1}^{(d-m)/2} \delta_i.$$

Define a regular subdivision of τ_d^2 to be the union of $\delta_1, \ldots, \delta_{(d-m)/2}$ with the subdivision from the previous construction restricted to τ. The arguments used in the first

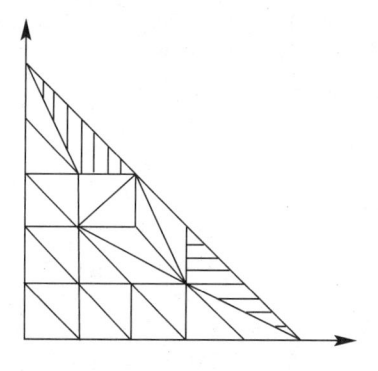

FIGURE 3

step show that the prescribed index distribution corresponding to the subdivision of τ can be provided independently of the signs at the vertices of $\delta_1, \ldots, \delta_{(d-m)/2}$. Now to any δ_i we assign the trinomial F_i from $\mathcal{P}(\delta_i)$ with unit coefficients. It is easy to show that F_i has two critical points in $(\mathbb{R}^*)^2$. For example, if $1 \leqslant i \leqslant (d+1)/4$, then

$$F_i = x^{2i-2} y^{d-2i+2} + x^{2i} y^{d-2i} + x^{2i-1} y^{d-2i}$$

has two real critical points

$$x = -\frac{d+2i-2}{2d}, \qquad y = \pm \frac{\sqrt{(d+2i-2)(d-2i)}}{2d}.$$

According to Proposition 2.2.4, both the critical points are extrema. Note that the change of variables $y \mapsto -y$ turns F_i into $-F_i$, hence one of the critical points is a maximum and the other is a minimum. Thereby we obtain any prescribed index distribution provided

$$\min\{c_0, c_2\} \geqslant (d-m)/2.$$

3.4. Third step. Assume that $d \geqslant 6$ is even, $m = d$, $p = 0$.

To the subdivision of τ_{d-1}^2 constructed in the first step, we add the following compatible regular triangulation of the quadrangle Q with vertices $(d; 0)$, $(0; d)$, $(0; d-1)$, $(d-1; 0)$: we divide Q into $2d - 1$ elementary triangles by segments, connecting the point $(1; d-1)$ with the points $(i; d-1-i)$, $i = 0, \ldots, d-1$, and by segments, connecting the point $(d-1; 0)$ with the points $(i; d-i)$, $i = 2, \ldots, d-1$. It is easy to see that any new triangle is either odd or even of type 2. So, repeating the sign determination procedure from the first step, we can get an arbitrary index distribution that satisfies (3.1.2), (3.1.3).

3.5. Fourth step. Assume that $d \geqslant 6$ is even, $0 < m < d$, $p = 0$.
Introduce the following triangles of area $V = 2$:

(3.5.1) $\quad \delta_i = \{(d-1; 0), (d-i+1; i-1), (d-i-1; i+1)\}, \qquad 1 \leqslant i \leqslant d-1.$

LEMMA 3.5.2. *The polynomial*

$$F_1(x, y) = x^d + a_1 x^{d-1} y + x^{d-2} y^2 + x^{d-1},$$

where $0 < a_1^2 < 4$, with Newton triangle δ_1 has a minimum as its only critical point in

$(\mathbb{R}^*)^2$. For $1 < i \leqslant d - 1$, the polynomial
(3.5.3)
$$F_i = x^{d-i+1}y^{i-1} + a_i x^{d-i}y^i + x^{d-i-1}y^{i+1} + x^{d-1}, \qquad 4(i^2 - 1)/i^2 < a_i^2 < 4,$$

has exactly two critical points in $(\mathbb{R}^*)^2$, namely a saddle and a minimum.

PROOF. Let $1 < i$. Let us add to δ_i the subdivision of $\tau_d^2 \setminus \delta_i$ constructed in the third step. For this triangulation we choose suitable polynomials compatible with F_i, and denote by c_0', c_1', c_2' the numbers of corresponding critical points provided by all the polynomials except F_i. Also put $\bar{c}(F_i) = (c_0'', c_1'', c_2'', p'')$. If we replace F_i by a gluing of two trinomials with elementary triangles, then from (3.1.2), (3.1.3) we obtain the equations

(3.5.4) $\qquad c_0' + c_1' + c_2' = (d-1)^2 - 2, \qquad c_0' - c_1' + c_2' = 1 - d.$

On the other hand, from (3.5.3), we obtain

$$c_0' + c_1' + c_2' + c_0'' + c_1'' + c_2'' + p'' = (d-1)^2,$$
$$c_0' + c_0'' - c_1' - c_1'' + c_2' + c_2'' = 3 - d.$$

These formulas and (3.5.4) imply that $p'' = 2$, $c_0'' = c_1'' = c_2'' = 0$, or

(3.5.5) $\qquad p'' = 0, \qquad c_0'' + c_2'' = c_1'' = 1.$

The equations $\partial F_i/\partial x = \partial F_i/\partial y = 0$ have two solutions

(3.5.6) $\qquad x_j = -\dfrac{d-1}{d(\lambda_j^2 + a_i \lambda_j + 1)\lambda_j^{i-1}}, \qquad y_j = \lambda_j x_j, \qquad j = 1, 2,$

where λ_1, λ_2 are the roots of the equation

(3.5.7) $\qquad (i+1)\lambda^2 + a_i i \lambda + (i-1) = 0.$

By (3.5.3), both λ_1 and λ_2 are real, and therefore (3.5.5) holds. If a_i^2 tends to $4(i^2-1)/i^2$, then the two critical points merge into one degenerate critical point $(x_0; y_0)$ of type $\pm x^2 + y^3$ (see [AGV]). It is well known [AGV] that the point $x^2 + y^3$ splits into a saddle and a minimum, and $-x^2 + y^3$ splits into a saddle and a maximum. In our situation, for $a = \pm 2\sqrt{i^2 - 1}/i$, we have

$$\lambda_1 = \lambda_2 = \lambda = \mp\sqrt{\dfrac{i-1}{i+1}}, \quad x_0 = -i(i+1)\dfrac{d-1}{2d}\left(\dfrac{i+1}{i-1}\right)^{(i-1)/2} < 0, \quad y_0 = \lambda x_0,$$

$$\dfrac{\partial^2 F_i}{\partial x^2}(x_0, y_0) = -(d-1)x_0^{d-3} > 0,$$

which implies that the point $(x_0; y_0)$ is of type $x^2 + y^3$, and we are done. The case $i = 1$ is similar. $\square$

Now, we substitute $d - m$ elementary triangles

(3.5.8) $\qquad \{(d-1;0), (d-i+1;i-1), (d-i;i)\}, \qquad i = 2,\ldots, d-m+1,$

from the subdivision of τ_d^2 constructed at the previous step for the triangles δ_{2i}, where $i = 1,\ldots, (d-m)/2$. Then gluing the polynomials F_{2i}, $i = 1,\ldots, (d-m)/2$, with suitable polynomials from the previous step, we obtain an arbitrary index distribution, except in the case $c_0 < (d - m)/2$. But in this case we have

$$c_2 = ((d - 1)^2 + 1 - n)/2 - c_0 \geqslant (d - m)/2,$$

hence we obtain the desired distribution (c_0, c_1, c_2) by multiplying one of the above polynomials by -1.

3.6. Fifth step. Assume that $d \geqslant 6$ is even, $m = p = 0$.

Let us take the following subdivision of τ_d^2. We subdivide the triangle τ_{d-1}^2 as in the first step, the triangle $\{(d - 1; 0), (d; 0), (0; d)\}$ into triangles δ_{2i-1}, $i = 1, \ldots, d/2$, given by (3.5.1), and the triangle $\{(d-1; 0), (0; d), (0; d-1)\}$ into elementary triangles

(3.6.1) $\qquad \delta_i' = \{(0; d), (d - i; i - 1), (d - i - 1; i)\}, \qquad i = 1, \ldots, d - 1.$

Here the polynomials $F_1, \ldots, F_{d/2}$ give us $d/2$ minima, and the corresponding polynomials with Newton triangles (3.6.1) give us $d-2$ minima according to Proposition 2.2.5. Therefore we can realize an arbitrary distribution (c_0, c_1, c_2) provided $c_0 \geqslant 3d/2 - 2$. If $c_0 < 3d/2 - 2$, then the condition $d \geqslant 6$ yields

$$c_2 = ((d - 1)^2 + 1)/2 - c_0 \geqslant 3d/2 - 2,$$

hence we obtain the desired vector (c_0, c_1, c_2) by multiplying one of the above polynomials by -1.

3.7. Sixth step. Assume that $d \geqslant 5$ is odd, $m = d$, $0 < 2p \leqslant (d - 2)(d - 3)$.
We need the following preliminary statement.

LEMMA 3.7.1. *For any $k \geqslant 2$ and any nonnegative $s \leqslant \min\{k/2, k(k - 1)/2\}$ there is a real NDP F' with Newton triangle*

$$T' = \{(0; 0), (0; k + 1), (k; 0)\}$$

with prescribed coefficients of the monomials x^k, y^{k+1} and such that

(3.7.2) $\qquad \overline{c}(F') = (s, s, 0, k(k - 1)/2 - s).$

PROOF. Let $k = 2$. Then $s = 0$, and the polynomial $F' = ay^3 + ay + 1 + bx + bx^2$ satisfies (3.7.2).

Let $k \geqslant 3$ be odd. Then $s \leqslant (k - 1)/2$. Denote by θ the triangle with vertices $(0; k + 1), (0; 2), (k; 0)$. It is not difficult to see that, for fixed $a, b \neq 0$ and almost all $\varepsilon > 0$, the polynomial

$$G = ax^k + axy^2 + axy^k + by^{k+1} + y^2 + \varepsilon x^{k-1}y$$

has Newton triangle θ and is nondegenerate. For sufficiently small ε, the polynomial G has only imaginary critical points, because we have

$$G_x = akx^{k-1} + ay^2 + ay^k + \varepsilon(k - 1)x^{k-2}y \neq 0$$

in $(\mathbb{R}^*)^2$. The number of critical points of G in $(\mathbb{R}^*)^2$ evidently equals $(k - 1)^2$. Note that the origin is a critical point of G of type A_{k-1} (in Arnold's notation [**AGV**]), which is equivalent to $y^2 + x^k$ over $\mathbb{R}$. According to [**AGV**], any deformation of such a point can be realized by adding monomials yx^r, $r < k/2$, and x^l, $l < k$. On the other hand, according to [**Chi**], there is a deformation splitting this point into s maxima, s saddles and $(k - 1)/2 - s$ pairs of imaginary critical points. Now we obtain the desired polynomial F' by gluing G with a suitable polynomial having Newton triangle $(0;2)$, $(0;0)$, $(k;0)$.

Let $k \geq 4$ be even. Then $s \leq k/2$. In this case we divide the initial triangle into the triangles

$$\theta = \{(0;k+1),(k;0),(2;0)\} \quad \text{and} \quad \{(0;k+1),(2;0),(0;0)\}.$$

Repeating the above arguments, we obtain the desired polynomial F'. □

Let $k(k-1)/2$ be the triangular number closest to p. Then

$$p = k(k-1)/2 + r, \quad 2 \leq k \leq d-2, \quad -k/2 \leq r \leq k/2.$$

Assume that $r \leq 0$. We search for a polynomial $F \in \mathcal{P}(\tau_d^2)$ with $\bar{c}(F) = (c_0, c_1, c_2, p)$, where c_0, c_1, c_2 satisfy (3.1.2), (3.1.3). Here

$$c_0 + c_2 = (d-1)(d-2)/2 - p \geq (d-1)(d-2)/2 - (d-2)(d-3)/2 - r$$
$$= d - 2 - r \geq -2r.$$

Hence $\max(c_0, c_2) \geq -r$. Without any loss of generality, we can assume that $c_0 \geq -r$. Take the polynomial F' from Lemma 3.7.1 corresponding to $s = -r$. Now we must find a polynomial F'' with Newton quadrangle

$$Q = \{(0;k+1),(0;d),(d;0),(k;0)\}$$

and $\bar{c}(F'') = (c_0 + r, c_1 + r, c_2, 0)$. Let us construct a regular subdivision of Q. Consider two possibilities. Let k be even. Then we restrict the regular subdivision of τ_d^2 constructed in the first step to the quadrangle $\{(0;k+1),(0;d),(d;0),(k+1;0)\}$, and add to it the elementary triangles

(3.7.3) $\quad \{(k;0),(k+1-i;i),(k-i;i+1)\}, \quad i = 0,\ldots,k.$

The arguments from the first step show that this subdivision of Q allows us to realize the desired index distribution. Let k be odd. Then we restrict the regular subdivision of τ_d^2 constructed in the first step to the quadrangle

$$\{(0;k+2),(0;d),(d;0),(k+2;0)\},$$

add to it the elementary triangles (3.7.3) and

$$\{(i;k+1-i),(i+1;k+1-i),(i;k+2-i)\}, \quad i = 0,\ldots,k+1,$$
$$\{(i;k+1-i),(i+1;k+1-i),(i+1;k-i)\}, \quad i = 0,\ldots,k.$$

It is easy to see that for any point on the segment $[(k+1;0),(0;k+1)]$ with both even coordinates there is at most one even triangle of type 1 containing this point. Therefore, using again the arguments from the first step we see that the given subdivision of Q allows us to realize the desired index distribution.

Assume that $r > 0$. Then $k \leq d-3$. Take the polynomial F' from Lemma 3.7.1 corresponding to $s = 0$, and the subdivision of Q constructed above. It is easy to see that this subdivision contains at least $(d-3)/2$ hexagons H_{tu}. For $r \leq k/2 \leq (d-3)/2$ of them, we change the construction of the polynomial from $\mathcal{P}(H_{tu})$ presented in the first step in the following way: to σ_{tu} we assign a trinomial with this Newton triangle. According to Proposition 2.2.5, for any signs of the coefficient, such a trinomial has a saddle and two imaginary critical points. Next, we construct the desired index distribution as explained above.

3.8. Seventh step. Assume that $d \geqslant 5$ is odd, $m = d$, $2p > (d-2)(d-3)$. In this situation, according to (3.1.2), (3.1.3), we have

$$p = (d-2)(d-3)/2 + r, \qquad 0 < r \leqslant d-1.$$

If $r = d - 1$, then the gluing of the polynomial F' corresponding to $k = d - 1$, $s = 0$, with trinomials corresponding to the odd elementary triangles

(3.8.1) $\qquad \{(d-1; 0), (i; d-i), (i+1; d-i-1)\}, \qquad i = 0, \ldots, d-1,$

gives us the desired polynomial.

Let $r < d - 1$, $r = 2r_1 + \varepsilon$, where ε is 0 or -1.

LEMMA 3.8.2. *For any $k \geqslant 2$ and any positive even $l \leqslant k$ there is a real NDP F'' with Newton triangle*

$$T'' = \{(k; 0), (k+2; 0), (k-l; l+1)\},$$

with prescribed coefficients of monomials x^k, x^{k+2}, $x^{k-l}y^{l+1}$, and with $\overline{c}(F'') = (0, 0, 0, l)$.

PROOF. It is easy to see that the polynomial

$$F'' = a_1 x^k + a_2 x^{k+2} + a_3 x^{k-l} y^{l+1} + a_3 x^k y$$

has no critical points in $(\mathbb{R}^*)^2$, because $F''_y \neq 0$ in $(\mathbb{R}^*)^2$. Hence it satisfies the requirements of this lemma. $\square$

Put $k = d - 2$, $s = -\varepsilon$, $l = 2r_1$, and take the corresponding compatible polynomials F', F'' from Lemmas 3.7.1 and 3.8.2. Then we divide the complement of $T' \cup T''$ in τ_d^2 into the following elementary triangles:

$$\{(d-2; 0), (d-2-l-i; l+1+i), (d-3-l-i; l+i+2)\},$$
$$0 \leqslant i \leqslant d-3-l,$$
$$\{(d-2-l; l+1), (d-i; i), (d-i-1; i+1)\}, \qquad 0 \leqslant i \leqslant l,$$
$$\{(d-2-l-i; l+1+i), (d-2-l-i; l+2+i), (d-1-l-i; l+1+i)\},$$
$$0 \leqslant i \leqslant d-2-l,$$
$$\{(d-2-l-i; l+1+i), (d-2-l-i; l+2+i), (d-3-l-i; l+2+i)\},$$
$$0 \leqslant i \leqslant d-3-l.$$

It is easy to see that to any integer point with both even coordinates there corresponds at most one even triangle with a vertex at this point. Therefore we can obtain the prescribed index distribution by setting suitable signs at the vertices of the regular subdivision constructed.

3.9. Eighth step. Assume that $d \geqslant 6$ is even, $m = d$, $p > 0$.

If $2p \leqslant (d-2)(d-3)$, then we obtain the desired polynomial by combining
(1) the construction of a suitable polynomial of degree $d - 1$ with $2p$ imaginary critical points given in the sixth and seventh steps;
(2) the construction of a polynomial with the Newton quadrangle

$$\{(d; 0), (0; d), (0; d-1), (d-1; 0)\}$$

given in the third step.

According to (3.1.2), (3.1.3), we have $2p \leqslant (d-1)(d-2)$. If $2p = (d-1)(d-2)$, then we glue the polynomial F' from Lemma 3.7.1 corresponding to $k = d-1$, $s = 0$, with trinomials corresponding to odd elementary triangles (3.8.1), thus obtaining the desired index distribution. Let

$$p = (d-2)(d-3)/2 + 2r + \varepsilon,$$

where $r \geqslant 1$, $2r + \varepsilon < d - 1$, and ε is 0 or -1. Put $k = d - 2$, $s = -\varepsilon$, $l = 2r$, and take the corresponding compatible polynomials F', F'' from Lemmas 3.7.1 and 3.8.2. Then we divide the complement of $T' \cup T''$ in τ_d^2 into the following elementary triangles:

$$\{(d-2;0),(d-2-l-i;l+1+i),(d-3-l-i;l+i+2)\},$$
$$0 \leqslant i \leqslant d-3-l,$$

$$\{(d-2-l;l+1),(d;0),(d-1;1)\},$$
$$\{(d-1;1),(d-2-l-i;l+1+i),(d-3-l-i;l+2+i)\},$$
$$0 \leqslant i \leqslant d-3-l,$$

$$\{(0;d-1),(d-1-i;i+1),(d-2-i;i+2)\}, \quad 0 \leqslant i \leqslant d-2.$$

Here we have no even triangles of type 1, hence we can provide the desired index distribution as shown above.

3.10. Ninth step. Assume $d \geqslant 6$ is even, $0 < m < d$, $2p > 0$.

In the previous regular subdivision of τ_d^2 we substituted the triangles (3.5.8) (or their versions obtained by permutation of axes) for the triangles δ_{2i}, $i = 1, \ldots, (d-m)/2$, defined by (3.5.1).

LEMMA 3.9.1. *Under the conditions of Lemma 3.5.2, for $1 < i \leqslant d - 1$ the polynomial F_i with $a_i = 0$ has two imaginary critical points in* $(\mathbb{C}^*)^2$.

PROOF. It is easy to see that $F_{i,y} \neq 0$ in $(\mathbb{R}^*)^2$. □

If $2p \leqslant d - m$, we put $a_{2i} = 0$ for $1 \leqslant i \leqslant p$, and take a_{2i}, satisfying (3.5.3), for $p < i \leqslant (d-m)/2$. Gluing the polynomials F_{2i}, $i = 1, \ldots, (d-m)/2$, with suitable polynomials constructed in the fourth step, we get an arbitrary index distribution except in the case

$$c_0 < (d-m)/2 - p.$$

But here

$$c_2 = ((d-1)^2 + 1 - m)/2 - p - c_0 \geqslant (d-m)/2,$$

hence we obtain the desired index distribution by multiplying the above polynomial by -1.

If $2p > d - m$, we put $a_{2i} = 0$ for any $i = 1, \ldots, (d-m)/2$. Then the procedure described in the previous steps gives the desired index distribution.

3.11. Tenth step. Assume that $d \geqslant 6$ is even, $m = 0$, $p > 0$.

Let us combine the constructions presented in the fifth and ninth steps. If $2p < d - 2$, then we obtain any prescribed index distribution as long as

$$\max\{c_0, c_2\} \geqslant 3d/2 - 2 - p.$$

But this inequality holds for $d \geqslant 6$ because of (3.1.2), (3.1.3). If $2p \geqslant d - 2$, then we obtain any prescribed index distribution satisfying $\max\{c_0, c_2\} \geqslant d - 1$, which is equivalent to $2p \leqslant (d-3)^2 - 1$.

3.12. Eleventh step. Assume that $d \geqslant 5$ is odd, $m < d$, $p > 0$.

In this case, combining the constructions presented in the second, sixth, and seventh steps, we can get any index distribution that satisfies

$$\min\{c_0, c_2\} \geqslant (d - m)/2.$$

3.13. Twelfth step. Assume that $d \leqslant 4$. Then all the index distributions except (3.1.7), (3.1.9) can be easily realized as explained above.

§4. New restrictions to the topology of T-hypersurfaces

The topological classification of nonsingular algebraic hypersurfaces in $\mathbb{R}P^n$ is known only for $n = 2$ [**Ha, V4**]. Here we consider this problem in connection with the index distribution of real polynomials. Let us fix $n \geqslant 3$. Let F be a nondegenerate real polynomial of degree d in n variables, and let the projective closure $P\{F\} \subset \mathbb{R}P^n$ of the affine hypersurface $\{F = 0\} \subset \mathbb{R}^n$ be nonsingular. A polynomial F is said to be a T-*polynomial* (or ET-*polynomial*) if it is glued by (1.2.1) from $(n+1)$-nomials (resp. from $(n+1)$-nomials with elementary Newton simplices). The corresponding hypersurface $P\{F\}$ will be called a T-*hypersurface*, or ET-*hypersurface*, respectively (cf. [**S3**]). The aim of this section is to give new restrictions on the topology of T-hypersurfaces by means of the results presented above.

Put

$$b_i(P\{F\}) = \dim H_i(P\{F\}, \mathbb{Z}/2\mathbb{Z}).$$

THEOREM 4.1. *Let $n \geqslant 3$ be fixed, and let $P\{F\}$ be a nonsingular hypersurface of degree d in $\mathbb{R}P^n$. If $P\{F\}$ is an ET-hypersurface, then*

(4.2)
$$b_0(P\{F\}) \leqslant \frac{d^n}{n!} + O(d^{n-1}),$$

(4.3)
$$b_i(P\{F\}) \leqslant \frac{1}{n-2i}\left(1 + \frac{n-4i}{n!}\right)d^n + O(d^{n-1}), \quad 1 \leqslant i \leqslant \frac{n-1}{2}.$$

If $P\{F\}$ is a T-hypersurface, then

(4.4)
$$b_0(P\{F\}) \leqslant \frac{2^{n-2}}{n!}d^n + O(d^{n-1}).$$

REMARK 4.5. Restrictions (4.2), (4.4) are new for $n \geqslant 3$. Restriction (4.3) is new for $n \geqslant 5$. For example, in the case $n = 3$, a standard application of the Smith and Comessatti inequalities (see [**Kh, Wi**]) gives only

$$b_0(P\{F\}) \leqslant 5d^3/12 + O(d^2),$$

while (4.2) gives

$$b_0(P\{F\}) \leqslant d^3/6 + O(d^2),$$

and (4.4) gives

$$b_0(P\{F\}) \leqslant d^3/3 + O(d^2).$$

Note also that Viro [**V1**] constructed surfaces with $b_0 = (7d^3 - 24d^2 + 32d)/24$ for any $d = 4k + 2$. This means, in particular, that these surfaces are not homeomorphic to ET-surfaces.

PROOF OF THEOREM 4.1. We use the following

PROPOSITION 4.6. *If* $\deg F = d$ *and* $\bar{c}(F) = (c_0, \ldots, c_n, p)$, *then*

(4.7) $\qquad b_i(P\{F\}) \leq c_i + c_{n-i} + O(d^{n-1}), \qquad 0 \leq i \leq (n-1)/2,$

and, for $n = 2k$,

(4.8) $\qquad b_k(P\{F\}) \leq c_k + O(d^{n-1}).$

This is a trivial consequence of Morse theory, Poincaré and Alexander–Pontryagin duality, and the fact that the total Betti number of a nonsingular hypersurface of degree d in $\mathbb{R}P^{n-1}$ is bounded from above by $O(d^{n-1})$ (see [**Fa, Ro**]).

We consider the Betti numbers of $T\{F\} = P\{F\} \cap (\mathbb{R}^*)^n$, which coincide with those for $P\{F\}$ up to terms of order $O(d^{n-1})$ and do not depend on the multiplication of F by any monomial.

Let F be T-polynomial with the underlying regular triangulation $\Delta_1, \ldots, \Delta_r$ of $\Delta = \tau_d^n$. Multiplying F by a suitable monomial, we can ensure $i_0(\Delta_1) = \cdots = i_0(\Delta_r) = 0$.

Now assume that F is ET-polynomial. According to Propositions 2.2.4 and 4.6, the inequalities (4.7), (4.8) yield

(4.9) $\qquad b_0(T\{F\}) \leq \sigma_n + O(d^{n-1}),$

(4.10) $\qquad b_j(T\{F\}) \leq \sigma_j + \sigma_{n-j} + O(d^{n-1}), \qquad 1 \leq j \leq (n-1)/2,$

(4.11) $\qquad b_k(T\{F\}) \leq \sigma_k + O(d^{n-1}), \qquad \text{if } n = 2k,$

where σ_j is the number of simplices with $i_+ = j$, $j = 1, \ldots, n$. To any simplex Δ_s with $i_+(\Delta_s) = n$ we assign its unique vertex belonging to $\partial K_+(\Delta_s) \setminus \partial K(\Delta_s)$. Thus we obtain a one-to-one correspondence between the simplices with $i_+ = n$ and integer points from Δ without some of its facets. This means that

$$\sigma_n = d^n/n! + O(d^{n-1}),$$

hence (4.9) implies (4.2). Now let us introduce the face vector $(f_0, \ldots, f_n)$ of the simplicial complex Σ defined by the triangulation $\Delta_1, \ldots, \Delta_r$, where f_j is the number of j-dimensional faces. If we compute, for any Δ_s, the number of j-dimensional faces intersecting with $\text{Int}(K_+(\Delta_s) \cup \Delta_s)$, then we obtain easily

$$f_j = \sum_{l=0}^{j} \binom{n-l}{j-l} \sigma_{n-l} + O(d^{n-1}),$$

or, conversely,

(4.12) $\qquad \sigma_{n+1-j} = \sum_{i=1}^{j} (-1)^{j-i} \binom{n+1-i}{n+1-j} f_{i-1} + O(d^{n-1}), \qquad j = 1, \ldots, n.$

Let us construct the following simplicial n-polytope (triangulated boundary of a convex $(n+1)$-dimensional polytope):
  (1) First we take the convex hull of the graph of the convex piecewise linear function v corresponding to the regular subdivision $\Delta_1, \ldots, \Delta_r$. It is clear that this polytope has $\leq 2^{n+1}$ nonsimplicial faces.
  (2) We add $\leq 2^{n+1}$ new vertices so as to replace the nonsimplicial faces by simplices.

Evidently, the face vector $(f'_0, \ldots, f'_n)$ of the obtained polytope coincides with the face vector of Σ up to terms of order $O(d^{n-1})$. On the other hand, for the so-called h-vector $(h_0, \ldots, h_{n+1})$ of the new polytope, we have (see [**Mc**])

$$h_j = \sum_{i=1}^{j} (-1)^{j-i} \binom{n+1-i}{n+1-j} f'_{i-1} + (-1)^j \binom{n+1}{n+1-j}, \qquad j = 1, \ldots, n,$$

$$h_0 = h_{n+1} = 1.$$

Hence (4.12) immediately yields

(4.13) $$\sigma_{n+1-j} = h_j + O(d^{n-1}), \qquad j = 1, \ldots, n.$$

The final step in proving (4.3) is to substitute (4.13) into the Dehn–Sommerville equations and the McMullen inequalities [**Gr, Mc**]

$$h_j = h_{n+1-j}, \quad h_{j-1} \leqslant h_j, \quad 1 \leqslant j \leqslant (n+1)/2,$$

and then to find in (4.10), (4.11) the maximin of several affine functions.

Let us pass to (4.4). According to Proposition 2.2.4 and (4.7), we must estimate the numbers $c^{(l)}$ of real critical points of $(n+1)$-nomials with Newton simplices Δ_l satisfying $i_+(\Delta) = n$. For such a simplex Δ_l, introduce the set S_l of integer points in Δ_l, except n vertices belonging to $\partial K(\Delta_l)$. Note that the sets S_l are nonempty and disjoint, which implies

(4.14) $$\operatorname{card} S_l \geqslant 1, \quad \sum \operatorname{card} S_l \leqslant f_0 \leqslant d^n/n! + O(d^{n-1}).$$

Let v_0 be the vertex of Δ_l in S_l. Consider the vectors $w_k = v_k - v_0$ introduced in the proof of Proposition 2.2.1. If all the coordinates of $w_1, \ldots, w_n$ are even, then S_l contains the midpoints of the edges $[v_0; v_1], \ldots, [v_0; v_n]$, hence card $S_l \geqslant n+1$, which means, according to Proposition 2.2.1,

(4.15) $$\frac{c^{(l)}}{\operatorname{card} S_l} \leqslant \frac{2^n}{n+1} \leqslant 2^{n-2}.$$

Assume that there is a vector w_k, say w_1, with at least one odd coordinate. Then the power d_1 in (2.2.3) is odd. If in addition d_2 is odd, then, clearly, (4.15) holds. Let $d_2, \ldots, d_n$ be even. Then $c^{(l)} \leqslant 2^{n-1}$. On the other hand, the last assumption implies that all the powers $p_{ij}, i \neq 1, i < j$, are even. In this case, if p_{12} or p_{13} is even, then the midpoint of the edge $[v_0; v_2]$ or of the edge $[v_0; v_3]$ belongs to S_l. Thus, card $S_l \geqslant 2$, and (4.15) holds. If p_{12}, p_{13} are odd, then the midpoint of the edge $[v_2; v_3]$ belongs to S_l, hence again we get (4.15). Finally, (4.14), (4.15) give

$$\sum c^{(l)} \leqslant \frac{2^{n-2}}{n!} d^n + O(d^{n-1}),$$

and the proof is completed. □

References

[Ar] V. I. Arnold, *On some problems in singularity theory*, Geometry and Analysis. Papers dedicated to the memory of V. K. Patodi, Springer-Verlag, Berlin, 1981, pp. 1–11.

[AGV] V. I. Arnold, S. M. Guseĭn-Zade, and A. N. Varchenko, *Singularities of differentiable maps.* I, "Nauka", Moscow, 1984; English transl., Birkhäuser, Basel, 1985.

[ChB] B. Chevallier, *Sur le courbes maximales de Harnack*, C. R. Acad. Sci. Paris Sér. I Math. **300** (1985), no. 4, 109–114.

[ChV] V. Chekanov, *Asymptotics of the number of maxima for a product of linear functions of two variables*, Vestnik Moskov. Univ. Ser. I Mat. Mekh.; English transl., Moscow Univ. Math. Bull. **41** (1986), 85–87.

[Chi] Yu. S. Chislenko, *Decompositions of simple singularities of real functions*, Funktsional. Anal. i Prilozhen. **22** (1988), no. 4, 52–67; English transl., Functional Anal. Appl. **22** (1988), no. 4, 297–310.

[Du] A. Durfee, N. Kronenfeld, H. Munson, J. Roy, and I. Westby, *Counting critical points of real polynomials in two variables*, Amer. Math. Monthly **100** (1993), no. 3, 255–271.

[Fa] I. Fary, *Cohomologie des variétés algébriques*, Ann. of Math. **65** (1957), no. 1, 21–73.

[Gr] B. Grünbaum, *Convex polytopes*, Wiley-Interscience, London, 1967.

[Gu] D. A. Gudkov, *The topology of real projective algebraic varieties*, Uspekhi Mat. Nauk **29** (1974), no. 4, 3–79; English transl., Russian Math. Surveys **29** (1974), no. 4, 1–79.

[Ha] A. Harnack, *Über die Vieltheiligkeit der ebenen algebraischen kurven*, Math. Ann. **10** (1876), 189–199.

[Kh] V. M. Kharlamov, *Real algebraic surfaces*, Proc. Intern. Congress Math. (Helsinki, 1978), vol. 1, Acad. Sci. Fennica, Helsinki, 1980, pp. 421–428. (Russian)

[Ko] A. G. Koushnirenko, *Polyèdres de Newton et nombres de Milnor*, Invent. Math. **32** (1976), 1–31.

[Mc] P. McMullen, *The number of faces of simplicial polytopes*, Israel J. Math. **9** (1971), 559–570.

[Ri] J.-J. Risler, *Construction d'hypersurfaces réelles [d'après Viro]*, Séminaire N. Bourbaki, vol. 763, 1992-93, Novembre 1992.

[Ro] V. A. Rokhlin, *Congruences modulo 16 in sixteenth Hilbert problem*, Funktsional. Anal. i Prilozhen. **6** (1972), no. 4, 58–64; English transl., Functional Anal. Appl. **6** (1972), no. 4, 301–306.

[S1] E. Shustin, *Gluing of singular algebraic curves*, Methods of Qualitative Theory, Gor'ky Univ. Press, Gor'ky, 1985, pp. 116–128. (Russian)

[S2] ———, *Hyperbolic and minimal smoothings of singular points*, Selecta Math. Soviet. **10** (1991), no. 1, 19–25.

[S3] ———, *Topology of real plane algebraic curves*, Proc. Intern. Conf. Real Algebraic Geometry, Rennes, June 24–29, 1991, Lecture Notes in Math., vol. 1524, Springer-Verlag, 1992, pp. 97–109.

[S4] ———, *Real plane algebraic curves with prescribed singularities*, Topology **32** (1993), no. 4, 845–856.

[St] B. Sturmfels, *Viro's theorem for complete intersections*, Preprint, Cornell Univ., Ithaca, 1992.

[V1] O. A. Viro, *Construction of multicomponent real algebraic surfaces*, Dokl. Akad. Nauk SSSR **248** (1979), no. 2, 279–282; English transl., Soviet Math. Dokl. **20** (1979), no. 5, 991–995.

[V2] ———, *Gluing of algebraic hypersurfaces, smoothing of singularities and construction of curves*, Proc. Leningrad Intern. Topological Conf. (Leningrad, Aug. 1983), "Nauka", Leningrad, 1983, pp. 149–197. (Russian)

[V3] ———, *Gluing of plane real algebraic curves and construction of curves of degree 6 and 7*, Lecture Notes in Math., vol. 1060, Springer-Verlag, Berlin, 1984, pp. 187–200.

[V4] ———, *Progress in the topology of real algebraic varieties over the last six years*, Uspekhi Mat. Nauk **41** (1986), no. 3, 45–67; English transl., Russian Math. Surveys **41** (1986), no. 3, 55–82.

[Wi] G. Wilson, *Hilbert's sixteenth problem*, Topology **17** (1978), no. 1, 53–74.

Translated by THE AUTHOR

SCHOOL OF MATHEMATICAL SCIENCES, TEL AVIV UNIVERSITY, RAMAT AVIV, TEL AVIV 69978, ISRAEL

On a Nonlinear Boundary Value Problem of Mathematical Physics

G. A. Utkin

In memory of Professor D. A. Gudkov

ABSTRACT. A nonlinear boundary value problem of mathematical physics describing the transverse vibrations of a string and a bead moving along the latter is considered. The theorem of existence and uniqueness of the solution to this problem is proved. A formula for the bead motion average velocity along the string under wave pressure, taking into account friction, is obtained.

The motion of a bead along a string subject to a wave pressure force was considered in [1]. The physical realization of the process is schematically pictured in Figure 1, where $U(x, t)$ is the transverse string displacement, α is a damping factor, k and M are stiffness and mass of fixation respectively, $F(t)$ is the external force, m is the mass of the bead, and A are rigid directing bars such that transverse motions of the bead are impossible and transverse waves of the string do not penetrate over the bead. In [1] some properties of the solution of such a problem were announced. This paper presents the proof of these properties.

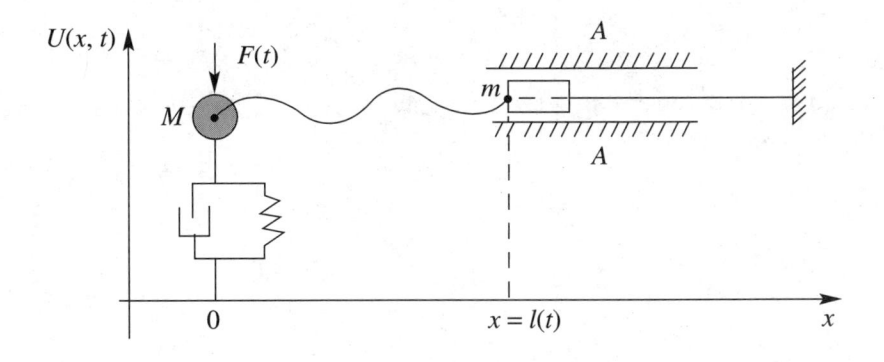

FIGURE 1

1991 *Mathematics Subject Classification*. Primary 35L70.

As shown in [2], the law of motion $l(t)$ of the bead along the string and the transverse motion $U(x, t)$ of the string are defined as solutions of the problem

$$\rho U_{tt} - N U_{xx} = 0, \tag{1}$$

$$MU_{tt} + \alpha U_t + kU - NU_x|_{x=0} = -F(t), \quad U(l(t), t) = 0, \tag{2}$$

$$m\ddot{l}(t) = F_{\text{pr}} - F_{\text{fr}}, \tag{3}$$

$$U(x, 0) = \varphi(x), \tag{4.1}$$

$$U(x, 0) = \psi(x), \tag{4.2}$$

$$l(0) = l_0, \quad \dot{l}(0) = v < c = \sqrt{N/\rho}, \tag{4.3}$$

$$\varphi(l_0) = 0, \tag{5.1}$$

$$\psi(l_0) + v\varphi'(l_0) = 0, \tag{5.2}$$

$$Mc^2\varphi''(0) + \alpha\psi(0) + k\varphi(0) - N\varphi'(0) = -F(0), \tag{5.3}$$

$$\varphi(l_0)F_{\text{fr}}|_{t=0} = -(\rho/2)(c^2 - v^2)(\varphi'(l_0))^3 + m[\varphi''(l_0)(c^2 + v^2) + 2v\psi'(l_0)], \tag{5.4}$$

where ρ is the density of the string, N is the tension of the string, l_0 and v are the initial state and velocity of the bead respectively, $F_{\text{fr}}(l, U_t, U_x)$ is the force of friction, $F_{\text{pr}} = (\rho/2)(c^2 - \dot{l}^2(t)) U_x^2(l(t), t)$ is the wave pressure force, and φ and ψ are the initial distributions of displacements and velocities of the string, satisfying the matching conditions (5.1)–(5.4). Let us assume that $\varphi(x), \psi(x) \in C^2_{[0,l_0]}$ and $F(t) \in C^2[0, +\infty]$. It is well known that the solution of equation (1) can be represented as the superposition of two traveling meeting waves

$$U(x, t) = g_1(t + x/c) - g_2(t - x/c). \tag{6}$$

Using (1)–(5), we obtain the following equations for the functions g_1, g_2, and l:

$$g_1(t + l(t)/c) = g_2(t - l(t)/c), \tag{7}$$

$$M(g_1''(t) - g_2''(t)) + (\alpha - N/c)g_1'(t) \\ - (\alpha + N/c)g_2'(t) + k(g_1(t) - g_2(t)) = -F(t), \tag{8}$$

$$m\ddot{l}(t) = 2\rho \frac{c - \dot{l}(t)}{c + \dot{l}(t)} \left(g_2'\left(t - \frac{l(t)}{c}\right) \right)^2 - F_{\text{fr}}\left(l, g_2'\left(t - \frac{l(t)}{c}\right)\right); \tag{9}$$

for the initial conditions for $g_1(y)$, $g_2(y)$, where $y \in [0, y_0]$, $y_0 = l_0/c$, and for $l(t)$ at $t = 0$, we have

$$g_1(y) - g_2(y) = \varphi(cy), \tag{10.1}$$

$$g_1'(y) = (1/2)(c\varphi'(cy) + \psi(cy)), \tag{10.2}$$

$$g_2'(y) = (1/2)(c\varphi'(cy) - \psi(cy)), \tag{10.3}$$

$$l(0) = l_0, \quad \dot{l}(0) = v. \tag{10.4}$$

It follows from (2) and (9) that the bead motion function $l(t)$ is defined by the falling wave pressure force F_{pr} and by the friction force F_{fr}. If $-c < \dot{l}(t) < 0$, then $F_{\text{fr}} < 0$, $F_{\text{pr}} - F_{\text{fr}} > 0$, so that $\dot{l}(t)$ increases. Thus if $\dot{l}(0) = v$ and $|v| < c$, then the equality $\dot{l}(t) = -c$ is not possible for any value of t. As $\dot{l}(t) \to c - 0$, the friction force F_{fr} increases and the wave pressure force tends to zero (if the velocity of the bead is equal to that of the wave, then they do not interact). So when $\dot{l}(t) \geqslant c - \varepsilon$, where

$\varepsilon > 0$ is defined by the external force $F(t)$ and by the friction force F_{fr}, we obtain $F_{\text{pr}} - F_{\text{fr}} \leqslant 0$ and the velocity $l(t)$ begins to fall. Thus for any $t \leqslant 0$, we have the inequality $|\dot{l}(t)| \leqslant c - \varepsilon < c$.

THEOREM 1. *If $\varphi(x), \psi(x) \in C^2[0, l_0]$, $F(t) \in C^2[0, \infty)$, and $F_{\text{fr}}(\dot{l}(t), U_t, U_x)$ $\in C^2_{\mathbb{R}^3}$, then there exists a unique solution of the nonlinear boundary problem (1)–(5) such that $U(x,t) \in C^2[0, l(t)]$ and $l(t) \in C^2[0, +\infty]$.*

PROOF. The relation
$$g_1(l_0/c) - g_2(-l_0/c) = U(l(0), 0) = 0$$
follows from (10.1), and conditions (10.2) and (10.3) define the functions $g_1(y)$ and $g_2(-y)$ on the segment $[0, y_0]$ up to a constant, which is the same both for g_1 and for g_2. So if the value $g_1(+0)$ is known, then $g_2(-0) = g_1(+0) - \varphi(0)$. Since the function $g_1(y)$ defined on the segment $[0, y_0]$ is known, we can use equation (8) and find a unique function $g_2(y)$ such that $g_2(y) \in C^2[0, y_0]$ and g_2 satisfies the initial conditions
$$g_2(0) = g_2(+0) = g_2(-0) \quad \text{and} \quad g_2'(0) = g_2'(+0) = (1/2)(c\varphi'(0) - \psi(0)).$$
Hence the function $g_2(y)$ is defined as $g_2(y) \in C^2[-y_0, 0]$ and also in $C^2[0, y_0]$ (this follows from (10.3)) and $g_2(-0) = g_2(+0)$, $g_2'(-0) = g_2'(+0)$. From conditions (5) and from equation (8), it follows that
$$M(g_1''(+0) - y) + (\alpha - N/c)g_1'(+0) - (\alpha + N/c)g_2'(+0) + k(g_1(+0) - g_2(+0)) = -F(0),$$
where y can be substituted both into $g_2(+0)$ and $g_2(-0)$. Since we obtain $g_2''(-0) = g_2''(+0)$, we see that $g_2(y)$ is defined on the segment $[-y_0, y_0]$.

If $g_2'(y)$ is known, then the function $l(t) \in C^2[0, t_1]$ satisfying the given initial conditions (10.4) can be found as a unique solution of equation (8), where t_1 is the root of the equation $t_1 - l(t_1)/c = y_0 = l/c$.

Since we know the function $l(t)$, $t \in [0, t_1]$, we can extend the function $g_1(y)$ to the segment $[y_0, y_1]$, where $y_1 = t_1 + l(t_1)/c$ follows from (7). From (4.1) and (6) it follows that $g_1(y_0 + 0) = g_2(-y_0 + 0)$, and equation (7), as $t \to +0$, implies the relationship $g_1(y_0 - 0) = g_2(-y_0 + 0)$. So we obtain $g_1(y_0 - 0) = g_1(y_0 + 0)$. From condition (10.2), we get
$$g_1'(y_0 - 0) = \frac{1}{2}(c\varphi'(0) + \psi(0)), \qquad g_1''(y_0 - 0) = \frac{1}{2}c(c\varphi''(0) + \psi'(0)),$$
and from equation (7), in accordance to (5.2), it follows that
$$g_1'(y_0 + 0) = \frac{c-v}{c+v}g_2'(-y_0 + 0) = \frac{c-v}{c+v}(c\varphi'(l_0) - \psi(l_0))$$
$$= \frac{1}{2}\frac{c-v}{c+v}(c\varphi'(l_0) + v\varphi'(l_0)) = \frac{1}{2}(c-v)\varphi'(l_0)$$
$$= \frac{1}{2}(c\varphi'(l_0) + \psi(l_0)) = g_1'(y_0 - 0).$$

Differentiating equation (7) twice and passing to the limit as $t \to +0$, we obtain
$$g_1''(y_0 + 0)\left(1 + \frac{v}{c}\right)^2 = g_1''(-y_0 + 0)\left(1 - \frac{v}{c}\right)^2 - \frac{\ddot{l}(0)}{c}(g_1'(y_0 + 0) + g_2'(-y_0 + 0))$$
$$= -\frac{c}{2}(c\varphi''(l_0) - \psi'(l_0))\left(1 - \frac{v}{c}\right)^2 - \varphi'(l_0)\ddot{l}(0).$$

In the last equation let us substitute the value

$$\ddot{l}(0) = \frac{1}{m}\left\{\frac{\rho}{2}(c^2 - v^2)[\varphi'(l_0)]^2 - F_{\text{fr}}|_{t=0}\right\}$$

found from equation (3) and take into account condition (5.4), obtaining

$$g_1''(y_0 + 0) = g_1''(y_0 - 0).$$

So we have proved that the function $g_1(y)$ has a continuous second derivative $g_1''(y)$ at the point $y = y_0$ and therefore on the segment $[0, y_1]$.

Since the function $g_1(y)$ is known on the segment $[0, y_1]$, it is possible to find a unique function $g_2(y) \in C^2[0, y_1]$ as the solution of the differential equation (8) with initial conditions (10.1)–(10.3). So this function $g_2(y)$ and the function g_2 found earlier are the same on the segment $[0, y_0]$.

Knowing the function $g_2(y)$ on the segment $[-y_0, y_0]$, we can find a unique function $l(t) \in C^2[0, t_2]$ as the solution of equation (9) with initial conditions (10.4), where t_2 is the root of the equation

$$t_2 - l(t_2)/c = y_1 = t_1 + l(t_1)/c.$$

Using equation (7), we can determine the function $g_1(y)$ on the segment $[0, y_2]$, where $y_2 = t_2 + l(t_2)/c$.

The process presented above allows us to construct the function $l(t)$ for any $t \in [0, \infty)$, the function $g_1(y)$ for $y \in [0, \infty)$ and the function $g_2(y)$ for $[-y_0, \infty)$. At each step of the construction, the continuations of the functions $g_1(y)$ and $g_2(y)$ are uniquely defined up to a constant, which is the same for both functions. Since $U(x, t) = g_1(t + x/c) - g_2(t - x/c)$, it follows that $U(x, t)$ is defined in a unique way and does not depend on the choice of the value of $g_1(+0)$. The right-hand side of equation (7) depends on $g_2'(y)$, but $g_2'(y)$ does not depend on the value $g_1(+0)$, hence the function $l(t)$ is defined in a unique way too. The theorem is proved.

Given the function $g_2(y)$ on the half line $[-y_0, \infty)$, we can define the corresponding functions $\varphi(x)$, $\psi(x)$, and $F(t)$ in a unique way. The friction force F_{fr} depends on $\dot{l}$, U_x, U_t and we assume this dependence to be given. So further in this paper, the falling wave is assumed to be given, while the external force and initial conditions are not given. Accordingly, we note, for example, that if $M = k = 0$, $\alpha = N/c$, then the following relation holds:

$$g_2'(y) = \begin{cases} (1/2)(c\varphi'(-cy) - \psi(-cy)), & y \in [-y_0, 0], \\ (1/2)(c/N)F(y), & y \in [0, +\infty). \end{cases}$$

If the wave pressure force is less than the friction force, the motion of the bead is decelerated and the bead stops within finite time. When these forces are comparable in value, the bead moves with stops. In practice, it is more interesting when the wave pressure force is sufficiently large and the bead moves without stops. So let us consider the case of a monotone motion of the bead ($\dot{l}(t) > 0$), when $F_{\text{fr}} = \delta_0 \operatorname{sgn} \dot{l}(t) + \delta_1 \dot{l}(t)$. The most interesting question is what can the attainable average velocity of the bead moving under the wave pressure force be when $F_{\text{fr}} = \delta_0 + \delta_1 \dot{l}(t)$? By "average velocity" we naturally mean the value $\dot{l}_m = \lim_{t \to \infty} l(t)/t$. The next theorem gives an answer to this question.

THEOREM 2. *If the limits*

$$A = \lim_{t \to \infty} \frac{1}{t} \int_0^t (g_2'(y))^2 dy \quad \text{and} \quad B = \lim_{t \to \infty} \frac{1}{t} \int_0^t \dot{l}^2(\tau) d\tau$$

exist, then the average velocity $\dot{l}_m$ *exists and is defined by the formula*

$$\dot{l}_m = \frac{c(2\rho A - \delta_0) - B\delta_1}{2\rho A + \delta_0 + c\delta_1}.$$

PROOF. Integrating equation (7), we obtain the relation

$$\frac{m}{2c}(c+\dot{l}(t))^2 = \frac{m(c+v)^2}{2c} + 2\rho \int_{-y_0}^{t-l(t)/c} (g_2'(y))^2 dy - \int_0^t \left(1 + \frac{l(\dot{y})}{c}\right)(\delta_0 + \delta_1 l(\dot{y})) dy.$$

Dividing this equation by t and passing to the limit as $t \to +\infty$, we obtain a linear equation for the average velocity

$$2\rho A(1 - \dot{l}_m/c) - \delta_0 - (\delta_0/c + \delta_1)\dot{l}_m - \delta_1 B/c = 0.$$

Solving this equation for $\dot{l}_m$, we get the required formula.

References

1. A. I. Vesnitsky and G. A. Utkin, *Body motion along string by wave pressure force*, Dokl. Akad. Nauk SSSR **302** (1988), no. 2, 278–280; English transl. in Soviet Math. Dokl. **38** (1989).
2. A. I. Vesnitsky, L. E. Kaplan, and G. A. Utkin, *Laws of variation of energy and momentum for one-dimensional system with moving fastenings and loads*, Prikl. Mat. Mekh. **47** (1983), no. 5, 863–866; English transl. in J. Appl. Math. Mech. **47** (1983).

Translated by THE AUTHOR

23, GAGARIN AVENUE, NIZHNY NOVGOROD 603600, RUSSIA

Generic Immersions of the Circle to Surfaces and the Complex Topology of Real Algebraic Curves

Oleg Viro

ABSTRACT. In a recent paper [1], Arnold introduced three new invariants of a generic immersion of the circle to the plane. These invariants are similar to Vassiliev invariants of classical knots. In a sense they are of degree one. In this paper an investigation based on similar ideas is made for real algebraic plane projective curves. In this more algebraic setting, Arnold's invariants have natural counterparts, two of which admit definitions in terms of the complexification of the curve. On the other hand, the Rokhlin complex orientation formula for a real algebraic curve bounding in its complexification suggests new combinatorial formulas for these two Arnold invariants. Using these formulas, a conjecture of Arnold is proved. Arnold's invariants are generalized to generic collections of immersions of the circle to the projective plane and other surfaces. Some invariants of high degrees admitting similar formulas are discussed.

§0. Introduction

This paper presents an interaction between theories of real algebraic curves and smooth immersions of the circle to the plane. I must acknowledge that this interaction exceeds my original expectations.

The initial point was Arnold's study [1] of analogs of Vassiliev invariants for immersions of the circle.[1] I started from the straightforward idea of applying the same approach to the theory of real plane projective algebraic curves. I hoped to get invariants that would be useful for the description of the topology of a real plane algebraic curve with singularities.

Almost immediately it became clear that two of the three Arnold invariants have the same behavior as the following two characteristics of a real plane algebraic curve separating its complexification: the number of imaginary self-intersection points of a half of the complexification and the number of imaginary intersection points of the halves. These numbers are involved in some versions of the Rokhlin complex orientation formulas.

1991 *Mathematics Subject Classification.* Primary 14P25.

[1] Also see Arnold's paper in this volume.

In the situations studied by Arnold, there is neither complexification nor hope of constructing its substitute: an arbitrary differentiable immersion of the circle to the plane does not admit a complexification.

Nonetheless the analogy started to work. The Rokhlin complex orientation formula suggested one look for its counterpart in the theory of immersions. The formula discovered in this way allowed one to prove Arnold's conjecture on the range of values of his invariants. It suggests generalizations of Arnold's invariants to the case of immersions of the circle and several copies of the circle to various other surfaces. A straightforward generalization of the formula provides an infinite series of invariants of finite degree for immersions of the circle to the plane.

§1. Arnold's work on immersed circles

1.1. Space of immersions. By a generic immersion of the circle S^1 into the plane $\mathbb{R}^2$, one means an immersion without triple points and without points of self-tangency. It has only ordinary double points of transversal self-intersection.

A triple point of an immersion is said to be *ordinary* if the branches at the point are transversal to each other. A self-tangency point of an immersion is said to be *ordinary* if the branches have distinct curvatures at the point. A self-tangency point of an immersion is called a point of *direct tangency* if the velocity vectors point in the same direction; otherwise it is called a point of *inverse tangency*.

The space of all immersions is an infinite-dimensional manifold. It consists of infinitely many connected components. The components are in a natural one-to-one correspondence, provided by the *Whitney index*, with the integers. The latter is an integer-valued characteristic of an immersion, which is also called its *winding number*, and may be defined as the rotation number of the velocity vector, as well as the degree of the Gauss map. It determines the immersion up to a regular homotopy, i.e., a path in the space of immersions.

In the space of immersions, all the nongeneric immersions form a hypersurface called the *discriminant hypersurface* or briefly the *discriminant*.

This hypersurface is stratified. There are three main strata (open in the discriminant):

1. The set of all immersions without triple points, with only one double point which is not transversal, and such that this point is an ordinary direct self-tangency point.
2. The set of all immersions without triple points, with only one double point which is not transversal, and such that this point is an ordinary inverse self-tangency point.
3. The set of immersions which have only one triple point, this point is ordinary, and besides this point there are only double points of transversal self-intersection.

A generic path in the space of immersions (i.e., a generic regular homotopy) intersects the discriminant hypersurface in a finite number of points, and these points belong to the main strata. Changes experienced by an immersion when it goes through the strata were called *perestroikas* by Arnold. They are shown in Figures 1.1, 1.2, and 1.3.

By a *coorientation* of a hypersurface one means a choice of one of the two parts separated by the hypersurface in a neighborhood of any of its points. Arnold [1] has constructed natural coorientations of the main strata of the discriminant hypersurface. The direction shown in Figures 1.1, 1.2, and 1.3 is positive for these coorientations.

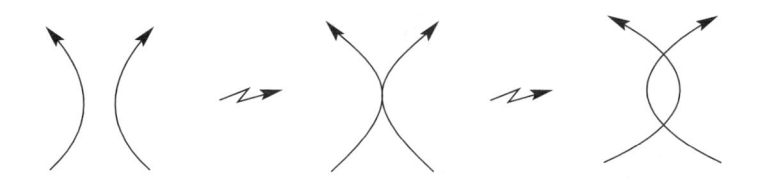

FIGURE 1.1. Direct self-tangency perestroika.

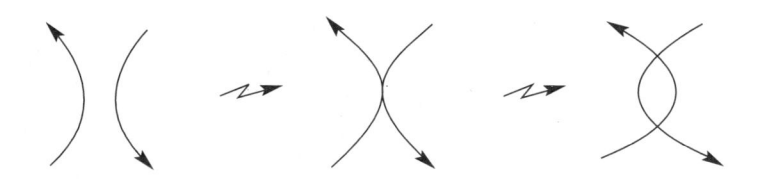

FIGURE 1.2. Inverse self-tangency perestroika.

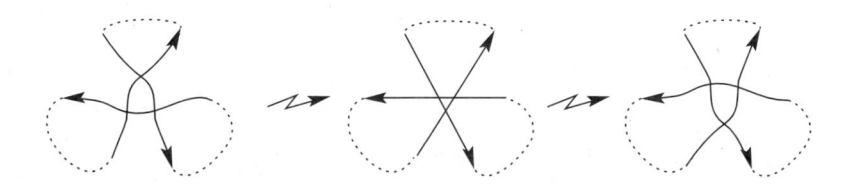

FIGURE 1.3. Triple point perestroika; dotted curves indicate the scheme of connection of the branches, which is needed to determine the positive direction of the perestroika; on the right-hand side we have $q = 0$, on the left-hand side we have $q = 3$, thus the positive direction is to the right.

In the case of the self-tangency strata, the positive direction is the one along which the number of double points increases. The coorientation of the triple point stratum is defined as follows. A transversal passage through this stratum is positive if the newborn vanishing triangle is positive. A vanishing triangle is the triangle formed by the three branches of a curve close to a curve with a triple point. The sign of a vanishing triangle is defined as follows. The orientation of the curve defines a cyclic ordering of the sides of the vanishing triangle, and hence an orientation of the triangle. Denote by q the number of sides of the vanishing triangle whose orientation as a part of the curve coincides with the orientation defined by the orientation of the triangle. The *sign* of a vanishing triangle is $(-1)^q$. See Figure 1.3.

1.2. The three Arnold invariants. For a generic immersion, Arnold [1] introduced numerical characteristics J^+, J^-, and St which[2] are defined by the following properties:

1.2.A. *J^+, J^-, and St are invariant under regular homotopy in the class of generic immersions.*

---

[2] The latter is denoted by St because Arnold called it *strangeness*. Indeed, St seems to be more subtle than $J^{\pm}$.

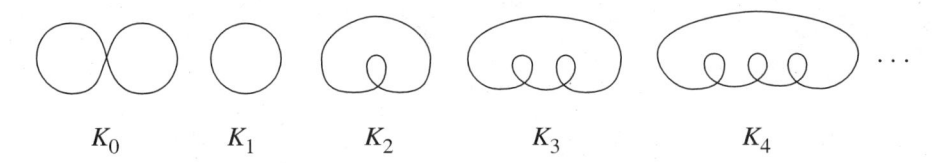

FIGURE 1.4. The standard curves of Whitney indices $0, \pm 1, \pm 2, \ldots$.

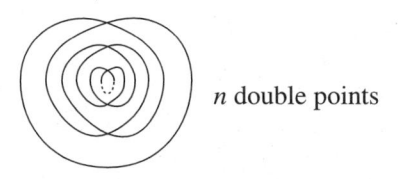

FIGURE 1.5. The curve A_{n+1}.

1.2.B. *J^+ does not change when the immersion experiences an inverse self-tangency perestroika or a triple point perestroika, but increases by two under a positive direct self-tangency perestroika.*

1.2.C. *J^- does not change when the immersion experiences a direct self-tangency perestroika or a triple point perestroika, but decreases by 2 under a positive (increasing the number of double points) inverse self-tangency perestroika.*

1.2.D. *St does not change when the immersion experiences a self-tangency perestroika, but increases by 1 under a positive triple point perestroika.*

1.2.E. *For the immersions K_i with $i = 0, 1, 2, \ldots$ shown in Figure 1.4,*

$$J^+(K_0) = 0, \quad J^+(K_{i+1}) = -2i \quad (i = 0, 1, \ldots),$$
$$J^-(K_0) = -1, \quad J^-(K_{i+1}) = -3i \quad (i = 0, 1, \ldots),$$
$$St(K_0) = 0, \quad St(K_{i+1}) = i \quad (i = 0, 1, \ldots).$$

At first glance the normalization provided by 1.2.E looks artificial. It is motivated by the desire to have invariants with nice properties: it is the only normalization giving invariants additive with respect to connected sum.

1.3. Arnold's conjecture. In [1] Arnold formulated several conjectures on the range of values of his invariants. In this paper I prove one of them. It was stated as follows.

1.3.A. CONJECTURE. *The minimal values of $J^\pm$ on all generic curves with n double points is attained only on the curve A_{n+1} of Figure* 1.5:

$$J^+ \geqslant -n^2 - n, \quad J^- \geqslant -n^2 - 2n.$$

§2. Real algebraic variations on the theme of $J^\pm$

2.1. The curves under consideration. The closest real algebraic counterparts of immersions $S^1 \to \mathbb{R}^2$ are real plane projective rational curves with an infinite set of real points. However, if one has in mind only $J^\pm$, it is not difficult to consider an essentially wider situation. (For a counterpart of St, see my preprint [9], where it is defined only for real plane projective rational curves with an infinite set of real points.)

Namely, consider irreducible plane projective curves of degree m and genus g. To distinguish direct and inverse self-tangencies, one needs an orientation. Especially if curves have several connected components, which may happen when $g > 1$. A natural orientation of the set of real points of an algebraic curve appears if the set of real points is zero homologous in the complexification. Curves with this property are called *curves of type* I (for example, see [5]).

If a curve of type I is irreducible, the real part of its normalization divides the set of complex points of the normalization into two halves. The images of the halves of the normalization in the set of points of the original curve may intersect each other. However, I shall call these images the *halves* of the curve.

Each of the halves of the normalization is oriented (as a piece of a complex curve) and induces an orientation on the real part as on its boundary. These two orientations are opposite to each other. They are called *complex orientations* of the real curve.

We shall consider irreducible plane projective curves of degree m, genus g, and type I with a distinguished complex orientation. The latter means that we shall consider curves with a selected half of their complexification.

Curves of this kind constitute a finite-dimensional stratified real algebraic variety. A curve all of whose singular points are ordinary double points will be called a *generic curve*. As is well known, generic curves constitute a Zariski open set in the space of all curves described above.

2.2. Singularities of a generic curve. A generic curve has only ordinary double singularities. These singularities are equivalent from the viewpoint of complex algebraic geometry. Real algebraic geometry distinguishes several types of such points.

First, a singular point may be real or imaginary.

Second, a real double point may belong to two real branches or to two imaginary branches conjugate to each other. I shall call a real ordinary double point a *crossing* if it is of the former type, and a *solitary double point* if it is of the latter type.

Third, an imaginary double point may be a self-intersection point of one of the halves, or an intersection point of different halves. Denote the number of points of the former type by τ and the number of points of the latter type by σ. (Certainly, both τ and σ are even.)

For a solitary ordinary double point, the choice of a half of the complexification determines a local orientation of $\mathbb{R}P^2$. It can be defined as the local orientation such that the imaginary branch of the curve belonging to the chosen half intersects $\mathbb{R}P^2$ (equipped with this local orientation) at this point with intersection number $+1$.

(Another, equivalent, definition is: one can perturb the curve keeping it of type I and converting the solitary point into an oval. The oval acquires the complex orientation. The latter induces an orientation of the disk bounded by the oval. This orientation coincides with the local orientation of $\mathbb{R}P^2$ above. To prove their coincidence, it suffices to consider a model example. Say, a conic $x^2 + y^2 = 0$ and its perturbation $x^2 + y^2 = \varepsilon$ with $\varepsilon > 0$.)

2.3. The Rokhlin formula. Curves of type I satisfy Rokhlin's complex orientation formula. For the sake of simplicity, I formulate it below only for generic curves. I preface it with several definitions.

For a generic curve A of type I, by a *smoothing* $\widetilde{\mathbb{R}A}$ of its real part $\mathbb{R}A$ we shall understand a smooth oriented 1-dimensional submanifold of $\mathbb{R}P^2$ obtained from $\mathbb{R}A$ by the modification at each real double point determined by the complex orientation as shown in Figure 2.1.

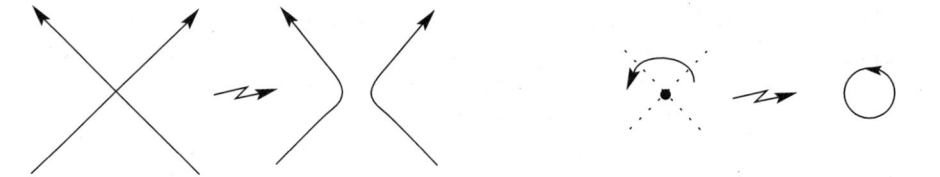

FIGURE 2.1. Smoothing of ordinary double points determined by an orientation; a solitary double point is shown on the right-hand side in an old-fashioned way (complex conjugate imaginary branches are indicated by dashed lines); at this point the local orientation of the real plane is shown by a circle arrow.

For an oriented closed 1-dimensional submanifold C of $\mathbb{R}P^2$ and a point $x \in \mathbb{R}P^2 \setminus C$, there is the index $\operatorname{ind}_C(x)$ of the point with respect to the curve. It is a nonnegative number defined as follows. Draw a line L on $\mathbb{R}P^2$ through x transversal to C. Equip it with a normal vector field vanishing only at x. For such a vector field one may take the velocity field of a rotation of the line around x. At each intersection point of L and C there are two directions transversal to L: the direction of the vector belonging to the normal vector field and the direction defined by the local orientation of C at the point. Denote the number of intersection points where both directions point to the same side of L by i_+ and the number of intersection points where they point to opposite sides of L by i_-. Then set[3] $\operatorname{ind}_C(x) = |i_+ - i_-|/2$. It is easy to check that ind_C is well defined: it depends neither on the choice of L nor on the choice of the normal vector field.

The second prerequisite notion is a sort of unusual integration: integration with respect to the Euler characteristic, in which the Euler characteristic plays the role of measure. It is well known that the Euler characteristic shares an important property of measures: it is additive in the sense that for any sets A, B such that the Euler characteristics $\chi(A)$, $\chi(B)$, $\chi(A \cap B)$, and $\chi(A \cup B)$ are defined,

$$\chi(A \cup B) = \chi(A) + \chi(B) - \chi(A \cap B).$$

However, the Euler characteristic is neither σ-additive nor positive. Thus the usual theory of integrals cannot be developed for it. This can be done, however, if one restricts to a very narrow class of functions. Namely, to functions that are finite linear combinations of characteristic functions of sets belonging to some algebra of subsets of a topological space such that each element of the algebra has a well-defined Euler characteristic. For a function $f = \sum_{i=1}^r \lambda_i \mathbf{1}_{S_i}$, set

$$\int f(x)\, d\chi(x) = \sum_{i=1}^r \lambda_i \chi(S_i).$$

For details and other applications of this notion see [8].

---

[3] Division by 2 appears here to ensure that this notion generalizes the corresponding well-known notion for an affine plane curve. In the definition for the affine situation, one uses a ray instead of an entire line. In the projective situation, there is no natural way to divide a line into rays, but we still have the opportunity of dividing the result by 2. Another distinction from the affine situation is that there the index may be negative. This is related to the fact that the affine plane is orientable, while the projective plane is not.

2.3.A. *Let A be a generic real plane projective algebraic curve of degree m and type I. Then*

$$\frac{m^2}{4} = \sigma + \int_{\mathbb{R}P^2 \setminus \widetilde{\mathbb{R}A}} (\mathrm{ind}_{\widetilde{\mathbb{R}A}}(x))^2 \, d\chi(x),$$

where σ is (as above) the number of imaginary double points of A at which different halves of the complexification meet each other.

This theorem has a rather long history. Its first special case was discovered by Rokhlin [4] in 1974. Then it was stated only for nonsingular plane projective curves of even degree m with a maximal number of ovals (equal to $(m-1)(m-2)/2 + 1$). The case of nonsingular curves of odd degrees with a maximal number of ovals was done by Mishachev [3]. For nonsingular curves of type I, it was stated by Rokhlin [5]. In terms of the integral against the Euler characteristic, the Rokhlin formula of [5] was rewritten in [8] (implicitly this was done earlier by Sharpe [6]). Double points and σ appeared in [7] and the most general formula in [10]. All versions of the formula are proved in the following way. Take a half of the complexification of the curve. Complete it with a chain contained in $\mathbb{R}P^2$ to a 2-cycle. Calculate the intersection number of this cycle with its image under the complex conjugation involution. This calculation may be done geometrically (putting the cycles in general position to each other and studying the intersection) and homologically (finding the homology classes of the cycles, which are in fact $m/2[\mathbb{C}P^1]$ and $-m/2[\mathbb{C}P^1]$). Comparison of the results gives rise to the formula.

Theorem 2.3.A provides a tool for understanding what happens with σ and τ when a curve experiences various perestroikas.

2.4. Perestroikas. The complement of the set of generic curves in the variety of all real plane projective algebraic curves of degree m, genus g, and type I is a sort of discriminant hypersurface. It contains six main strata. Each of them consists of curves having only one singular point which is not an ordinary double point. The type of that singular point defines the stratum. Here is the list:
1. real cusp;
2. real point of direct ordinary tangency;
3. real point of inverse ordinary tangency;
4. real point of ordinary tangency of two imaginary branches;
5. real ordinary triple point of intersection of three real branches;
6. real ordinary triple point of intersection of a real branch and two conjugate imaginary branches.

These singularities and the corresponding perestroikas are shown in Figures 2.2, 1.1, 1.2, 2.3, 1.3, and 2.4, respectively. The behavior of the local orientations at solitary double points under the cusp perestroika and solitary self-tangency perestroika shown in Figures 2.2 and 2.3 follows from Rokhlin's formula 2.3.A.

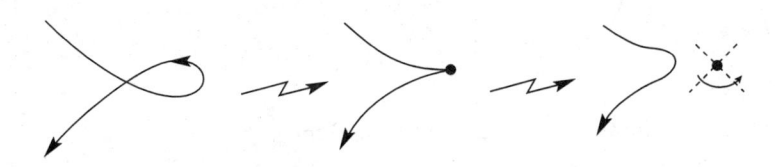

FIGURE 2.2. Cusp perestroika: σ and τ do not change.

FIGURE 2.3. In this picture everything but one point in the center and two points in the right-hand side is not real; this is a solitary self-tangency perestroika: two solitary double points come from the world of imaginary; at the moment of their arrival they show up as a single solitary self-tangency point. σ decreases by 2, while τ is constant.

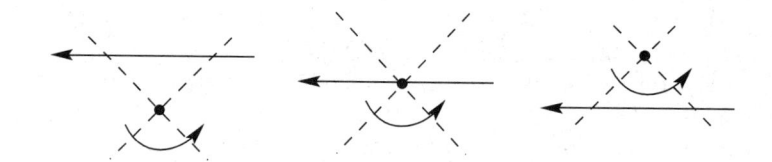

FIGURE 2.4. Perestroika of a triple point with imaginary branches: σ increases by 2, while τ decreases by 2.

In self-tangency perestroikas and in the perestroika of Figure 2.4 imaginary double points are involved. Theorem 2.3.A and the fact that $\sigma+\tau$ is the number of all imaginary double points and that the total number of double points (real and imaginary) is constant ($= (m-1)(m-2)/2 - g$) imply that σ and τ change under the six perestroikas as follows:

1. cusp perestroika (Figure 2.2): σ and τ do not change;
2. direct self-tangency perestroika (Figure 1.1): σ is constant, while τ decreases by 2;
3. inverse self-tangency perestroika (Figure 1.2): σ decreases by 2, while τ is constant;
4. solitary self-tangency perestroika (Figure 2.3): σ decreases by 2, while τ is constant;
5. triple point perestroika (Figure 1.3): σ and τ do not change;
6. triple point with imaginary branches perestroika (Figure 2.4): σ increases by 2, while τ decreases by 2.

This suggests that *σ is a counterpart of J^-, while $-\tau$ is a counterpart of J^+.*

§3. Back to immersed circles

3.1. Rokhlin-type formula. By 2.3.A, for generic real algebraic curves we have

$$\sigma = \frac{m^2}{4} - \int_{\mathbb{R}P^2 \setminus \widetilde{\mathbb{R}A}} (\mathrm{ind}_{\widetilde{\mathbb{R}A}}(x))^2 \, d\chi(x),$$

and $m^2/4$ does not change under perestroikas. Hence, the integral

$$-\int_{\mathbb{R}P^2 \setminus \widetilde{\mathbb{R}A}} (\mathrm{ind}_{\widetilde{\mathbb{R}A}}(x))^2 \, d\chi(x)$$

has the same behavior under direct and inverse self-tangency perestroikas and triple point perestroikas as σ and J^-.

This suggests one compare $J^-(C)$ with

$$-\int_{\mathbb{R}^2\setminus\widetilde{C}} (\operatorname{ind}_{\widetilde{C}}(x))^2 \, d\chi(x)$$

in Arnold's original situation: for a generic immersed circle C. Here $\widetilde{C}$ means smoothing of C defined exactly as in 2.3. The integral is defined in the same way, too.

3.1.A. *For any generic immersed circle C,*

$$J^-(C) = 1 - \int_{\mathbb{R}^2\setminus\widetilde{C}} (\operatorname{ind}_{\widetilde{C}}(x))^2 \, d\chi(x).$$

PROOF. It is easy to see that in this nonalgebraic situation

$$-\int_{\mathbb{R}^2\setminus\widetilde{C}} (\operatorname{ind}_{\widetilde{C}}(x))^2 \, d\chi(x)$$

changes under perestroikas of C as J^-. See Figure 3.1, where smoothings of the fragments involved into Arnold's perestroikas are shown.

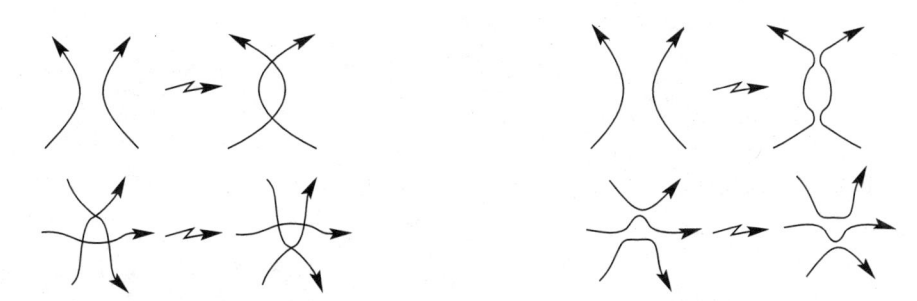

The two perestroikas above do not change smoothing

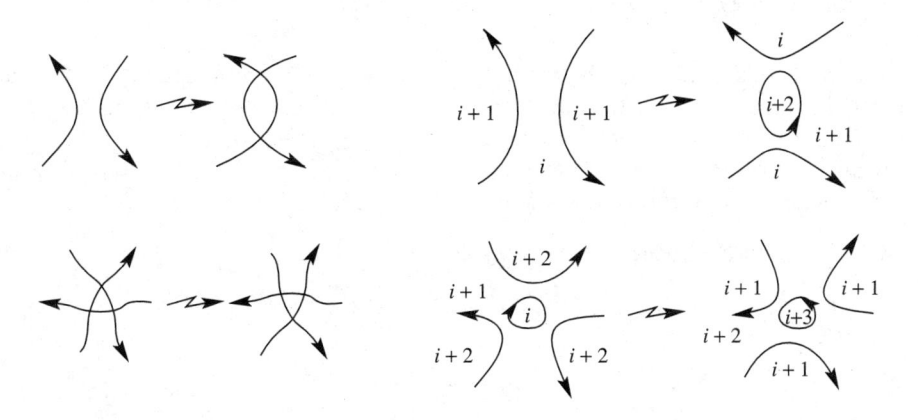

FIGURE 3.1. Smoothings of perestroikas.

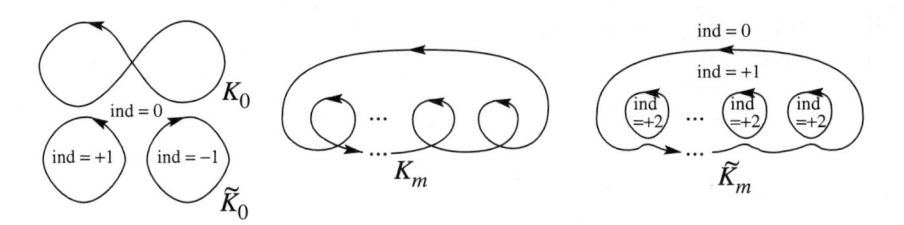

FIGURE 3.2. Calculation of $\int_{\mathbb{R}^2\setminus\widetilde{K}_m}(\operatorname{ind}_{\widetilde{K}_m}(x))^2\,d\chi(x)$; on the left-hand side one can see that $\int_{\mathbb{R}^2\setminus\widetilde{K}_0}(\operatorname{ind}_{\widetilde{K}_0}(x))^2\,d\chi(x) = 1+1 = 2$; on the right-hand side after smoothing there are $m-1$ interior ovals; therefore $\int_{\mathbb{R}^2\setminus\widetilde{K}_m}(\operatorname{ind}_{\widetilde{K}_m}(x))^2\,d\chi(x) = 1(1-(m-1))+4(m-1) = 3m-2$.

Furthermore,

$$\int_{\mathbb{R}P^2\setminus\widetilde{K}_m}(\operatorname{ind}_{\widetilde{K}_m}(x))^2\,d\chi(x)$$
$$= \begin{cases} 2, & \text{if } m=0, \\ (1-(m-1))+4(m-1) = 3m-2, & \text{if } m>0. \end{cases}$$

See Figure 3.2.

On the other hand, by 1.2.E

$$J^-(K_0) = -1, \qquad J^-(K_m) = -3m+3 \quad (m=1,2,\ldots).$$

Comparing, we obtain the desired result. □

3.1.B. COROLLARY. *For any generic immersed circle C with n double points,*

$$J^+(C) = 1 + n - \int_{\mathbb{R}^2\setminus\widetilde{C}}(\operatorname{ind}_{\widetilde{C}}(x))^2\,d\chi(x).$$

PROOF. Indeed, $J^+ = J^- + n$. □

3.2. A version of 3.1.A without $\widetilde{C}$. Let C be a generic immersed circle. On each connected component E of the complement $R^2\setminus C$, the function assigning to a point $x\in E$ its index with respect to C is constant. Denote its value by $\operatorname{ind}_C(E)$.

A double point V of C is adjacent to four angles of $\mathbb{R}^2\setminus C$. Denote by $\operatorname{ind}_C(V)$ the arithmetic mean of the values taken by ind_C on these four angles.

3.2.A. OBVIOUS LEMMA. *For a generic immersed circle C,*

$$\int_{\mathbb{R}^2\setminus\widetilde{C}}(\operatorname{ind}_{\widetilde{C}}(x))^2\,d\chi(x)$$
$$= \sum_{E \text{ a component of } \mathbb{R}^2\setminus C}(\operatorname{ind}_C(E))^2 - \sum_{V \text{ a double point of } C}(\operatorname{ind}_C(V))^2.$$

3.2.B. COROLLARY. *For a generic immersed circle, C*

$$J^-(C) = 1 - \sum_{E \text{ a component of } \mathbb{R}^2 \setminus C} (\text{ind}_C(E))^2$$
$$+ \sum_{V \text{ a double point of } C} (\text{ind}_C(V))^2,$$
$$J^+(C) = 1 - \sum_{E \text{ a component of } \mathbb{R}^2 \setminus C} (\text{ind}_C(E))^2$$
$$+ \sum_{V \text{ a double point of } C} (1 + (\text{ind}_C(V))^2).$$

3.3. A version of 3.1.A in terms of the mutual position of the components of $\widetilde{C}$. Two disjoint circles embedded into $\mathbb{R}^2$ constitute an *injective pair* if one of them is contained in a disk bounded by the other. An injective pair of oriented circles is said to be *positive* if the orientations of the circles are induced by an orientation of the annulus bounded by the circles. Otherwise it is said to be *negative*. See Figure 3.3.

positive injective pair   negative injective pair

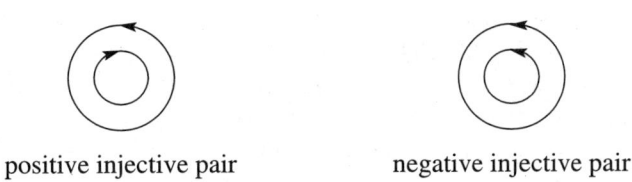

FIGURE 3.3. Positive and negative injective pairs of circles.

Given a generic immersed circle C, denote by l the number of components of its smoothing $\widetilde{C}$, by Π the number of injective pairs of components of $\widetilde{C}$, by Π^+ the number of positive injective pairs of components of $\widetilde{C}$, and by Π^- the number of negative injective pairs of components of $\widetilde{C}$.

3.3.A. EASY LEMMA.

$$\int_{\mathbb{R}^2 \setminus \widetilde{C}} (\text{ind}_{\widetilde{C}}(x))^2 \, d\chi(x) = l - 2\Pi^+ + 2\Pi^-. \quad \square$$

In fact this presentation of $\int_{\mathbb{R}^2 \setminus \widetilde{C}} (\text{ind}_{\widetilde{C}}(x))^2 \, d\chi(x)$ is due to Rokhlin [5]. He used it instead of the integral in the original expression of his formula.

3.3.B. COROLLARY. *For a generic immersed circle C,*

$$J^- = 1 - l + 2\Pi^+ - 2\Pi^-. \quad \square$$

3.4. Proof of Arnold's Conjecture 1.3.A. By 3.3.B

$$J^- = 1 - l + 2\Pi^+ - 2\Pi^-.$$

Since $\Pi = \Pi^+ + \Pi^-$, one has $2\Pi^+ - 2\Pi^- \geq -2\Pi$ and therefore

$$J^- \geq 1 - l - 2\Pi.$$

The number of injective pairs Π is not greater than the number of all pairs of components of $\widetilde{C}$, which is equal to $(l^2 - l)/2$. Therefore

$$J^- \geqslant 1 - l - l^2 + l = 1 - l^2.$$

On the other hand, obviously $l \leqslant n + 1$. Consequently,

$$J^- \geqslant 1 - (n+1)^2 = -n^2 - 2n.$$

If the equality $J^- = -n^2 - 2n$ holds, then any two components of $\widetilde{C}$ constitute an injective pair, this pair is negative, and the number of double points of C is equal to $l - 1$. These conditions imply that $C = A_{n+1}$.

Since $J^+ = J^- + n$, the statement about J^+ follows from the one about J^-. □

§4. Generalizations of $J^\pm$

Theorem 3.1.A can be used other than just as a tool for calculations. It gives a new proof of the existence of J^- (i.e., of the existence of a generic immersed curve characteristic satisfying 1.2.A and 1.2.C).

In this section the problem of generalizing $J^\pm$ to new situations is considered. The main object is a generic collection C of k circles immersed into a surface F. Here by genericity I mean basically the same as in the case of a single immersed circle: intersections and self-intersections are transversal and at each point of F there are at most two branches of C. By a generalization of J^+ (respectively, J^-) I mean a numerical characteristic of a generic collection which is invariant under regular homotopy in the class of generic collections of immersed circles and satisfies 1.2.B (respectively, 1.2.C).

The most interesting question about this concerns the existence of such a characteristic for collections of curves of a given regular homotopy class. It can be solved by studying the stratification of the space of such collections, as was done by Arnold [1] for single immersions of the circle to the plane. In this section another approach suggested by Theorem 3.1.A is used.

The question of uniqueness has an obvious solution: to a characteristic of a generic collection of immersed circles one can add any function of a regular homotopy class to get another characteristic satisfying the same conditions, and any characteristic satisfying these conditions can be obtained in this way.

It is sufficient to consider the problems only for J^-, since for a characteristic J^- satisfying the definition above for a generalization of J^-, the characteristic

$$J^-(C) + \text{the number of double points of } C$$

satisfies our definition for a generalization of J^+.

I make special efforts to construct generalizations of J^- with the most interesting additional properties. There are indications, e.g. Theorem 4.4.A, that sometimes this goal is achieved.

4.1. In the affine plane. For a collection C of k circles immersed to $\mathbb{R}^2$, set

$$J^-(C) = k - \int_{\mathbb{R}^2 \setminus \widetilde{C}} (\operatorname{ind}_{\widetilde{C}}(x))^2 \, d\chi(x).$$

It has the properties of J^- and is additive under both connected sum and disjoint sum.

4.2. In the projective plane.
The same approach can be applied to more general situations. For example, we can extend this definition to the class of generic collections of circles immersed to the projective plane. Given a generic collection C of k circles immersed to $\mathbb{R}P^2$, denote by $J^-(C)$ the number

$$1 - \int_{\mathbb{R}P^2 \setminus \widetilde{C}} (\operatorname{ind}_{\widetilde{C}}(x))^2 \, d\chi(x).$$

It is easy to see that $J^-(C)$ is invariant under regular homotopy in the class of generic immersions and 1.2.C holds for it. It is a generalization of Arnold's J^- in the sense that if C is a composition of an immersion C' of S^1 to $\mathbb{R}^2$ and an embedding $\mathbb{R}^2 \to \mathbb{R}P^2$, then $J^-(C) = J^-(C')$.

4.3. A straightforward generalization: zero-homologous curves in a surface with trivial 2-homology.
Generalizations of J^- given in Sections 4.1 and 4.2 are based on the notion of index of a point with respect to a curve. In both cases it has a simple homological meaning.

In the case of $\mathbb{R}^2$, a curve C is zero homologous and it bounds a chain. Since $H_2(\mathbb{R}^2) = 0$, the homology class of the chain in the relative homology $H_2(\mathbb{R}^2, C)$ is unique. The index of a point $x \in \mathbb{R}^2 \setminus C$ is the image of this class in the local homology group $H_2(\mathbb{R}^2, \mathbb{R}^2 \setminus x) = \mathbb{Z}$. It is really a well-defined integer, since the isomorphism $H_2(\mathbb{R}^2, \mathbb{R}^2 \setminus x) \to \mathbb{Z}$ is determined by the orientation of $\mathbb{R}^2$.

In the case of $\mathbb{R}P^2$, all these arguments are still valid except for the following two points: first, C may be nonhomologous to zero; second, the projective plane is not orientable, therefore the index is defined up to multiplication by -1. Fortunately, both obstructions are easy to overcome. To overcome the first one, it suffices to take rational homology instead of integer homology, or even homology with $\mathbb{Z}[\frac{1}{2}]$ coefficients. As for the second one, we need only the square of the index.

This approach is easy to apply to generic immersions of a collection of circles into a connected surface F with $H_2(F) = 0$ realizing the zero rational homology class.

Indeed, for a generic collection C of k circles immersed in a connected surface F and realizing the zero element of $H_1(F; \mathbb{Q})$, we construct a smoothing $\widetilde{C}$, then find a homology class ξ in $H_2(F, \widetilde{C}; \mathbb{Q})$ whose image under the boundary homomorphism $\partial : H_2(F, \widetilde{C}; \mathbb{Q}) \to H_1(\widetilde{C}; \mathbb{Q})$ is the fundamental class of $\widetilde{C}$. Since $H_2(F; \mathbb{Q}) = 0$, such a class ξ is unique. For each $x \in F \setminus \widetilde{C}$, we take the image of ξ under the relativization homomorphism $H_2(F, \widetilde{C}; \mathbb{Q}) \to H_2(F, F \setminus x; \mathbb{Q})$. Denote the absolute value of this image under an isomorphism $H_2(F, F \setminus x; \mathbb{Q}) \to \mathbb{Q}$ by $\operatorname{ind}_{\widetilde{C}}(x)$ and denote by $J^-(C)$ the number

$$k - \int_{F \setminus \widetilde{C}} (\operatorname{ind}_{\widetilde{C}}(x))^2 \, d\chi(x).$$

It is easy to see that $J^-(C)$ is invariant under regular homotopy in the class of generic collections of immersions and 1.2.C holds for it. The number $J^-(C)$ is a generalization of Arnold's J^- in the sense that if C is a composition of an immersion C' of S^1 to $\mathbb{R}^2$ and an embedding $\mathbb{R}^2 \to F$, then $J^-(C) = J^-(C')$.

4.4. In the sphere.
In the case of the sphere, although any curve C is zero homologous, the homology class in $H_2(S^2, C)$ of a chain bounded by C is not unique since one can add to such a class the fundamental class of S^2 multiplied by any integer.

Thus, contrary to the case of affine and projective planes, in the sphere, at first glance, there is no index of a point with respect to a 1-cycle. However, for a 1-cycle which is a collection of immersed circles, there is a nice replacement for the index.

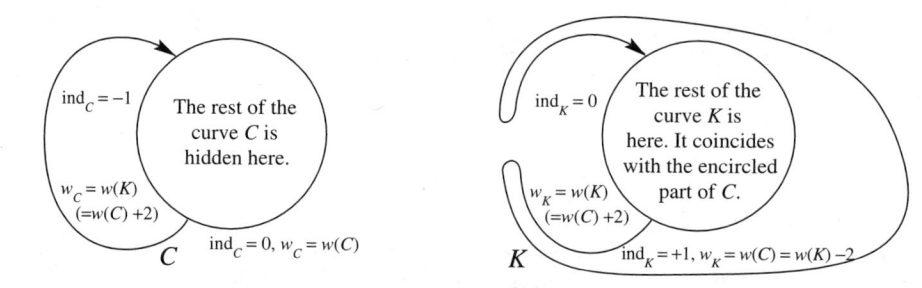

FIGURE 4.1. Behavior of $\operatorname{ind}_C(x)$ and $w_C(x)$ when x jumps over a branch of C; to visualize it, we move the piece of C which is jumped over around the sphere (through the point at infinity of the plane of picture); the winding number of the curve K obtained differs by 2 from the winding number of C.

Recall that for a collection C of circles immersed to $\mathbb{R}^2$ there is a well-defined number $w(C)$ called the *winding number*[4] and, in the case of a single immersed circle, this number defines the immersion up to regular isotopy.

Given a point $x \in S^2$ and a collection of immersed circles $C \subset S^2$ with $x \notin C$, set $w_C(x)$ to be the winding number of C in $S^2 \setminus x$. The local behavior of $w_C(x)$ and $\operatorname{ind}_C(x)$ are similar: when x jumps over a branch of C from the left-hand side to the right-hand side, $\operatorname{ind}_C(x)$ decreases by 1, whereas $w_C(x)$ increases by 2, see Figure 4.1.

Therefore
$$-\int_{S^2 \setminus \tilde{C}} \left(\frac{w_{\tilde{C}}(x)}{2}\right)^2 d\chi(x)$$
changes under self-tangency and triple point perestroikas exactly like J^-. This suggests the following definition.

For a generic collection C of k circles immersed to S^2, set

(4.1) $$J^-(C) = k - \frac{1}{4}\int_{S^2 \setminus \tilde{C}} (w_{\tilde{C}}(x))^2 \, d\chi(x).$$

Note that this is not a generalization of J^- of plane curves in the sense above. It is a counterpart rather than a generalization. It is related to J^- of plane curves, but the relation is a bit more complicated, namely:

4.4.A. *Let C be a collection of circles immersed in the plane, $i : \mathbb{R}^2 \to S^2$ be an embedding and C' be $i(C)$. Then $J^-(C') = J^-(C) + (w(C))^2/2$.*

Arnold [2] observed that $J^-(C) + (w(C))^2/2$ is an invariant of C'. He called $J^-(C) + (w(C))^2/2$ a *conformal invariant* of C (it is invariant under conformal transformations). Theorem 4.4.A means that

$$J^-(C) = k - \frac{1}{4}\int_{S^2 \setminus \tilde{C}} (w_{\tilde{C}}(x))^2 \, d\chi(x)$$

coincides with the conformal invariant.

To prove 4.4.A, we need a more explicit relation between ind_C and w_C.

---

[4]It is also called the Whitney index, it is the rotation number of the velocity vector as well as the degree of the Gauss map.

4.4.B. LEMMA 1. *Let C be a collection of circles immersed in the plane, $i: \mathbb{R}^2 \to S^2$ be an embedding and C' be $i(C)$. Then for any $x \in \mathbb{R}^2 \setminus C$,*

$$\operatorname{ind}_C(x) = \frac{1}{2}(w(C) - w_{C'}(i(x))).$$

PROOF. Observe that the formula is correct for x belonging to the outer component of $\mathbb{R}^2 \setminus C$ (where both sides are equal to zero) and that both sides change by the same quantity when x jumps over a branch of C. □

4.4.C. LEMMA 2. *Let K be a smooth closed oriented 1-manifold in S^2. Then*

$$\int_{S^2 \setminus K} w_K(x)\, d\chi(x) = 0.$$

PROOF. In the case of one circle this is obvious, since the complement $S^2 \setminus K$ consists of 2 open disks, on one of them w_K is equal to 1, while on the other it is -1. The general case follows, because $w_{A \cup B}(x) = w_A(x) + w_B(x)$ and the integral is additive. □

PROOF OF 4.4.A. By the definition given in Section 4.1,

$$J^-(C) = k - \int_{\mathbb{R}^2 \setminus \widetilde{C}} (\operatorname{ind}_{\widetilde{C}}(x))^2\, d\chi(x).$$

By Lemma 1, the right-hand side is equal to

$$k - \int_{\mathbb{R}^2 \setminus \widetilde{C}} \frac{1}{4}(w_{\widetilde{C}'}(i(x)))^2\, d\chi(x) + \int_{\mathbb{R}^2 \setminus \widetilde{C}} \frac{1}{2} w(C) w_{\widetilde{C}'}(i(x))\, d\chi(x) - \frac{1}{4}(w(C))^2.$$

Since for any function $\varphi: S^2 \to \mathbb{R}$ which is constant on $S^2 \setminus i(\mathbb{R}^2)$ we have

$$\int_{\mathbb{R}^2 \setminus \widetilde{C}} \varphi(x)\, d\chi(x) = \int_{S^2 \setminus \widetilde{C}'} \varphi(i(x))\, d\chi(x) - \varphi(S^2 \setminus i(\mathbb{R}^2)),$$

we can rewrite the expression of $J^-(C)$ obtained above as follows:

$$J^-(C) = k - \int_{S^2 \setminus \widetilde{C}'} \frac{1}{4}(w_{\widetilde{C}'}(i(x)))^2\, d\chi(x) + \frac{1}{4}(w(C))^2$$

$$+ \frac{1}{2}w(C) \int_{S^2 \setminus \widetilde{C}'} w_{\widetilde{C}'}(i(x))\, d\chi(x) - \frac{1}{2}(w(C))^2 - \frac{1}{4}(w(C))^2$$

$$= k - \int_{S^2 \setminus \widetilde{C}'} \frac{1}{4}(w_{\widetilde{C}'}(i(x)))^2\, d\chi(x)$$

$$+ \frac{1}{2}w(C) \int_{S^2 \setminus \widetilde{C}'} w_{\widetilde{C}'}(i(x))\, d\chi(x) - \frac{1}{2}(w(C))^2.$$

By Lemma 2, it is equal to

$$k - \int_{S^2 \setminus \widetilde{C}'} \frac{1}{4}(w_{\widetilde{C}'}(i(x)))^2\, d\chi(x) - \frac{1}{2}(w(C))^2.$$

By the definition of $J^-(C')$, it equals

$$J^-(C') - \frac{1}{2}(w(C))^2. \quad \square$$

Lemma 2 suggests a way to characterize $w_K(x)/2$ as a reasonable choice for the index without appealing to differential topology: consider all functions ind_K obtained as above from chains with boundary K and select one of them satisfying the identity

$$\int_{S^2\setminus K} \mathrm{ind}_K(x)\, d\chi(x) = 0.$$

This rule assumes that K is a collection of disjoint embedded circles. In the general case, one must either smooth the singularities, or change the integral by extending the function ind_K over K and taking the integral over the whole S^2.

The same natural chain in S^2 with a given boundary has been constructed by Arnold [2] in a different way.

Consider now any chain in S^2 with boundary K and the function ind_K related to this chain. From the natural chain related to $w_K(x)/2$, this one differs by $c[S^2]$ for some $c \in \mathbb{Z}$, therefore $\mathrm{ind}_K(x) = w_K(x)/2 + c$. The number c can be found as follows: integrating the latter relation

$$\int_{S^2\setminus K} \mathrm{ind}_K(x)\, d\chi(x) = \frac{1}{2}\int_{S^2\setminus K} w_K(x)\, d\chi(x) + c\chi(S^2\setminus K)$$

and taking into account Lemma 2, one obtains

$$c = \frac{1}{2}\int_{S^2\setminus K} \mathrm{ind}_K(x)\, d\chi(x).$$

Therefore

$$\frac{1}{2}w_K(x) = \mathrm{ind}_K(x) - \frac{1}{2}\int_{S^2\setminus K} \mathrm{ind}_K(u)\, d\chi(u).$$

The latter formula allows us to rewrite the definition (4.1) of $J^-(C)$ for a spherical curve in terms of an arbitrary chain with boundary $\widetilde{C}$:

(4.2)
$$J^-(C) = k - \int_{S^2\setminus \widetilde{C}} \left(\mathrm{ind}_{\widetilde{C}}(x) - \frac{1}{2}\int_{S^2\setminus \widetilde{C}} \mathrm{ind}_{\widetilde{C}}(u)\, d\chi(u) \right)^2 d\chi(u)$$
$$= k - \int_{S^2\setminus \widetilde{C}} (\mathrm{ind}_{\widetilde{C}}(x))^2\, d\chi(x) + \frac{1}{2}\left(\int_{S^2\setminus \widetilde{C}} \mathrm{ind}_{\widetilde{C}}(x)\, d\chi(x) \right)^2.$$

This formula has a real algebraic counterpart. Namely, for a generic real algebraic curve A of type I on the sphere S^2, where A is the intersection of S^2 and a surface of degree d, we have

$$\frac{d^2}{2} = \int_{S^2\setminus \widetilde{\mathbb{R}A}} (\mathrm{ind}_{\widetilde{\mathbb{R}A}}(x))^2\, d\chi(x) - \frac{1}{2}\left(\int_{S^2\setminus \widetilde{\mathbb{R}A}} \mathrm{ind}_{\widetilde{\mathbb{R}A}}(x)\, d\chi(x) \right)^2 + \sigma.$$

4.5. Zero-homologous curves in orientable closed surfaces with nonzero Euler characteristic. Consider now a generic collection of k immersed circles in an orientable closed connected surface F with $\chi(F) \neq 0$ realizing the zero of $H_1(F)$.

As before, construct a smoothing $\widetilde{C}$ of C, then consider homology classes $\xi \in H_2(F, \widetilde{C}; \mathbb{Q})$ whose image under the boundary homomorphism $\partial \colon H_2(F, \widetilde{C}; \mathbb{Q}) \to H_1(\widetilde{C}; \mathbb{Q})$ is the fundamental class of $\widetilde{C}$. Any two ξ's of this sort can be obtained one from another one by adding the image of $q[F]$ with $q \in \mathbb{Q}$ under the relativization homomorphism $\rho \colon H_2(F; \mathbb{Q}) \to H_2(F, \widetilde{C}; \mathbb{Q})$. For each $x \in F \setminus \widetilde{C}$, take the image of

ξ under the relativization homomorphism $H_2(F, \widetilde{C}; \mathbb{Q}) \to H_2(F, F \setminus x; \mathbb{Q})$. Denote the absolute value of this image under the isomorphism $H_2(F, F \setminus x; \mathbb{Q}) \to \mathbb{Q}$ by $\mathrm{ind}_\xi(x)$. Consider

$$\int_{F \setminus \widetilde{C}} \mathrm{ind}_\xi(x) \, d\chi(x).$$

This number is rational and if we replace ξ by $\xi + p(q[F])$, it changes by $q\chi(F)$. Since by assumption $\chi(F) \neq 0$, there exists a unique ξ_0 such that

$$\int_{F \setminus \widetilde{C}} \mathrm{ind}_{\xi_0}(x) \, d\chi(x) = 0.$$

Pick this ξ_0 and denote by $J^-(C)$ the number

$$k - \int_{F \setminus \widetilde{C}} (\mathrm{ind}_{\xi_0}(x))^2 \, d\chi(x).$$

This generalizes the construction of the previous subsection. It is easy to prove that this is invariant under regular isotopy in the class of generic collections of immersed circles and satisfies 1.2.C.

4.6. Curves in the torus and collections of contractible curves. Since the Euler characteristic of torus is zero, it is impossible to apply the construction of the previous section to a generic collection of circles immersed in a torus. This "unfortunate" property of torus can be partially compensated by the commutativity of its fundamental group. Partially means that it allows us to deal with generic collections of immersed circles in which each circle is zero-homologous, but not with generic collections with the sum of the classes realized by all the immersed circles equal to zero.

Consider a generic collection C of circles $C_1, \ldots, C_k$ immersed into torus T such that each C_i realizes $0 \in H_1(T)$. Since the fundamental group is commutative, each C_i is homotopic to zero. Therefore C_i can be presented as the composition of a generic immersion C_i^* of the circle into the plane $\mathbb{R}^2$ and the universal covering $\pi \colon \mathbb{R}^2 \to T$.

Take the function

$$x \mapsto \mathrm{ind}_{\bigcup_{i=1}^k C_i^*}(x) \quad \text{on } \mathbb{R}^2 \setminus \bigcup_{i=1}^k C_i^*.$$

Since for any $x \in T \setminus \bigcup_{i=1}^k C_i$ this function does not vanish only at a finite number of points of $\pi^{-1}(x)$, there exists the direct image

$$T \setminus C \to \mathbb{Z} \colon x \mapsto \sum_{y \in \pi^{-1}(x)} \mathrm{ind}_{\bigcup_{i=1}^k C_i^*}(y).$$

Denote the latter function by ind_C. It is easy to see that ind_C does not depend on the choice of the liftings C_i^*.

For a smoothing $\widetilde{C}$ of C, there is a unique locally constant function on $T \setminus \widetilde{C}$ coinciding with ind_C outside the neighborhoods of double points of C where the smoothing takes place. Denote this function by $\mathrm{ind}_{\widetilde{C}, C}$. It actually depends not only on $\widetilde{C}$, but on C.

Set

$$J^-(C) = k - \int_{T \setminus \widetilde{C}} (\mathrm{ind}_{\widetilde{C}, C}(x))^2 \, d\chi(x).$$

This number is invariant under regular isotopy in the class of generic collections of immersed circles and satisfies 1.2.C.

The same construction can be applied to a generic collection of circles immersed in any surface (not only the torus) such that each of the circles is contractible in the surface. If the surface is orientable, but not T, then one can apply the construction of the previous section. The results differ.

§5. Some high degree invariants of plane generic curves

5.1. Momenta of the index. From formula (4.2) and Theorem 4.4.A it follows that for a generic plane curve C, we have

$$w(C) = \int_{\mathbb{R}^2 \setminus \tilde{C}} \operatorname{ind}_{\tilde{C}}(x)\, d\chi(x).$$

This formula is easy to prove *ab ovo*, too. It suggests that one should consider all "momenta" $\int_{\mathbb{R}^2 \setminus \tilde{C}} (\operatorname{ind}_{\tilde{C}}(x))^r \, d\chi(x)$ of $\operatorname{ind}_{\tilde{C}}$.

Given a generic plane curve C, denote by $M_r(C)$ the integral

$$\int_{\mathbb{R}^2 \setminus \tilde{C}} (\operatorname{ind}_{\tilde{C}}(x))^r \, d\chi(x).$$

By (4.2) $M_1(C) = w(C)$. This number is invariant under regular homotopy. Thus it can be regarded as an invariant of degree 0.

It follows from the definition of 4.1 that $M_2(C) = k - J^-(C)$, where k is the number of components of the immersed curve. Thus M_2 is an invariant of degree 1. This suggests that M_r may be an invariant of degree $r - 1$. Below this is proved to be the case.

Following the general scheme of definition of finite degree invariants, for any characteristic of generic immersions of the circle in the plane which is locally constant on the space of generic immersions, one defines its *first derivative*. It is a characteristic of immersions having only one double point which is not ordinary and this point is either an ordinary triple or ordinary self-tangency point. On such a curve, the first derivative of the original characteristic is defined to be the difference between the values of the original characteristic on the adjacent generic immersions. In other words, it is the jump of the original characteristic happening at the corresponding perestroika. Of course, to define it, one has to specify a direction of the perestroika (a coorientation of the stratum of the discriminant hypersurface). In the case of self-tangency there is a natural direction: from curves with less double points to curves with more double points. In the case of triple points a coorientation was defined by Arnold [1], see 1.1 above. However, we shall need another local coorientation.

5.1.A. *For any r, a direct self-tangency perestroika of C does not change $M_r(C)$.*

This is obvious: $\tilde{C}$ does not change under a direct self-tangency perestroika; see Figure 4.1. □

Theorem 5.1.A says that the first derivative of any M_r vanishes on curves with a direct self-tangency point.

Studying M_r, one must distinguish two kinds of the triple point perestroikas. At the moment of perestroika, take vectors at the triple point tangent to the branches and directed according to their orientations. If one of the vectors can be presented as a

FIGURE 5.1. Weak (on the left) and strong (on the right) triple point perestroikas.

FIGURE 5.2. Inverse self-tangency point with index i of the two types.

linear combination of the other two vectors with positive coefficients, the perestroika is said to be *weak*, otherwise it is said to be *strong*. See Figure 5.1.

5.1.B. *For any r, a weak triple point perestroika of C does not change $M_r(C)$.*

This holds for the same reason as 5.1.A: a weak triple point perestroika does not change $\widetilde{C}$. See Figure 4.1. □

Theorem 5.1.B says that the first derivative of any M_r vanishes on curves with a weak triple point.

$M_r(C)$ changes under an inverse self-tangency perestroika if $r > 1$, and under a strong triple point perestroika if $r > 2$. To describe the change, let me introduce the following notion.

For a multiple point p of a circle C immersed to the plane, let the *index* of p be the minimal number i such that there exists a small perturbation C' of C and a point p' in $\mathbb{R}^2 \setminus C'$ arbitrarily close to p and having index i with respect to C'.

For example, the index of an inverse self-tangency point is equal to the index of the narrow adjacent domains if the latter is smaller than the index of the adjacent wide domains. Otherwise it is smaller by 2 than the index of the adjacent narrow domains. See Figure 5.2.

5.1.C. *For any r an inverse self-tangency perestroika of C changes $M_r(C)$ by $(i+2)^r - 2(i+1)^r + i^r$, where i is the index of the self-tangency point.*

PROOF. The corresponding perestroika of $\widetilde{C}$ replaces a vanishing arc by two arcs and the disk bounded by the vanishing newborn oval. See Figure 5.3. This means that there is a homeomorphism mapping the complement of the arc onto the complement of the two arcs and the disk and mapping $\widetilde{C}$ before the perestroika to $\widetilde{C}$ after it. This homeomorphism preserves the index. Thus the difference between the integrals is the integral over the newborn disk and two arcs minus the integral over the vanishing arc. It is easy to see that it equals $(i+2)^r - 2(i+1)^r + i^r$. □

In the case of a strong triple point, its index is i if the adjacent domains have indices $i+1$ and $i+2$, because then there is a perturbation with a vanishing triangle

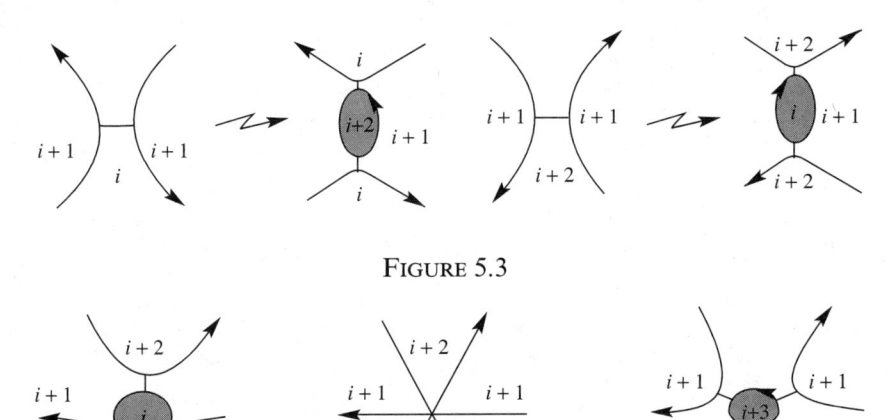

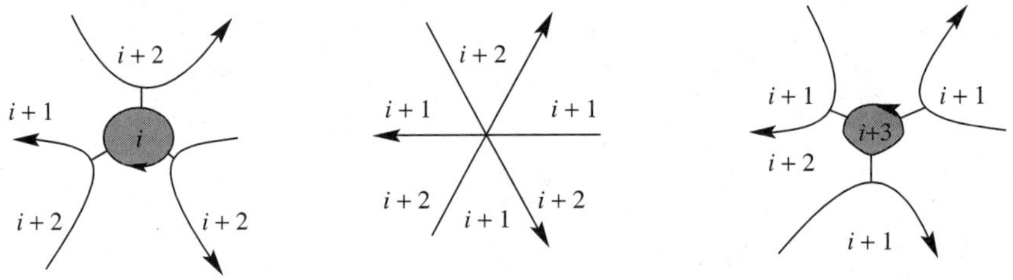

FIGURE 5.3

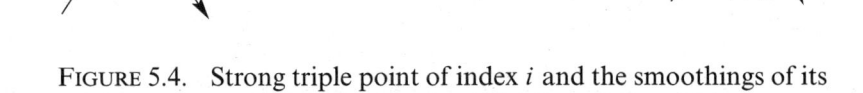

FIGURE 5.4. Strong triple point of index i and the smoothings of its perturbations; the vanishing arcs and disks are shown thicker.

born in place of the point with index i. There are only two topologically distinct perturbations. In the other perturbation, the vanishing triangle has index $i + 3$. See Figure 5.4.

By the *positive* direction of a strong triple point perestroika, we mean the direction in which points from the area inside the newborn triangle have index $i + 3$, where i is the index of the triple point.

5.1.D. *For any r, a strong triple point perestroika of C changes $M_r(C)$ by $(i+3)^r - 3(i+2)^r + 3(i+1)^r - i^r$, where i is the index of the triple point.*

The proof is similar to that of 5.1.C above. □

On the set of plane curves with a single nonordinary multiple point, the index of this point is an invariant of degree 1: a generic perestroika changes it by a constant depending only on the local structure of the perestroika. Therefore polynomial functions of the index of this point are invariants of finite degree. Thus the theorems of this section imply that $M_r(C)$ are invariants of C of finite degree.

5.2. A polynomial of immersions. For a generic plane curve C consider the formal power series

$$(5.1) \qquad P_C(h) = \sum_{r=0}^{\infty} \frac{M_r(C) h^r}{r!}.$$

It allows us to reformulate the results of the preceding section in the following concise form.

5.2.A. *Direct self-tangency and weak triple point perestroikas of C do not change P_C. An inverse self-tangency perestroika at a point of index i adds $e^{ih}(e^h - 1)^2$ to $P_C(h)$. A strong triple point perestroika at a point of index i adds $e^{ih}(e^h - 1)^3$ to $P_C(h)$.* □

There is a more compact formula for $P_C(h)$:

5.2.B. *For a generic plane curve C, we have*

$$P_C(h) = \int_{\mathbb{R}^2 \setminus \widetilde{C}} e^{\operatorname{ind} \widetilde{c}(x) h} \, d\chi(x). \tag{5.2}$$

PROOF. Indeed,

$$\begin{aligned}
\int_{\mathbb{R}^2 \setminus \widetilde{C}} e^{\operatorname{ind} \widetilde{c}(x) h} \, d\chi(x) &= \int_{\mathbb{R}^2 \setminus \widetilde{C}} \sum_{r=0}^{\infty} \frac{(\operatorname{ind} \widetilde{c}(x))^r h^r}{r!} \, d\chi(x) \\
&= \sum_{r=0}^{\infty} \frac{h^r}{r!} \int_{\mathbb{R}^2 \setminus \widetilde{C}} (\operatorname{ind} \widetilde{c}(x))^r \, d\chi(x) \\
&= \sum_{r=0}^{\infty} \frac{h^r}{r!} M_r(C) = P_C(h). \quad \square
\end{aligned} \tag{5.3}$$

Proposition 5.2.B suggests that we introduce the variable $q = e^h$. The power series $P_C(h)$ turns into a Laurent polynomial $P_C(q)$ in q defined by

$$\int_{\mathbb{R}^2 \setminus \widetilde{C}} q^{\operatorname{ind} \widetilde{c}(x)} \, d\chi(x). \tag{5.4}$$

The coefficients of $P_C(q)$ have a simple geometric meaning: if

$$P_C(q) = \sum_{r=-\infty}^{\infty} p_r q^r,$$

then p_r is equal to the Euler characteristic of the subset of $\mathbb{R}^2 \setminus \widetilde{C}$ in which $\operatorname{ind}_{\widetilde{c}}(x) = r$. The changes of P_C under perestroikas also look simpler: an inverse self-tangency perestroika at a point of index i adds $q^i(q-1)^2$ and a strong triple point perestroika at a point of index i adds $q^i(q-1)^3$.

5.3. Analogy with knot polynomial invariants. Thus the polynomial $P_C(q)$ is very similar to the quantum knot polynomial invariants like the Jones polynomial. Replacing q by the exponent e^h we obtain a power series in h whose coefficients are invariants of finite degree. The behavior of $P_C(q)$ under perestroikas of C is similar to the skein relations. It allows us to calculate $P_C(q)$ inductively, using any regular homotopy connecting C with an immersion whose polynomial is known. Formula (5.4) can be viewed as an analog of the face state sum formulas for knot quantum polynomials.

References

1. V. I. Arnold, *Plane curves, their invariants, perestroikas and classifications*, Singularities and Bifurcations (V. I. Arnold, ed.), Adv. Sov. Math., vol. 21, 1994, pp. 33–91.
2. _____, *Conformally invariant version of J^+ and the Bennequin invariant on S^2*, Letter 3–1994 (March 3, 1994).
3. N. M. Mishachev, *Complex orientations of plane M-curves of odd degree*, Funktsional. Anal. i Prilozhen. **9** (1975), no. 4, 77–78; English transl. in Functional Anal. Appl. **9** (1975), no. 4.
4. V. A. Rokhlin, *Complex orientations of real algebraic curves*, Funktsional. Anal. i Prilozhen. **8** (1974), no. 4, 71–75; English transl. in Functional Anal. Appl. **8** (1974), no. 4.
5. _____, *Complex topological characteristics of real algebraic curves*, Uspekhi Mat. Nauk **33** (1978), no. 5, 77–89; English transl., Russian Math. Surveys **33** (1978), no. 5, 85–98.

6. R. W. Sharpe, *On the ovals of even degree plane curves*, Michigan Math. J. **22** (1976), no. 3, 285–288.
7. O. Ya. Viro, *Progress in the topology of real algebraic varieties over the last six years*, Uspekhi Mat. Nauk **41** (1986), no. 3, 45–67; English transl., Russian Math. Surveys **41** (1986), no. 3, 55–82.
8. _____, *Some integral calculus based on Euler characteristic*, Lecture Notes in Math., vol. 1346, Springer-Verlag, Heidelberg and Berlin, 1988, pp. 127–138.
9. _____, *First degree invariants of generic curves on surfaces*, Preprint Uppsala Univ., U.U.D.M. Report 1994:21.
10. V. I. Zvonilov, *Complex orientations of real algebraic curves with singularities*, Dokl. Akad. Nauk SSSR **268** (1983), no. 1, 22-26; English transl., Soviet Math. Dokl. **27** (1983), 14–17.

Translated by THE AUTHOR

DEPARTMENT OF MATHEMATICS, UPPSALA UNIVERSITY, BOX 480, S-751 06 UPPSALA, SWEDEN
E-mail address: Viro@mat.uu.se

DEPARTMENT OF MATHEMATICS, UNIVERSITY OF CALIFORNIA, RIVERSIDE, CALIFORNIA 92521
E-mail address: Viro@math.ucr.edu

POMI, FONTANKA 27, ST. PETERSBURG, 191011, RUSSIA

Stratified Spaces of Real Algebraic Curves of Bidegree $(m, 1)$ and $(m, 2)$ on a Hyperboloid

V. I. Zvonilov

ABSTRACT. The rigid isotopies of real algebraic curves on a hyperboloid, i.e., the isotopies in the class of algebraic curves with a given collection of singularities, are studied. We obtain the rigid isotopy classification of curves of bidegree $(m, 1)$ and of nonsingular curves of bidegree $(m, 2)$. Sufficient conditions for singular curves of bidegree $(m, 2)$ to be rigidly isotopic are found.

Introduction

Real algebraic curves on a hyperboloid are classical objects of investigation. The problem of real isotopic classification of such curves was studied at Gudkov's school in Nizhniĭ Novgorod [1–3] and by the author [4].

On the other hand, papers devoted to the study of the space of plane curves of given degree, both in the real and in the complex case, have appeared [5, 6]. In these papers the curves have been classified up to rigid isotopies. For nonsingular real algebraic curves, the notion of rigid isotopy was introduced by Rokhlin [8]. For complex curves with singularities, the definition of rigid isotopy was given by Degtyarev [7]. The present paper lies in the course of these investigations and is devoted to the study of rigid isotopies of real algebraic curves on a hyperboloid. For the real case, we modify the definition of rigid isotopy given in [7]. This definition is adjusted to the topological classification of singularities obtained from the smooth classification of [9] by ignoring the moduli (they do not appear in the situation that we are studying).

The lemma in 3.3, which is similar to the one of [7, §2.2], is the technical basis of the investigation.

§1. General definitions

In this paper *hyperboloid* means the product of two real projective lines. The complex part of the hyperboloid is $\mathbb{C}P^1 \times \mathbb{C}P^1$, the real one is $\mathbb{R}P^1 \times \mathbb{R}P^1$. The involution of complex conjugation in $\mathbb{C}P^1 \times \mathbb{C}P^1$ is denoted by conj. Let x_0, x_1 and y_0, y_1 be the homogeneous coordinates on the factors of the product. A *curve of bidegree* (m, n) is a polynomial in x_0, x_1, y_0, y_1 considered up to a constant factor

---

1991 *Mathematics Subject Classification.* Primary 14P25, 14H99.
Key words and phrases. Real algebraic curves on a hyperboloid, rigid isotopies of the curves.

©1996, American Mathematical Society

and homogeneous both in x_0, x_1 and y_0, y_1 of degrees m and n respectively. The complex (real) part of a real curve F is denoted by $\mathbb{C}F$ ($\mathbb{R}F$). Curves of bidegree $(0, 1)$ and $(1, 0)$ are called *horizontal* and *vertical lines* respectively. The two-sheeted covering $S^1 \to \mathbb{R}P^1$ takes the standard orientation of S^1 to an orientation of $\mathbb{R}P^1$, so all the horizontal and vertical lines are provided with orientations. The classes in $H_1(\mathbb{R}P^1 \times \mathbb{R}P^1)$ realized by the real lines of bidegree $(0, 1)$ and $(1, 0)$ are denoted by e_1 and e_2 respectively.

The set of real curves of bidegree (m, n) is a real projective space of dimension $mn + m + n$, denoted by $C_{m,n}$.

A *rigid isotopy* of curves $F_0, F_1 \in C_{m,n}$ is a pair consisting of a path F_t in $C_{m,n}$ joining F_0 with F_1 and a 1-parameter family of homeomorphisms $\varphi_t \colon \mathbb{C}P^1 \times \mathbb{C}P^1 \to \mathbb{C}P^1 \times \mathbb{C}P^1$, commuting with conj, such that $\varphi_0 = \text{id}$, $\varphi_t(\mathbb{C}F_0) = \mathbb{C}F_t$. For example, a pair $\pi = (\alpha, \beta)$ of projective transformations of coordinates x_0, x_1 and y_0, y_1 determines, obviously, a rigid isotopy joining the curves F and $\pi(F)$ if α and β preserve the orientation of $\mathbb{R}P^1$.

It is clear that rigidly isotopic curves have (topologically) the same collections of singularities and that rigid isotopy is an equivalence relation in $C_{m,n}$. The equivalence classes are called *strata* of $C_{m,n}$.

§2. Curves of bidegree $(m, 1)$

Let F be a real curve of bidegree $(m, 1)$ given by the polynomial

$$w(x_0, x_1)(a(x_0, x_1)y_0 + b(x_0, x_1)y_1),$$

where a, b are coprime. It is clear that the roots of w determine the components of F that are the vertical lines. The curve $\widetilde{F}$ generated by $ay_0 + by_1$ is the graph of the rational function $y_1/y_0 = -a/b$ on $\mathbb{C}P^1$. So $\widetilde{F}$ is nonsingular (hence every singular curve of bidegree $(m, 1)$ is reducible) and $\mathbb{R}\widetilde{F}$ is connected. Let $e_1 + \mu e_2$ be the class in $H_1(\mathbb{R}P^1 \times \mathbb{R}P^1)$ determined by $\mathbb{R}\widetilde{F}$ endowed with some orientation.

First, we show that the *rigid isotopy class of $\widetilde{F}$ is given by μ*. We take advantage of the fact that the rational function $y_1/y_0 = -a/b$ is determined by the collection of roots of a, b and by the value at a point different from these roots. If all the roots of a, b are simple and the real roots of these polynomials alternate, then the number of real roots of each of a, b is equal to $|\mu|$ and the sign of μ is given by the value of the above-mentioned function at the chosen point. It is clear that in this case the rigid isotopy class of $\widetilde{F}$ is determined by μ. The general case is reduced to this special one in the following way. By a slight change of the coefficients of a, b, we make the roots of these polynomials simple. Then moving two adjacent real roots of a (of b) that are not separated by the roots of b (of a), we merge them into one double root and divide it into two conjugate imaginary roots.

Let $\Delta_{k,1}$, or simply Δ, denote the set of all singular (i.e., reduced) curves of bidegree $(k, 1)$ and Str $C_{k,l}$ the set of strata of $C_{k,l}$.

We consider the inclusion $C_{r,0} \times (C_{s,1} \setminus \Delta) \to C_{r+s,1}$ determined by the multiplication of the polynomials defining the curves. It is clear that the map Str $C_{r,0} \times$ Str $(C_{s,1} \setminus \Delta) \to$ Str $C_{r+s,1}$ induced by the inclusion is injective. Obviously, the rigid isotopy class of a curve in $C_{r,0}$ is given by the collection of multiplicities of real and imaginary components into which the curve splits. So we have completed the proof of the following theorem.

THEOREM. *The rigid isotopy class of F is given by the collection of multiplicities of real and imaginary roots of w (i.e., the multiplicities of the corresponding components of F) and by μ. In particular, if F is nonsingular, then its rigid isotopy class is determined by its real scheme, i.e., by the diagram of the disposition of $\mathbb{R}F$ in $\mathbb{R}P^1 \times \mathbb{R}P^1$.*

Let r be the number of distinct roots of w and $(s, 1)$ be the bidegree of $\widetilde{F}$. If we fix the multiplicities of the roots of w, then this polynomial is determined, up to a constant factor, by the collection of its distinct roots. Besides, the set of the curves $\widetilde{F}$ with μ fixed is open in $C_{s,1}$. So *the dimension of the stratum containing F is equal to $r + 2s + 1$.*

§3. Curves of bidegree $(m, 2)$

3.1. The discriminant. Let F be the real curve of bidegree $(m, 2)$ defined by the polynomial

(1) $\qquad f = w(x_0, x_1)(a(x_0, x_1) y_0^2 + b(x_0, x_1) y_0 y_1 + c(x_0, x_1) y_1^2),$

where a, b, c are coprime polynomials of degree n and w is a nonzero polynomial of degree $m - n$. Let $\widetilde{F}$ denote the curve determined by the polynomial $\tilde{f} = f/w$ and let d be the discriminant $b^2 - 4ac$ of $\widetilde{F}$ (so that $w^2 d$ is the one of F). It is not difficult to verify that a point $(x, y) \in \mathbb{C}F$ is singular if and only if x is a multiple root of $w^2 d$. It is clear that the roots of w give the components of F that are vertical lines.

The curves with zero discriminant are studied below in 3.7.2. So in 3.1–3.6 we assume that d is nonzero.

Since the discriminant of a curve is defined up to a positive constant factor, we require that the sum of squares of the coefficients of the discriminant be equal to 1. Such polynomials form the unit sphere S_{2n} in the space of real homogeneous polynomials of degree $2n$ in x_0, x_1.

If we free $d \in S_{2m}$ from squares, i.e., divide it by the square of the polynomial of maximal degree, then we get the polynomial r with simple roots which coincide with the roots of odd degree of d. The polynomial r determines the triple (Z, p, u), where Z is the two-sheeted covering space of $\mathbb{C}P^1$ branched over the roots of r, the map $p\colon Z \to \mathbb{C}P^1$ is the covering projection, and $u\colon Z \to Z$ is an antiholomorphic involution which covers the involution conj$\colon \mathbb{C}P^1 \to \mathbb{C}P^1$ of complex conjugation and has the set fix u of fixed points lying over the set $\{x \in \mathbb{R}P^1 \mid r(x) \geq 0\}$ (we note that the degree of r is even and so the sign of r on $\mathbb{R}P^1$ is defined). In turn, such a triple determines r up to a positive constant factor: its roots are the branch points of p and the sign is connected with fix u as stated above. In this way we can identify the set $\mathcal{H}_g$ of triples (Z, p, u), where Z is a curve of genus g, with the subspace of those $r \in S_{2g+2}$ that have only simple roots. It is clear that if d is the discriminant of F, then the corresponding triple (Z, p, t) is the real normalization of $\widetilde{F}$.

Let $\mathcal{D}_n \mathcal{H}_g$ denote the union, over all $(Z, p, t) \in \mathcal{H}_g$, of sets of effective divisors on Z that have degree n and are invariant under u. The topology of $\mathcal{D}_n \mathcal{H}_g$ can be defined as follows. We consider an effective divisor on Z as an unordered collection of n points of Z, i.e., as an element of the symmetric product $Z^{(n)}$ (here $Z^{(n)}$ is the quotient space of the product of n copies of Z under the action of the permutation group). Since, in addition, Z can be defined by the equation $x_2^2 = r(x_0, x_1)$ in the weighted homogeneous projective plane $P(1, 1, g + 1)$ (see, e.g., [10]), the set $\mathcal{D}_n \mathcal{H}_g$ lies in $\mathcal{H}_g \times (P(1, 1, g + 1))^{(n)}$ and is endowed with the induced topology.

3.2. Algebraic remark: interpolation problem for homogeneous polynomials. Let $\mathbb{C}H_l[x_0, x_1]$ denote the space of complex homogeneous polynomials of degree l in x_0, x_1. First, we shall show that there exists a polynomial in $\mathbb{C}H_l[x_0, x_1]$ with given partial derivatives of order $k - 1$ at a point $a \in \mathbb{C}^2 \setminus 0$. For this purpose we define $V_{ka} \colon \mathbb{C}H_l[x_0, x_1] \to \mathbb{C}^k$, where $1 \leqslant k \leqslant l + 1$, which maps f to the collection of the values of its partial derivatives of order $k-1$ at the point a. It is clear that V_{ka} is linear. It follows from the Euler theorem for homogeneous polynomials that $\ker V_{ka}$ consists of polynomials f with $\mu_f(a) \geqslant k$, where $\mu_f(a)$ is the multiplicity of the root a of f. So $\dim \ker V_{ka} = l - k + 1$. Hence, V_{ka} is surjective and the required polynomial exists.

Now, let $a_1, \ldots, a_s \in \mathbb{C}^2 \setminus 0$ be placed so that their projections on $\mathbb{C}P^1$ are distinct. We seek $f \in \mathbb{C}H_l[x_0, x_1]$ having for $i = 1, \ldots, s$ given values of partial derivatives of order $k-1$ in a_i. Let $D = \sum k_i a_i$ be an effective divisor. The arguments above establish that if $\deg D \leqslant l + 1$, then the map $V_D = (V_{k_1 a_1}, \ldots, V_{k_s a_s}) \colon \mathbb{C}H_l[x_0, x_1] \to \mathbb{C}^{\deg D}$ is surjective and $\dim \ker V_D = l + 1 - \deg D$. Hence, the polynomial f exists and such polynomials form an affine subspace of dimension $l + 1 - \deg D$.

It is clear that if D is invariant under conj, then f is real.

Since $V_{k_i \lambda a_i} = \lambda^{l-k_i+1} V_{k_i a_i}$, it follows that $\ker V_{k_i \lambda a_i}$ coincides with $\ker V_{k_i a_i}$. We denote this kernel by $\ker k_i[a_i]$, where $[a_i]$ is the image of a_i under the projection $\mathbb{C}^2 \setminus 0 \to \mathbb{C}P^1$. In a similar manner we can define $\ker[D]$, where $[D] = \sum k_i[a_i]$.

Let D_1, D_2 be effective divisors on $\mathbb{C}^2 \setminus 0$, and let D be the least common multiple of $[D_1]$, $[D_2]$. It is not difficult to check that $\ker V_{D_1} \cap \ker V_{D_2} = \ker D$. So for any $v_i \in \operatorname{im} V_{D_i}$, $i = 1, 2$, the intersection $V_{D_1}^{-1}(v_1) \cap V_{D_2}^{-1}(v)$ is either empty or is of dimension $l + 1 - \deg D$.

3.3. The map $N_{n,g}$. Curves with horizontal components are considered in 3.7.3. In 3.3–3.6 we eliminate them from our considerations, so we study the space $C^*_{m,2}$ of curves that have a nonzero discriminant and have no horizontal line as a component.

Let $C(n, g)$ be the set of $F \in C^*_{m,2}$ such that $\widetilde{F}$ is the curve of bidegree $(n, 2)$ and of genus g. For $F \in C(n, g)$, let F^ν be the real normalization of $\widetilde{F}$, $p^\nu \colon \mathbb{C}F^\nu \to \mathbb{C}\widetilde{F}$ the corresponding projection and $\operatorname{Pr}_i \colon \mathbb{C}P^1 \times \mathbb{C}P^1 \to \mathbb{C}P^1$ the projection on the ith factor ($i = 1, 2$). Let the divisor A_F on $\mathbb{C}F^\nu$ be equal to $(\operatorname{Pr}_2 p^\nu)^*[1:0]$, where $[1:0]$ is the point in $\mathbb{C}P^1$ with the coordinates $y_0 = 1$, $y_1 = 0$. We denote $\operatorname{Pr}_2 p^\nu$ by p_F. It is clear that A_F lies in $\mathcal{D}_n \mathcal{H}_g$ and that $(p_F)_* A_F$ is the divisor of the polynomial a.

Let $\mathcal{D}_k$ denote the space of effective divisors of degree k on $\mathbb{C}P^1$ which are invariant under conj. If F is defined by the polynomial (1), then the relation $N_{n,g}(F) = ((w), d, A_F)$, where (w) is the divisor of the polynomial w and $d \in S_{2n}$ is the normalized discriminant of $\widetilde{F}$, gives the map $N_{n,g} \colon C(n, g) \to \mathcal{D}_{m-n} \times S_{2n} \times \mathcal{D}_n \mathcal{H}_g$.

LEMMA. (i) *The image of $N_{n,g}$ ($1 \leqslant n \leqslant m$, $0 \leqslant g \leqslant n - 1$) consists of all the triples $(W, d, A) \in \mathcal{D}_{m-n} \times S_{2n} \times \mathcal{D}_n \mathcal{H}_g$ that satisfy the following conditions*:
   1) *The triple $(Z, p, t) \in \mathcal{H}_g$ such that A lies on Z is determined by the polynomial $r \in S_{2g+2}$ (see 3.1), which is discriminant d free from squares.*
   2) *For $x \in \mathbb{C}P^1$, if $\mu_d(x)$ is odd, then $p^{-1}(x)$ appears in A either with the coefficient $\mu_d(x)$, or with an even one which is less than $\mu_d(x)$. If $\mu_d(x)$ is even (in particular, is equal to zero), then either both of the points of $p^{-1}(x)$ appear in A with the same coefficient less than $\mu_d(x)/2$, or the smallest of the coefficients of these points is equal to $\mu_d(x)/2$.*

(ii) Let $(W, d, A) \in \operatorname{im} N_{n,g}$. We denote by Q' the maximal divisor on $\mathbb{C}P^1$ obeying the inequality $2Q' \leqslant \gcd(p_*A, (d))$. Then $N_{n,g}^{-1}(W, d, A)$ is a quasiprojective variety of dimension $2 + \deg Q'$.

(iii) If a path in $\operatorname{im} N_{n,g}$ beginning at (W, d, A) leaves W and d unchanged, and keeps the coefficients of A and Q' constant, then for any curve $F \in N_{n,g}^{-1}(W, d, A)$, this path is covered by a rigid isotopy of F.

PROOF. The fact that condition 1) of (i) is necessary for (W, d, A) to belong to the image of $N_{k,g}$ is evident from the definition of the map. Let us prove the necessity of 2). By the relation $b^2 = d + 4ac$ and because a, b, c are coprime, if $\mu_d(x)$ is odd, then either $\mu_a(x) = \mu_d(x)$, or $\mu_a(x)$ is even, and $\mu_a(x) < \mu_d(x)$. If $\mu_d(x)$ is even, then the branches of $\bar{F}$ lying over a neighborhood of x are given by the equations $y_0/y_1 = (-b \pm \sqrt{d})/(2a)$. Using the relation $b^2 = d + 4ac$, we see that $\mu_a(x) \leqslant \mu_d(x)$ implies $\mu_a(x)$ is even and the multiplicities of intersection of these branches over x with the line $y_1 = 0$ are equal to $\mu_a(x)/2$, and if $\mu_a(x) > \mu_d(x)$, then the multiplicities are equal to $\mu_d(x)/2$ and $\mu_a(x) - \mu_d(x)/2$.

Now we prove that 1), 2) are sufficient for (W, d, A) to belong to $\operatorname{im} N_{n,g}$.

Let $\gcd(p_*A, (d)) = \sum k_i t_i$ and $r_i = [k_i/2]$, where $[\cdots]$ denotes the integer part. Let Q be the divisor on Z that is made up from the points $p^{-1}(t_i)$; if $p^{-1}(t_i)$ is a single point, then it has the coefficient $2r_i$ in Q, and if $p^{-1}(t_i)$ consists of two points, then both of them have the coefficient r_i. It is clear that $p_*Q = 2Q'$ (see (ii)). Let $k = \deg Q'$. We denote $A - Q$ by $\bar{A}$. Let us choose $a_i = (a_{i0}, a_{i1})$, $q_j = (q_{j0}, q_{j1})$ in $\mathbb{C}^2 \setminus 0$ so that $[a_i], [q_j] \in \mathbb{C}P^1$ run over the support of $p_*\bar{A}$, Q' respectively. We define $\bar{a}(x_0, x_1)$, $q(x_0, x_1)$ as $\prod(a_{i1}x_0 - a_{i0}x_1)$, $\prod(q_{j1}x_0 - q_{j0}x_1)$. Denote d/q^2 by $\bar{d}$.

The standard projection $\operatorname{Pr}: \mathbb{C}^2 \setminus 0 \to \mathbb{C}P^1$ and the covering p determine a two-sheeted branched covering $\tilde{p}: \tilde{Z} \to \mathbb{C}^2 \setminus 0$ appearing in the commutative diagram

$$\begin{array}{ccc} \tilde{Z} & \xrightarrow{\tilde{p}} & \mathbb{C}^2 \setminus 0 \\ \operatorname{Pr} \downarrow & & \downarrow \operatorname{Pr} \\ Z & \xrightarrow{p} & \mathbb{C}P^1. \end{array}$$

Let $\sqrt{\bar{d}}$ denote one of two regular functions on $\tilde{Z}$ whose square is $\bar{d} \circ \tilde{p}$.

We must find a real homogeneous polynomial $\bar{b}$ of degree $n - k$ such that $\bar{b}^2 - \bar{d}$ is divided by $\bar{a}$. This means that $\mu_{\bar{b}^2 - \bar{d}}(a_i) \geqslant \mu_{\bar{a}}(a_i)$ for any i. Let $k_i = \mu_{\bar{a}}(a_i)$ and $\alpha = \sum k_i a_i$. Then the last inequality is equivalent to the condition $V_\alpha(\bar{b}^2) = V_\alpha(\bar{d})$, from which we shall find $V_\alpha(\bar{b})$, i.e., $V_{k_i a_i}(\bar{b})$ for any i. From the definitions of $\bar{a}, \bar{d}$ and from 2) of (i), it follows that either $\mu_{\bar{d}}(a_i) = k_i = 1$, or $\mu_{\bar{d}}(a_i) = 0$ and $k_i \geqslant 1$. In the former case, $\bar{b}(a_i) = 0$. In the latter case, $\bar{b}(a_i)$ can have two opposite values. We exclude this nonuniqueness in the following way. Since $\mu_d(a_i)$ is even, the set $p^{-1}(\operatorname{Pr} a_i)$ consists of two points. Due to the definition of $\bar{A}$ and by 2) of (i), exactly one of them appears in $\bar{A}$. We denote it by z_i. It is clear that there exists a unique point $\tilde{z}_i \in \tilde{Z}$ such that $\operatorname{Pr} \tilde{z}_i = z_i$ and $\tilde{p}\tilde{z}_i = a_i$. Let $\bar{b}(a_i) = \sqrt{\bar{d}}(\tilde{z}_i)$. If $k_i > 1$, we get $V_{k_i a_i}(\bar{b})$ by equating to zero the partial derivatives of $\bar{b}^2 - \bar{d}$ in a_i and considering that $\bar{b}(a_i) \neq 0$.

To put this another way, if $\tilde{\alpha}$ is a divisor on $\tilde{Z}$ such that $\operatorname{Pr}_* \tilde{\alpha} = \bar{A}$ and $\tilde{p}_*\tilde{\alpha} = \alpha$, then, as in 3.2, a regular function h on $\tilde{Z}$ can be mapped into the collection $V_{\tilde{\alpha}}(h)$ of its partial derivatives of the corresponding orders at the points of $\tilde{\alpha}$. Then $V_\alpha(\bar{b}) = V_{\tilde{\alpha}}(\sqrt{\bar{d}})$. It is clear that $V_{\tilde{\alpha}}(-\sqrt{\bar{d}}) = -V_{\tilde{\alpha}}(\sqrt{\bar{d}})$.

Since V_α is surjective, the desired polynomial $\bar{b}$ exists. Such polynomials constitute $V_\alpha^{-1}(V_{\tilde{\alpha}}(\sqrt{\bar{d}}))$.

We choose a polynomial w with $(w) = W$, define $\bar{c}$ as $(\bar{b}^2 - \bar{d})/(4\bar{a})$ and determine the curve F by the polynomial

$$(2) \qquad w(\bar{a}q^2 y_0^2 + \bar{b}q y_0 y_1 + \bar{c} y_1^2).$$

Let us clarify when $\bar{a}q^2$, $\bar{b}q$, $\bar{c}$ are not coprime or, equivalently, $\bar{a}q^2$, $\bar{d}q^2$, $\bar{c}$ have a common root at a point $x \in \mathbb{C}^2 \setminus 0$. Due to the definition of $\bar{c}$, this implies $\mu_{\bar{b}^2 - \bar{d}}(x) \geq \mu_{\bar{a}}(x) + 1$. If $\mu_{\bar{a}}(x) = \mu_{\bar{d}}(x) = 1$, this inequality is impossible. Otherwise x is a root of q by 2) of (i). So $\bar{a}q^2$, $\bar{b}q$, $\bar{c}$ have a common root if and only if $V_\kappa(\bar{b}^2 - \bar{d}) = 0$, where $\kappa = \sum(\mu_{\bar{a}}(q_i) + 1) q_i$, or, as above, $\bar{b} \in V_\kappa^{-1}(V_{\tilde{\kappa}}(\sqrt{\bar{d}}))$.

The dimension of

$$V_\alpha^{-1}(V_{\tilde{\alpha}}(\sqrt{\bar{d}})) \cap V_\kappa^{-1}(V_{\tilde{\kappa}}(\sqrt{\bar{d}}))$$

is, by 3.2, less than $\dim V_\alpha^{-1}(V_{\tilde{\alpha}}(\sqrt{\bar{d}}))$ because the degree of the least common multiple of $[\alpha]$, $[\kappa]$ is evidently greater than $\deg \alpha$. So, if

$$\bar{b} \in B(\sqrt{\bar{d}}) = V_\alpha^{-1}(V_{\tilde{\alpha}}(\sqrt{\bar{d}})) \setminus V_\kappa^{-1}(V_{\tilde{\kappa}}(\sqrt{\bar{d}})),$$

then $N_{n,g}(F) = (W, d, A)$, and (i) is proved.

Now we prove (ii). Let the curve $F \in N_{n,g}^{-1}(W, d, A)$ be given by the polynomial (1) so that $b_2 - 4ac = d \in S_{2k}$. Then the pair of polynomials (a, b) is determined up to sign. Let $\bar{a}$, q be the above-mentioned polynomials. It is clear that $a/(\bar{a}q^2) \in \mathbb{R} \setminus 0$ and $b/q \in B(\sqrt{\bar{d}}) \cup B(-\sqrt{\bar{d}})$. Identifying (λ, β) with $(-\lambda, -\beta)$ for any $(\lambda, \beta) \in (\mathbb{R} \setminus 0) \times (B(\sqrt{\bar{d}}) \cup B(-\sqrt{\bar{d}}))$ defines the quotient space $\mathcal{AB}$. So the map $ab : N_{n,g}^{-1}(W, d, A) \to \mathcal{AB}$ with $ab(F) = (a/(\bar{a}q^2), b/q)$ is given. The inverse map is determined by the relation

$$ab^{-1}(\lambda, \beta) = w(\lambda \bar{a} q^2 y_0^2 + \beta q y_0 y_1 + \gamma y_1^2), \quad \text{where } \gamma = (\beta^2 - \bar{d})/(4\lambda \bar{a}).$$

Hence, $N_{n,g}^{-1}(W, d, A)$ is isomorphic to a quasiprojective variety of dimension $k + 2$, due to 3.2.

To prove (iii), it is sufficient to notice that if a path in $\operatorname{im} N_{n,g}$ satisfies the conditions of (iii), then the arguments in the proof of (ii) imply that the map $N_{n,g}$ considered over the image of the path is a trivial bundle.

REMARKS. (i) Let the points z_1, z_2 appear in $\bar{A}$ with the coefficient 1 and let $p(z_i) = [1 : \alpha_i]$. Then by the interpolation formula, for any $\bar{b} \in V_\alpha^{-1}(V_{\tilde{\alpha}}(\sqrt{\bar{d}}))$ we have

$$(3) \qquad b(1, x) = \frac{\sqrt{\bar{d}}(\tilde{z}_1)(x - \alpha_2) b_1(x)}{(\alpha_1 - \alpha_2) b_1(\alpha_1)} + \frac{\sqrt{\bar{d}}(\tilde{z}_2)(x - \alpha_1) b_1(x)}{(\alpha_2 - \alpha_1) b_1(\alpha_2)} + (x - \alpha_1)(x - \alpha_2) b_2(\alpha_1, \alpha_2; x),$$

where $\tilde{z}_i$ is a point of $\tilde{Z}$ with $\tilde{p}\tilde{z}_i = (1, \alpha_i)$ and $\operatorname{Pr} \tilde{z}_i = z_i$ (see the proof of the lemma), b_1 is a polynomial which is nonzero in α_1, α_2, while b_2 is a polynomial in x and is a continuous function in α_1, α_2 if z_1, z_2 do not coincide with the other points of $\bar{A}$.

(ii) Let (W, d, A) be in im $N_{n,g}$ and let all the coefficients of A be equal to 1. Then, by the proof of the lemma, $N_{n,g}^{-1}(W, d, A)$ is connected if all the points of p_*A are roots of d, and consists of two connected components otherwise. The component containing the curve determined by the polynomial (1) is given by the function b/a; curves with opposite signs of the functions lie in different components. Let Y denote the line $y_1 = 0$. Due to the equality $y_1/y_0 = 2a/(-b \pm \sqrt{d})$, the component is determined by the intersection of $\widetilde{F}$ and Y in a neighborhood of $p^v(z)$, where z is a real point of A with $p(z)$ different from the roots of d. The intersection is called *increasing* if $\widetilde{F}$ meets Y from left bottom to right top and *decreasing* if this happens from left top to right bottom.

3.4. The code of a curve. Let $F \in C^*_{m,2}$ be given by the polynomial (1) and let the roots of a be simple and different from those of d. In this case we shall define the code of F.

First, we divide the set of real points of A_F (of real roots of a) into two classes, upper and lower, in the following way.

If d has roots of odd multiplicity, let F_i be a connected component of $\mathbb{R}\widetilde{F}$ and let l_i, p_i be its extreme left and extreme right points. Then the set $(p^v)^{-1}(F_i \setminus \{l_i, p_i\})$ has two connected components. We consider their images under the map p^v in a neighborhood of l_i. If the image of one component lies higher than the image of the other one, the first component is called *upper* and the second is *lower*. The points of A_F which lie in these components, as well as the real roots of a, are also called *upper* and *lower*.

Let d have no real roots of odd multiplicity and $\mathbb{R}F^v$ consist of two components. If $\widetilde{F}$ splits into two curves of bidegree $(m_1, 1)$, $(m_2, 1)$ with $m_1 \neq m_2$, then the *upper component* of $\mathbb{R}F^v$ corresponds to the largest of the numbers m_1, m_2. If $\widetilde{F}$ splits and $m_1 = m_2$ or $\widetilde{F}$ does not split and $\mathbb{R}F^v$ consists of two components, let $e_1 + \mu_1 e_2$, $e_1 + \mu_2 e_2 \in H_1(\mathbb{R}P^1 \times \mathbb{R}P^1)$ be the classes defined by the components of $\mathbb{R}F^v$. If $\mu_1 \neq \mu_2$, then the *upper component* of $\mathbb{R}F^v$ corresponds to the largest of the numbers μ_1, μ_2. If $\mu_1 = \mu_2$, we choose any of the components of $\mathbb{R}F^v$ and call it the *upper component*. Now the upper and lower points of A_F (roots of a) are determined as above. So the upper and lower points of A_F (roots of a) are always uniquely defined, except for the last case (with $\mu_1 = \mu_2$), when the real points of A_F (real roots of a) are divided into only two classes.

Finally, let d have no root of odd multiplicity and $\mathbb{R}F^v$ be connected. We say that a segment on $\mathbb{R}P^1$ bounded by adjacent roots of a can be *lifted* if a component of the inverse image of the segment under the map p_F is bounded by points of A_F. We assign two adjacent roots of a to one of two classes if the segment bounded by them can be lifted, and to the opposite class otherwise. It is clear that the real points of A_F are also divided into two classes. We choose any of the classes and call its elements *upper points*.

The orientation of a horizontal line defines a cyclic order on the set of real roots of d. To the set $[d_i, d_{i+1})$ we assign the pair $(\delta_i; \omega_i)$, where δ_i is the multiplicity of d_i as a root of d and ω_i is the collection of multiplicities of the roots of w lying on $[d_i, d_{i+1})$ (the collection begins with zero if d_i is not a root of w). In addition, if $d > 0$ on the interval (d_i, d_{i+1}), we write $1, -1, 0$ after $(\delta_i; \omega_i)$: plus one is for every upper root of a lying on (d_i, d_{i+1}), minus one is for a lower root, zero is for the absence of roots. To any imaginary root of the product wd, we assign the pair of its multiplicities as a root of d and of w. We put the obtained collection of pairs $(\delta_i; \omega_i)$ and the numbers

0, ±1 between the angle brackets ⟨ ⟩ and call it the *code* of the curve F. Besides, we separate the real part of the code from the imaginary one by ";".

Let us point out that if $d > 0$ on $\mathbb{R}P^1$, then the real part of the code involves merely ±1 or 0, but if $d < 0$ on $\mathbb{R}P^1$, the real part is empty.

Hence, if either d has roots of odd multiplicity, or $\widetilde{F}$ is reduced and $m_1 \neq m_2$ or $\mu_1 \neq \mu_2$, then the code is uniquely defined. Otherwise the code is defined up to change of sign in all the ones.

3.5. Elementary isotopies. A translation of $F \in C^*_{m,2}$ along a vertical line is, evidently, a rigid isotopy. This allows us to assume that the line $Y: y_1 = 0$ is not tangent to $\widetilde{F}$ and does not meet $\widetilde{F}$ in the inverse images of the roots of d. Then all the roots of a are simple and different from the ones of d. Hence, the code of F is defined.

As mentioned in 3.1, the singularities of F are determined by the pair (W, d), where $W = (w)$, which is in turn determined by the code of F. Since, in addition, any isotopy of $\mathbb{C}P^1$ equivariant under conj evidently preserves the code, we assume (W, d) is fixed in the subsequent study of rigid isotopies.

We suggest some types of rigid isotopies with (W, d) fixed. In the notation for the isotopies, we put the code of the initial curve to the left of the arrow and the code of the resulting one to the right.

THEOREM. *Let L and M be parts of a code, ω be the collection of numbers defined above, and $\varepsilon = \pm 1$, $l \geq 0$. Then the following isotopies $I_1 - I_6$ exist and, if a has imaginary roots, then the isotopies $I_7 - I_9$ exist as well.*

$$I_1: L\varepsilon, \varepsilon, M \to LM,$$
$$I_2: L(4l, \omega), \varepsilon, -\varepsilon M \to L\varepsilon, -\varepsilon, (4l, \omega) M,$$
$$I_3: L(4l+2, \omega), \varepsilon, -\varepsilon M \to L-\varepsilon, \varepsilon, (4l+2, \omega) M,$$
$$I_4: L\varepsilon, (2, \omega), -\varepsilon M \to L-\varepsilon, (2, \omega), \varepsilon M,$$
$$I_5: \begin{matrix} L(2l+1, \omega), \varepsilon, -\varepsilon M \to L(2l+1, \omega) M, \\ L\varepsilon, -\varepsilon, (2l+1, \omega) M \to L(2l+1, \omega) M, \end{matrix}$$
$$I_6: \begin{matrix} L(1, \omega), \varepsilon M \to L(1, \omega), -\varepsilon M, \\ L\varepsilon, (1, \omega) M \to L-\varepsilon, (1, \omega) M, \end{matrix}$$
$$I_7: L\varepsilon, (4l, \omega), \varepsilon M \to L-\varepsilon, (4l, \omega)-\varepsilon M,$$
$$I_8: L\varepsilon, (4l+2, \omega), -\varepsilon M \to L-\varepsilon, (4l+2, \omega)\varepsilon M,$$
$$I_9: \begin{matrix} L(2l+1, \omega), \varepsilon M \to L(2l+1, \omega), -\varepsilon M, \\ L\varepsilon, (2l+1, \omega) M \to L-\varepsilon, (2l+1, \omega) M. \end{matrix}$$

PROOF OF THE EXISTENCE OF I_1. Suppose F has the code $L\varepsilon, \varepsilon, M$, the points z_1, z_2 of A_F correspond to the indicated pair ε, ε of the code and $p_F(z_i) = [1 : \alpha_i]$. Let α_2 tend to α_1. Then it is not difficult to calculate that the polynomial (3) tends to $b_0(1, x)$ with $b_0(1, \alpha_1) = \sqrt{\bar{d}}(\tilde{z}_1)$ and $b'_0(1, \alpha_1) = (\sqrt{\bar{d}})'(\tilde{z}_1)$. So F is rigidly isotopic to the curve F_0 corresponding to b_o. The line Y is tangent to F_0 in $p^v(z_1)$. Hence, moving the line to the appropriate side, we obtain a curve with the code LM.

REMARK. A similar argument shows that *if a has imaginary roots, then the isotopy inverse to I_1 exists.*

PROOF OF THE EXISTENCE OF $I_2 - I_9$. To construct $I_2 - I_5$, we use Remark (i) of 3.3. Let F be the initial curve, z_1, z_2 be the points of A_F corresponding to the indicated pair ε, $-\varepsilon$ of the code, and let $\delta \in \mathbb{R}P^1$ correspond to the root of d indicated in the code. We may assume that $\delta = [1:0]$. Let $p_F(z_i) = [1:\alpha_i]$ and $n = \mu_d(\delta)/2$. For $t \in [0, 1]$, we put

(4) $\quad\quad\quad\quad \alpha_1(t) = t\alpha_1, \quad\quad \alpha_2(t) = t\alpha_2 + (\alpha_2 - \alpha_1)t^n.$

For $t \neq 0$ these equalities define continuous families of divisors A_t, Q_t, where $A_1 = A_F$, $Q_1 = Q$, with coefficients independent of t. So, due to assertion (iii) of the Lemma, (4) gives a continuous family of curves F_t which are rigidly isotopic to F. If, in addition, as $t \to 0$ the limit curve F_0 lies in $N_{n,g}^{-1}(W, d, A_0)$, then F_0 is also rigidly isotopic to F. The condition $F_0 \in N_{n,g}^{-1}(W, d, A_0)$ is fulfilled if the coefficients a_0, b_0, c_0 of the limit polynomial $w_0(a_0 y_0^2 + b_0 y_0 y_1 + c_0 y_1^2)$ defining F_0 are coprime. It is clear that the common root of a_0, b_0, c_0 may be δ only. So it will suffice to show that $c_0(1, 0) \neq 0$. The polynomial $c_0(1, x)$ is determined by the equality $b_0^2 = d + a_0 c_0$. So, since $\mu_a(\delta) = 2$, the equality $c_0(1, 0) = 0$ is equivalent to $\mu_b(\delta) \geq 2$ when $\mu_d(\delta) > 2$ and to $(b_0'(1,0))^2 = d''(1,0)$ when $\mu_d(\delta) = 2$. Using (3) and taking into account the fact that z_1, z_2 are not in the same class (upper or lower), we obtain that $b_0'(1,0) = 2\alpha_1^n/(\alpha_1 - \alpha_2)$ when $n > 1$ and $b_0'(1,0) = (\alpha_1 + \alpha_2)/(\alpha_1 - \alpha_2)$ when $n = 1$. Hence, by a small change of α_1, α_2 which is, evidently, covered by a rigid isotopy of the initial curve, we can obtain $c_0(1, 0) \neq 0$.

Considering (4) for $t \in [-1, 1]$ and moving, if necessary, z_1, z_2 to satisfy the inequality $\alpha_1 < \alpha_2 < 2\alpha_1$, we obtain I_2, I_3. To get I_4, it is also sufficient to take (4) with $t \in [-1, 1]$ after a prior move of z_1, z_2 so that $\alpha_1 = -\alpha_2$, because in this case $d''(1, 0) \neq 0$.

To obtain I_5 (the upper row), we notice that because of the relation $\mu_a(\delta) = 2$, the line Y is not tangent to the curve F_0 over δ. So, moving the line to the appropriate side, we get a curve with the code $L(2l + 1, \omega)M$. The lower line of I_5 is constructed similarly.

The isotopy I_6 is easy to construct. Indeed, it is sufficient to move the corresponding point of A_F along fix u across $p_F^{-1}(\delta)$, where δ is the root of d indicated in the code, to a point of the opposite name (upper or lower). Since the coefficients of A_t, Q_t do not change during this move, we may apply assertion (iii) of the Lemma.

To construct I_7, we use the condition that A_F has a pair of conjugate imaginary points. So, due to the Remark in this subsection, there exists an isotopy inverse of I_1 which takes $L\varepsilon, (4l, \omega), \varepsilon M$ to $L\varepsilon, (4l, \omega), \varepsilon, -\varepsilon, -\varepsilon M$. Now I_2 gives $L\varepsilon, \varepsilon, -\varepsilon, (4l, \omega), -\varepsilon M$ and I_1 gives the desired code $L - \varepsilon, (4l, \omega), -\varepsilon M$.

The isotopies I_8, I_9 are built in a similar way:

$$L\varepsilon, (4l + 2, \omega), -\varepsilon M \xrightarrow{I_1^{-1}} L\varepsilon, (4l + 2, \omega), -\varepsilon, \varepsilon, \varepsilon M$$
$$\xrightarrow{I_3} L\varepsilon, \varepsilon, -\varepsilon, (4l + 2, \omega), \varepsilon M \xrightarrow{I_1} L - \varepsilon, (4l + 2, \omega), \varepsilon M$$

for I_8,

$$L(2l + 1, \omega), \varepsilon M \xrightarrow{I_1^{-1}} L(2l + 1, \omega), \varepsilon, -\varepsilon, -\varepsilon M \xrightarrow{I_4} L(2l + 1, \omega), -\varepsilon M$$

for I_9 (the upper row).

3.6. Rigid isotopic classification of nonsingular curves. The real isotopic classification of nonsingular curves is well known (see [4, §3.9]). The real scheme of such a curve consists of ovals (an *oval* is a component of the real part of the curve which bounds a topological disk in $\mathbb{R}P^1 \times \mathbb{R}P^1$) and nonovals. If d has real roots, then any nonoval realizes the class e_2 and all the ovals are outside of each other. If d has no real roots, then there are no ovals. In this case when m is even, the curve has either no nonovals (empty scheme) or two of them and every nonoval realizes the same class $e_1 + \mu e_2$ with $|\mu| \leqslant m/2$. When m is odd, there is a single nonoval; it realizes the class $2e_2 + (2\mu + 1)e_2$ with $|2\mu + 1| \leqslant m$.[1]

THEOREM. *The rigid isotopic classification of nonsingular curves of bidegree $(m, 2)$ coincides with the real isotopic classification.*

PROOF. Let F be the initial curve and let d have real roots. The isotopies I_1, I_6 give the curve F_1 rigidly isotopic to F and such that any component of $\mathbb{R}F_1$ has the code $(1, 0), \varepsilon, (1, 0)$, where $\varepsilon = 0$ when the curve is an oval and $\varepsilon = 1$ when it is a nonoval. It is clear that every nonoval realizes the class e_2. We may simplify F_1 by moving all the points of A_F into the inverse images of the roots of d under the map p_F. The coefficient b of the polynomial (1) defining the obtained curve may obviously be assumed equal to zero. So in this case by Remark (ii) of 3.3, all the curves with the same real scheme form a stratum of $C_{m,2}$.

All the curves with empty real scheme also form a stratum, because for such a curve we can, as above, bring the points of A_F into coincidence with the inverse images of some roots of d.

Finally, if d has no real roots but the curve has real points, then I_1 gives a curve with a code in which ones alternate with minus ones. In this case, by Remark (ii) of 3.3, the stratum of F is determined by the number N of real points of A_F and by the type T of the intersection (increasing or decreasing) of the curve F and the line Y, i.e., by the class $2e_1 + \mu e_2 \in H_1(\mathbb{R}P^1 \times \mathbb{R}P^1)$ realized by $\mathbb{R}F$ endowed with an orientation (N gives $|\mu|$ and T gives the sign of μ).

OBVIOUS COROLLARY. *Let F be nonsingular. Then F is rigidly isotopic to the curve given, if d is positive on $\mathbb{R}P^1$, by the polynomial $a_1 y_0^2 + b_1 y_0 y_1 - a_1 y_1^2$, where a_1, b_1 are coprime, their real roots are simple and alternate, while the polynomial $b_1^2 + 4a_1^2$ has no multiple roots; otherwise it is given by the polynomial $a_1 y_0^2 + c_1 y_1^2$, where a_1, c_1 are coprime and all their roots are simple.*

3.7. Rigid isotopies of singular curves. In this section we assume F to be the curve determined by the polynomial (1).

3.7.1. Curves without real singularities. Let all the singularities of $\widetilde{F}$ be imaginary. Using the reasoning of 3.6, we see that the corollary from 3.6 is valid in this case as well, except for the condition of simplicity of the imaginary roots of the discriminant.

3.7.2. Curves with zero discriminant. Let d be equal to zero. Since a, b, c are coprime, they are squares and, hence, so is $\tilde{f}$. Then the curve F_1 defined by the polynomial $w(\tilde{f})^{1/2}$ is of bidegree $(m - n/2, 1)$. It is clear that the rigid isotopy class of F_1 described in §2 determines the one of F.

---

[1] In [4] all the real schemes are listed without considering the orientations of horizontal and vertical lines of the hyperboloid, i.e., without considering the sign of μ. We note, in addition, that in the present paper we have interchanged the notation of e_1, e_2 as compared to [4].

3.7.3. *Curves containing a horizontal line.* If F splits into a real horizontal line and a curve of bidegree $(m, 1)$, then, using the results of §2, we can easily describe the rigid isotopy class of F. The case when F is a pair of imaginary horizontal lines is much simpler.

3.7.4. *Curves in $C_{m,2}^*$.* Curves with negative discriminant (empty curves) have been considered in 3.7.1. If d is nonpositive, then $\mathbb{R}\widetilde{F}$ is a set of isolated points. It is clear that in this case we can also move all the roots of a into the roots of d. So, up to a rigid isotopy, the curve F is given by the polynomial $w(ay_0^2 + cy_1^2)$.

If d changes sign, it has roots of odd multiplicity. Then the isotopies $I_1 - I_3$, I_5 give, evidently, a curve with the code

$$\langle k_1, k_2, \ldots, k_r; (\delta_1, \omega_1'), \ldots, (\delta_s, \omega_s')\rangle, \tag{5}$$

where k_i is the collection

$$(2l_{i1}+1, \omega_{i1}), \varepsilon_{i1}, (2l_{i2}, \omega_{i2}), \varepsilon_{i2}, \ldots, (2l_{in-1}, \omega_{in-1}), \varepsilon_{in-1}, (2l_{in}+1, \omega_{in})$$

with $\varepsilon_{ij} = 0, \pm 1$. If $l_{i1} = 0$ ($l_{in} = 0$), then I_6 allows ε_{i1} (ε_{in-1}) to be equal to 0 or 1. We show that *this is also the case when $l_{i1} > 0$ ($l_{in} > 0$)*. To do this, it will suffice to show that a has imaginary roots and to use I_9 if necessary. Let all the roots of a be real. Since any two roots of a are separated by a root of d, they cannot, by (i) of the Lemma of 3.3, arrive at an odd-multiple root of d while F_1 moves along the vertical line. On the other hand, there is a moment in the motion when the line Y intersects F_1 at a point over such a root of d. This contradiction completes the proof.

If d is nonnegative, then $I_1 - I_3$ gives, evidently, a curve with the code

$$\langle (2l_1, \omega_1), \underbrace{-\varepsilon_1, \varepsilon_1, \ldots, (-1)^\mu \varepsilon_1}_{\mu}, (2l_2, \omega_2), \varepsilon_2, \ldots, (2l_n, \omega_n), \varepsilon_n;$$
$$(\delta_1, \omega_1'), \ldots, (\delta_s, \omega_s')\rangle, \tag{6}$$

where $\varepsilon_i = 0, \pm 1$.

The following theorem now obviously follows from Remark (ii) of 3.3.

THEOREM. *Let $F_1, F_2 \in C_{m,2}^*$ have the same code* (5) *or* (6). *If the code is not uniquely defined* (see 3.4), *let either $\varepsilon_i = 0$ for all i, or the intersection of $\widetilde{F}_1$ with Y and the one of $\widetilde{F}_2$ with Y at the point corresponding to some nonzero ε_i be of the same type* (increasing or decreasing). *Then F_1, F_2 are rigidly isotopic.*

REMARK. The hypothesis of the theorem is not necessary because of the existence of the isotopies I_4, $I_7 - I_9$ and also because of the following example:

$$\langle (1,0), 0, (2,0), 1, (2,0), 0, (1,0) \rangle$$

$$\xrightarrow{I_1^{-1}} \langle (1,0), -1, -1, (2,0), 1, (2,0), 0, (1,0) \rangle$$

$$\xrightarrow{I_4, I_6} \langle (1,0), 1, 1, (2,0), -1, (2,0), 0, (1,0) \rangle$$

$$\xrightarrow{I_1} \langle (1,0), 0, (2,0), -1, (2,0), 0, (1,0) \rangle.$$

References

1. D. A. Gudkov, *On the topology of algebraic curves on a hyperboloid*, Uspekhi Mat. Nauk **34** (1979), no. 6, 26–32; English transl. in Russian Math. Surveys **34** (1979).
2. D. A. Gudkov and A. E. Usachev, *Nonsingular curves of lowest orders on a hyperboloid*, Methods of the Qualitative Theory of Differential Equations (E. A. Leontovich-Andronova, ed.), Gor'ky State Univ., Gor'ky, 1980, pp. 96–103. (Russian)
3. D. A. Gudkov and E. I. Shustin, *A classification of nonsingular curves of order eight on an ellipsoid*, Methods of the Qualitative Theory of Differential Equations (E. A. Leontovich-Andronova, ed.), Gor'ky State Univ., Gor'ky, 1980, pp. 104–107. (Russian)
4. V. I. Zvonilov, *Complex topological invariants of real algebraic curves on a hyperboloid and on an ellipsoid*, Algebra i Analiz **3** (1991), no. 5, 88–108; English transl. in Leningrad Math. J. **3** (1992).
5. V. M. Kharlamov, *Rigid isotopic classification of real plane curves of degree* 5, Funktsional. Anal. i Prilozhen. **15** (1981), no. 1, 88–89; English transl. in Functional Anal. Appl. **15** (1981).
6. D. A. Gudkov and G. M. Polotovskiĭ, *Stratification of the space of real plane algebraic curves of* 4*th order on algebraic-topological types*, I, II, Deposited in VINITI No. 5600-B87 and No. 6331-B90. (Russian)
7. A. I. Degtyarev, *Isotopic classification of complex plane projective curves of degree* 5, Algebra i Analiz **1** (1989), no. 4, 78–101; English transl. in Leningrad Math. J. **1** (1990).
8. V. A. Rokhlin, *Complex topological characteristics of real algebraic curves*, Uspekhi Mat. Nauk **33** (1978), no. 5, 77–89; English transl., Russian Math. Surveys **33** (1978), no. 5, 85–98.
9. V. I. Arnold, S. M. Guseĭn-Zade, and A. N. Varchenko, *Singularities of differentiable maps*. I, "Nauka", Moscow, 1984; English transl., Birkhäuser, Basel, 1985.
10. I. Dolgachev, *Weighted projective varieties*, Lecture Notes in Math., vol. 956, Springer-Verlag, Heidelberg and Berlin, 1982, pp. 34–71.

Translated by THE AUTHOR

Other Titles in This Series

(Continued from the front of this publication)

135 S. N. Artemov et al., Six Papers in Logic
134 A. Ya. Aĭzenshtat et al., Fourteen Papers Translated from the Russian
133 R. R. Suncheleev et al., Thirteen Papers in Analysis
132 I. G. Dmitriev et al., Thirteen Papers in Algebra
131 V. A. Zmorovich et al., Ten Papers in Analysis
130 M. M. Lavrent′ev, K. G. Reznitskaya, and V. G. Yakhno, One-dimensional Inverse Problems of Mathematical Physics
129 S. Ya. Khavinson, Two Papers on Extremal Problems in Complex Analysis
128 I. K. Zhuk et al., Thirteen Papers in Algebra and Number Theory
127 P. L. Shabalin et al., Eleven Papers in Analysis
126 S. A. Akhmedov et al., Eleven Papers on Differential Equations
125 D. V. Anosov et al., Seven Papers in Applied Mathematics
124 B. P. Allakhverdiev et al., Fifteen Papers on Functional Analysis
123 V. G. Maz′ya et al., Elliptic Boundary Value Problems
122 N. U. Arakelyan et al., Ten Papers on Complex Analysis
121 V. D. Mazurov, Yu. I. Merzlyakov, and V. A. Churkin, Editors, The Kourovka Notebook: Unsolved Problems in Group Theory
120 M. G. Kreĭn and V. A. Jakubovič, Four Papers on Ordinary Differential Equations
119 V. A. Dem′janenko et al., Twelve Papers in Algebra
118 Ju. V. Egorov et al., Sixteen Papers on Differential Equations
117 S. V. Bočkarev et al., Eight Lectures Delivered at the International Congress of Mathematicians in Helsinki, 1978
116 A. G. Kušnirenko, A. B. Katok, and V. M. Alekseev, Three Papers on Dynamical Systems
115 I. S. Belov et al., Twelve Papers in Analysis
114 M. Š. Birman and M. Z. Solomjak, Quantitative Analysis in Sobolev Imbedding Theorems and Applications to Spectral Theory
113 A. F. Lavrik et al., Twelve Papers in Logic and Algebra
112 D. A. Gudkov and G. A. Utkin, Nine Papers on Hilbert's 16th Problem
111 V. M. Adamjan et al., Nine Papers on Analysis
110 M. S. Budjanu et al., Nine Papers on Analysis
109 D. V. Anosov et al., Twenty Lectures Delivered at the International Congress of Mathematicians in Vancouver, 1974
108 Ja. L. Geronimus and Gábor Szegő, Two Papers on Special Functions
107 A. P. Mišina and L. A. Skornjakov, Abelian Groups and Modules
106 M. Ja. Antonovskiĭ, V. G. Boltjanskiĭ, and T. A. Sarymsakov, Topological Semifields and Their Applications to General Topology
105 R. A. Aleksandrjan et al., Partial Differential Equations, Proceedings of a Symposium Dedicated to Academician S. L. Sobolev
104 L. V. Ahlfors et al., Some Problems on Mathematics and Mechanics, On the Occasion of the Seventieth Birthday of Academician M. A. Lavrent′ev
103 M. S. Brodskiĭ et al., Nine Papers in Analysis
102 M. S. Budjanu et al., Ten Papers in Analysis
101 B. M. Levitan, V. A. Marčenko, and B. L. Roždestvenskiĭ, Six Papers in Analysis
100 G. S. Ceĭtin et al., Fourteen Papers on Logic, Geometry, Topology and Algebra
99 G. S. Ceĭtin et al., Five Papers on Logic and Foundations
98 G. S. Ceĭtin et al., Five Papers on Logic and Foundations
97 B. M. Budak et al., Eleven Papers on Logic, Algebra, Analysis and Topology
96 N. D. Filippov et al., Ten Papers on Algebra and Functional Analysis

(See the AMS catalog for earlier titles)